STUDENT'S SOLUTIONS MANUAL

Nancy S. Boudreau

Bowling Green State University

STATISTICS

Ninth Edition

McClave
Sincich

Prentice
Hall

Upper Saddle River, NJ 07458

Editor in Chief: Sally Yagan
Acquisitions Editor: Quincy McDonald
Supplement Editor: Joanne Wendelken
Assistant Managing Editor: John Matthews
Production Editor: Donna Crilly
Supplement Cover Manager: Paul Gourhan
Supplement Cover Designer: PM Workshop Inc.
Manufacturing Buyer: Ilene Kahn

© 2003 by Prentice Hall
Prentice-Hall, Inc.
Upper Saddle River, NJ 07458

The author and publisher of this book have used their best efforts in preparing this book.
These efforts include the development, research, and testing of the theories and programs
to determine their effectiveness. The author and publisher make no warranty of any kind,
expressed or implied, with regard to these programs or the documentation contained in
this book. The author and publisher shall not be liable in any event for incidental or
consequential damages in connection with, or arising out of, the furnishing,
performance, or use of these programs.

Printed in the United States of America

10 9 8 7 6 5 4 3

ISBN 0-13-066068-X

Pearson Education Ltd., *London*
Pearson Education Australia Pty. Ltd., *Sydney*
Pearson Education Singapore, Pte. Ltd.
Pearson Education North Asia Ltd., *Hong Kong*
Pearson Education Canada, Inc., *Toronto*
Pearson Educacíon de Mexico, S.A. de C.V.
Pearson Education—Japan, *Tokyo*
Pearson Education Malaysia, Pte. Ltd.
Pearson Education, *Upper Saddle River, New Jersey*

Contents

Preface

This solutions manual is designed to accompany the text, *Statistics*, Ninth Edition, by James T. McClave and Terry Sincich. It provides answers to most odd-numbered exercises for each chapter in the text. Other methods of solution may also be appropriate; however, the author has presented one that she believes to be most instructive to the beginning Statistics student. The student should first attempt to solve the assigned exercises without help from this manual. Then, if unsuccessful, the solution in the manual will clarify points necessary to the solution. The student who successfully solves an exercise should still refer to the manual's solution. Many points are clarified and expanded upon to provide maximum insight into and benefit from each exercise.

Instructors will also benefit from the use of this manual. It will save time in preparing presentations of the solutions and possibly provide another point of view regarding their meaning.

Some of the exercises are subjective in nature and thus omitted from the Answer Key at the end of *Statistics*, Ninth Edition. The subjective decisions regarding these exercises have been made and are explained by the author. Solutions based on these decisions are presented; the solution to this type of exercise is often most instructive. When an alternative interpretation of an exercise may occur, the author has often addressed it and given justification for the approach taken.

I would like to thank Kelly Barber for creating the art work and for typing this work.

Nancy S. Boudreau
Bowling Green State University
Bowling Green, Ohio

Statistics, Data, and Statistical Thinking

1.1   Statistics is a science that deals with the collection, classification, analysis, and interpretation of information or data. It is a meaningful, useful science with a broad, almost limitless scope of applications to business, government, and the physical and social sciences.

1.3   The first element of inferential statistics is the population of interest. The population is a set of existing units. The second element is one or more variables that are to be investigated. A variable is a characteristic or property of an individual population unit. The third element is the sample. A sample is a subset of the units of a population. The fourth element is the inference about the population based on information contained in the sample. A statistical inference is an estimate, prediction, or generalization about a population based on information contained in a sample. The fifth and final element of inferential statistics is the measure of reliability for the inference. The reliability of an inference is how confident one is that the inference is correct.

1.5   Quantitative data are measurements that are recorded on a meaningful numerical scale. Qualitative data are measurements that are not numerical in nature; they can only be classified into one of a group of categories.

1.7   A population is a set of existing units such as people, objects, transactions, or events. A sample is a subset of the units of a population.

1.9   An inference without a measure of reliability is nothing more than a guess. A measure of reliability separates statistical inference from fortune telling or guessing. Reliability gives a measure of how confident one is that the inference is correct.

1.11   The data consisting of the classifications A, B, C, and D are qualitative. These data are nominal and thus are qualitative. After the data are input as 1, 2, 3, and 4, they are still nominal and thus qualitative. The only differences between the two data sets are the names of the categories. The numbers associated with the four groups are meaningless.

1.13   a.   The bacteria count is a number. Therefore, it is quantitative.

        b.   Occupations take on values such as doctor, lawyer, carpenter, etc., which are not numeric. Therefore, it is qualitative.

        c.   Marital status can take on values such as married, single, divorced, etc., which are not numeric. Therefore, it is qualitative.

        d.   The time in months can take on values such as 1, 2, 3, etc. Therefore, it is quantitative.

1.15   a.   The population of interest is all citizens of the United States.

        b.   The variable of interest is the view of each citizen as to whether the impeachment trial of President Clinton was conducted fairly. It is qualitative.

c. The sample is the 2000 individuals selected for the poll.

d. The inference of interest is to estimate the proportion of all citizens who believe the impeachment trial of President Clinton was conducted fairly.

e. The method of data collection is a survey.

f. It is not very likely that the sample will be representative of the population of all citizens of the United States. By selecting phone numbers at random, the sample will be limited to only those people who have telephones. Also, many people share the same phone number, so each person would not have an equal chance of being contacted. Another possible problem is the time of day the calls are made. If the calls are made in the evening, those people who work in the evening would not be represented.

1.17   a. For question 1, the variable of interest is the e-commerce strategy status. Since the responses to this variable are "yes" or "no", the variable is Qualitative.

For question 2, the variable of interest is the time of implementation of an e-commerce plan. Depending on exactly how the question was asked, the variable could be either Qualitative or Quantitative. If the question was written such that the respondent had to pick from possible answers like "within the next year", "from one to 3 years", and "over 3 years", the variable would be Qualitative. On the other hand, if the respondent was requested to give a time to the closest month, then the variable could be considered Quantitative.

For question 3, the variable of interest is the status of delivering products over the internet. Since the responses to this variable are "yes" or "no", the variable is Qualitative.

For question 4, the variable of interest is company's total revenue in the last fiscal year. Since the responses to this variable are numbers such as $1 million, $700,000, etc., the variable is Quantitative.

b. The data collected from the 154 companies represent a sample. The Cutter Consortium surveyed only 154 out of many more U.S. companies. The Consortium did not survey all possible companies.

1.19   a. Length of maximum span can take on values such as 15 feet, 50 feet, 75 feet, etc. Therefore, it is quantitative.

b. The number of vehicle lanes can take on values such as 2, 4, etc. Therefore, it is quantitative.

c. The answer to this item is "yes" or "no," which are not numeric. Therefore, it is qualitative.

d. Average daily traffic could take on values such as 150 vehicles, 3,579 vehicles, 53,295 vehicles, etc. Therefore, it is quantitative.

e. Condition can take on values "good," "fair," or "poor," which are not numeric. Therefore, it is qualitative.

f.   The length of the bypass or detour could take on values such as 1 mile, 4 miles, etc. Therefore, it is quantitative.

g.   Route type can take on values "interstate," U.S.," "state," "county," or "city," which are not numeric. Therefore, it is qualitative.

1.21   a.   The data collection method used by the researchers was a designed experiment. Half of the boxers received the massage and half did not.

b.   The experimental units are the amateur boxers.

c.   There are two variables measured on the boxers – heart rate and blood lactate level. Both of these variables are measured on a numeric scale and thus, are quantitative.

d.   The inferences drawn from the analysis are: there is no difference in the mean heart rates between the two groups of boxers (those receiving massage and those not receiving massage) and there is no difference in the mean blood lactate levels between the two groups of boxers. Thus, massage did not affect the recovery rate of the boxers.

e.   No. Only amateur boxers were used in the experiment. Thus, all inferences relate only to boxers.

1.23   a.   The data for this study were collected from a designed experiment. The subjects in the study were randomly divided into two groups - one group received the drug and the other group received a placebo.

b.   This study involves interential statistics. The goal of the research was to see if the drug was effective in reducing blood loss of burn patients who undergo skin replacement surgery. The researchers are not particularly interested in the outcomes for the patients in the study, but rather the outcomes for all possible burn patients.

c.   The population of interest to the researchers is all possible burn patients. The sample is the 14 burn patients who participated in the study.

1.25   a.   The population of interest to the psychologist is the set of all male soccer players between the ages of 14 and 29 who played up to five times per week.

b.   The variables of interest are the frequency of "headers" and the IQ of each soccer player.

c.   The average number of headers per game per player can take on values such as 0, 1, 2, etc. Therefore, it is quantitative. IQ can take on values such as 85, 100, 103, etc. Therefore, it is quantitative.

d.   The sample is the set of 60 male soccer players between the ages of 14 and 29 who played up to five times per week.

e.   The inference made by the psychologist is that "heading" the ball in soccer lowers players' IQs.

f.　There could be several reasons why the inference could be misleading. First, "heading" the ball is a skill. It is possible that those players who are more athletically inclined are not as intellectually gifted as those players who are not as athletically inclined. Also, it appears that the data were collected only once per player. The only way one could conclude that "heading" the ball lowers players' IQs is by collecting data on each player over a period of time and seeing if each player's IQ decreases.

1.27　a.　The population of interest to the pollsters is the set of all Americans.

b.　The variable of interest is whether or not the person believes the Nazi extermination of the Jews happened. Since the views are not numeric, the variable is qualitative.

c.　The sample is the set of 1000 adults and high school students who were asked the question.

d.　The method of data collection is by survey.

e.　The pollsters inferred that one in five Americans believe that it is possible that the Holocaust never happened.

f.　The reliability of this inference is very suspect. The question presented to the sample is not a "yes," "no" question. It asks one to decide if it seems possible or if it seems impossible. Therefore, an answer of "yes" to this question is totally meaningless

Methods for Describing
Set of Data

2.1   First, we find the frequency of the grade A. The sum of the frequencies for all 5 grades must be 200. Therefore, subtract the sum of the frequencies of the other 4 grades from 200. The frequency for grade A is:

$$200 - (36 + 90 + 30 + 28) = 200 - 184 = 16$$

To find the relative frequency for each grade, divide the frequency by the total sample size, 200. The relative frequency for the grade B is 36/200 = .18. The rest of the relative frequencies are found in a similar manner and appear in the table:

Grade on Statistics Exam	Frequency	Relative Frequency
A: 90–100	16	.08
B: 80– 89	36	.18
C: 65– 79	90	.45
D: 50– 64	30	.15
F: Below 50	28	.14
Total	200	1.00

2.3   a.   To construct a relative frequency table for data, we must find the relative frequency for each species of rhinos. To find the relative frequency, divide the frequency by the total population size, 13,585. The relative frequency for African Black rhinos is 2,600/13,585 = .191. The rest of the relative frequencies are found in a similar manner and are reported in the table.

Rhino Species	Population Estimate	Relative Frequency
African Black	2,600	2600/13585 = .191
African White	8,465	8465/13585 = .623
(Asian) Sumatran	400	400/13585 = .029
(Asian) Javan	70	70/13585 = .005
(Asian) Indian	2,050	2050/13585 = .151
Total	13,585	.999

b.   The relative frequency bar chart is:

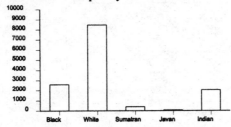

c.   The proportion of rhinos that are African is (.191 + .623) = .814.

The proportion of rhinos that are Asian is (.029 + .005 + .151) = .185.

2.5   a.   The variable "reason" is qualitative because the responses are categories. There are 7 values for the variable – Infant, Child, Medical, Infant & Medical, Child & Medical, Infant & Child, and Infant, Child & Medical.

b.   To compute the relative frequencies, divide the frequencies (or number of requests) by the total sample size of 30,337. The relative frequency for Infant is 1,852 / 30,337 = .061. The rest of the relative frequencies are computed in the table.

Reason	Number of Requests	Computation	Relative Frequency
Infant	1,852	1,852 / 30,337	.061
Child	17,148	17,148 / 30,337	.565
Medical	8,377	8,377 / 30,337	.276
Infant & Medical	44	44 / 30,337	.001
Child & Medical	903	903 / 30,337	.030
Infant & Child	1,878	1,878 / 30,337	.062
Infant & Child & Medical	135	135 / 30,337	.004
Total	30,337		.999

c.   A bar chart for the data is:

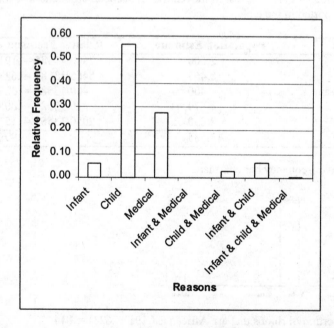

d.   The proportion of car owners who requested on-off bag switches who gave Medical as one of the reasons is $(8,377 + 44 + 903 + 135) / 30,337 = 9,459 / 30,337 = .312$.

2.7   a.   The type of graph used in this problem is a pie graph.

        b.   The type of cancer treated most often at the Moffitt Cancer Center is Breast Cancer. Of all the patients treated, 19% were treated for Breast Cancer, higher than any other form of cancer.

        c.   The percentage of Moffitt's patients treated for melanoma, lymphoma, or leukemia is the sum of the percentages for these three categories: 12% + 4% + 3% = 19%. Of all the treated patients, 19% were treated for melanoma, lymphoma, or leukemia.

2.9   First, construct a frequency table. There are 6 categories of highest degree obtained. The relative frequencies are found by dividing the frequencies by the total number of CEO's in the sample, which is 25. The table is:

Highest Degree Obtained	Frequency	Computation	Relative Frequency
None	1	1 / 25	.04
Bachelors	7	7 / 25	.28
Masters	11	11 / 25	.44
Doctorate	2	2 / 25	.08
JD	2	2 / 25	.08
LLB (law)	2	2 / 25	.08
Total	25		1.00

A relative frequency bar chart of the data is:

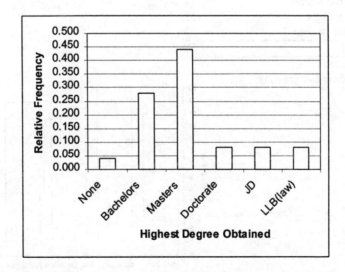

It appears that most CEOs have some type of advanced degrees. Of the 25 CEOs studied, 17 or 68% had some type of advanced degree.

2.11   a.   The opinion that occurred most often was "favorable/recommended" with 238 responses. The total number of responses was $19 + 37 + 35 + 238 + 46 = 375$. The proportion of books receiving a "favorable/recommended" opinion is $238/375 = .635$.

       b.   Books receiving either a 4 (favorable/recommended or a 5 (outstanding/significant) were reviewed as favorable and recommended for purchase. The total number of books receiving a rating of 4 or 5 is $238 + 46 = 284$. The proportion of books receiving these ratings is $284/375 = .757$. This proportion is more than .75 of 75%. Thus, the statement made is correct.

2.13   A relative frequency bar graph is used to depict the data:

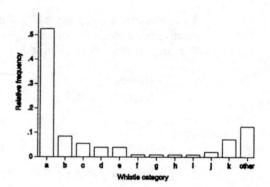

From the bar graph, over half of the whistle types were "Type a." The next most frequent category was the "Other types" with a relative frequency of about .14. Whistle types b and k were the next most frequent. None of the other whistle types had relative frequencies higher than .05.

2.15   a.   We will use a frequency bar graph to describe the data. First, we must add up the number of spills's under each category. These values are summarized in the following table:

Cause of Spillage	Frequency
Collision	11
Grounding	13
Fire/Explosion	12
Hull Failure	12
Unknown	2
Total	50

The frequency bar graph is:

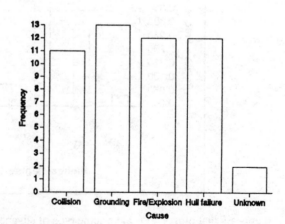

b.   Because each of the bars are about the same height, it does not appear that one cause is more likely to occur than any other.

2.17   To find the number of measurements for each measurement class, multiply the relative frequency by the total number of observations, $n = 500$. The frequency table is:

Measurement Class	Relative Frequency	Frequency
.5 – 2.5	.10	500(.10) = 50
2.5 – 4.5	.15	500(.15) = 75
4.5 – 6.5	.25	500(.25) = 125
6.5 – 8.5	.20	500(.20) = 100
8.5 – 10.5	.05	500(.05) = 25
10.5 – 12.5	.10	500(.10) = 50
12.5 – 14.5	.10	500(.10) = 50
14.5 – 16.5	.05	500(.05) = 25
		500

The frequency histogram is:

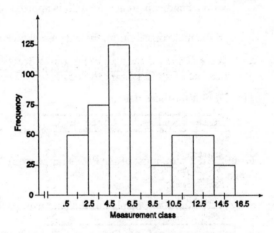

2.19   a.   This is a frequency histogram because the number of observations are displayed rather than the relative frequencies.

b.   There are 14 measurement classes used in this histogram.

c.   The total number of measurements in the data set is 49.

2.21   We will first combine the categories 0 and .1 – .9.

The frequency histogram for the data is:

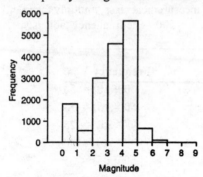

2.23   a.   Summing the percents in the bars above 20, the percentage of male USGA golfers with a handicap greater than 20 is approximately 28.5%.

b.   Summing the percents in the bars above 20, the percentage of female USGA golfers with a handicap greater than 20 is approximately 71%.

c.   In the male distribution, the data are skewed slightly to the right.

d.   In the female distribution, the data are skewed to the left.

2.25   The dot plot for these data is:

From the dot plot, most of the PMI's range from 3 to 7.5 (16 of the 22). The rest of the PMI's range from 10 to 15.

2.27   a.   A stem-and-leaf display of the data using MINITAB is:

```
Stem-and-leaf of FNE      N  = 25
Leaf Unit = 1.0

   2      0 67
   3      0 8
   6      1 001
  10      1 3333
  12      1 45
  (2)     1 66
  11      1 8999
   7      2 0011
   3      2 3
   2      2 45
```

b.   The numbers in bold in the stem-and-leaf display represent the bulimic students. Those numbers tend to be the larger numbers. The larger numbers indicate a greater fear of negative evaluation. Thus, the bulimic students tend to have a greater fear of negative evaluation.

c.   A measure of reliability indicates how certain one is that the conclusion drawn is correct. Without a measure of reliability, anyone could just guess at a conclusion.

2.29   a.   Using Minitab, the stem-and-leaf display for the data is:

```
Stem-and-Leaf of LOSS       N = 19
Leaf Unit = 1.0

   8        0  11234559
  (3)       1  123
   8        2  00
   6        3  9
   5        4  ⑥
   4        5  ⑥
   3        6  ① ⑤
   1        7
   1        8
   1        9
   1       10  ⓪
```

b.   The numbers circled on the display in part **a** are associated with the eclipses of Saturnian satellites.

c.   Since the five largest numbers are associated with eclipses of Saturnian satellites, it is much more likely that the greater light loss is associated with eclipses rather than occults.

2.31   a.   From the three dot plots, it appears that there is not much difference in the response rate between the Control group and the Treatment group. Both range from approximately 165 to 275. (The Treatment group did have a response of 139.) The response rate of the Familiarity group appears to be higher than for the other two groups. The response rate ranges from approximately 190 to 290.

b.   First, we will construct frequency tables for each of the groups. We will use the following classes: 20.5–25.05, 25.05–30.05, 30.05–35.05, 35.05–40.05, 40.0545.05, 45.05–50.05, and 50.05–55.05. The frequencies and relative frequencies (found by dividing the frequencies by the sample size) are listed in the following tables:

Control Group:

Measurement Class	Frequency	Relative Frequency
20.05–25.05	1	.1
25.05–30.05	6	.6
30.05–35.05	2	.2
35.05–40.05	1	.1
Totals	10	1.0

Treatment Group:

Measurement Class	Frequency	Relative Frequency
25.05–30.05	1	.1
30.05–35.05	4	.4
35.05–40.05	3	.3
40.05–45.05	2	.2
Totals	10	1.0

Familiarity Group:

Measurement Class	Frequency	Relative Frequency
35.05–40.05	1	.1
40.05–45.05	4	.4
45.05–50.05	4	.4
50.05–55.05	1	.1
Totals	10	1.0

The relative frequency histograms for the three groups are as follows:

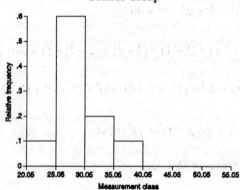

Control Group

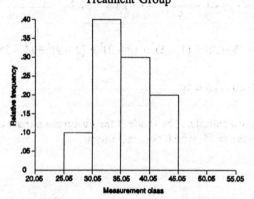

Treatment Group

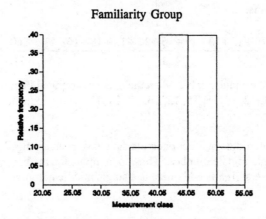

Familiarity Group

The three relative frequency histograms of the accuracy scores for the three groups indicate that the Familiarity group had the highest accuracy rate, the Treatment group had the second highest accuracy rate, and the Control group had the lowest accuracy rate.

2.33 a. $\sum x = 5 + 1 + 3 + 2 + 1 = 12$

  b. $\sum x^2 = 5^2 + 1^2 + 3^2 + 2^2 + 1^2 = 40$

  c. $\sum (x-1) = (5-1) + (1-1) + (3-1) + (2-1) + (1-1) = 7$

  d. $\sum (x-1)^2 = (5-1)^2 + (1-1)^2 + (3-1)^2 + (2-1)^2 + (1-1)^2 = 21$

  e. $\left(\sum x\right)^2 = (5+1+3+2+1)^2 = 12^2 = 144$

2.35 Using the results from Exercise 2.33,

  a. $\sum x^2 - \dfrac{\left(\sum x\right)^2}{5} = 40 - \dfrac{144}{5} = 40 - 28.8 = 11.2$

  b. $\sum (x-2)^2 = (5-2)^2 + (1-2)^2 + (3-2)^2 + (2-2)^2 + (1-2)^2 = 12$

  c. $\sum x^2 - 10 = 40 - 10 = 30$

2.37 Assume the data are a sample. The mode is the observation that occurs most frequently. For this sample, the mode is 15, which occurs 3 times.

The sample mean is:

$$\bar{x} - \frac{\sum x}{n} = \frac{18+10+15+13+17+15+12+15+18+16+11}{11} = \frac{160}{11} = 14.545$$

The median is the middle number when the data are arranged in order. The data arranged in order are: 10, 11, 12, 13, 15, 15, 15, 16, 17, 18, 18. The middle number is the 6th number, which is 15.

2.39   The median is the middle number once the data have been arranged in order. If n is even, there is not a single middle number. Thus, to compute the median, we take the average of the middle two numbers. If n is odd, there is a single middle number. The median is this middle number.

A data set with 5 measurements arranged in order is 1, 3, 5, 6, 8. The median is the middle number, which is 5.

A data set with 6 measurements arranged in order is 1, 3, 5, 5, 6, 8. The median is the average of the middle two numbers which is $\frac{5+5}{2} = \frac{10}{2} = 5$.

2.41   a. $\bar{x} = \frac{\sum x}{n} = \frac{85}{10} = 8.5$

b. $\bar{x} = \frac{400}{16} = 25$

c. $\bar{x} = \frac{35}{45} = .78$

d. $\bar{x} = \frac{242}{18} = 13.44$

2.43   a.   For a distribution that is skewed to the left, the mean is less than the median.

b.   For a distribution that is skewed to the right, the mean is greater than the median.

c.   For a symmetric distribution, the mean and median are equal.

2.45   a.   The mean of the data is $\bar{x} = \frac{\sum x}{n} = \frac{11.77}{8} = 1.4713$.

b.   The median is the average of the middle two numbers once the data are arranged in order. The data arranged in order are:

1.37, 1.41, 1.42, 1.48, 1.50, 1.51, 1.53, 1.55

The middle two numbers are 1.48 and 1.50. The median is $\frac{1.48+1.50}{2} = 1.49$

c.   Since the mean is less than the median, the data are somewhat skewed to the left.

2.47   Of all the applications, there could be some with 0 letters, 1 letter, 2 letters, 3 letters, etc. Of all of the possibilities, more applications had 3 letters than had any other number (mode). Since the median was also 3, at least half of the applications had 3 or more letters. The mean was 2.28. Since this is quite a bit smaller than the median, several of the applications must have had 0 or 1 letter.

2.49   a.   The mean response was 5.87. The average rating (across the 15 respondents) of the statement "The Training Game is a great way for students to understand the animal's perspective during training" was 5.87. Since the highest rating the statement could receive was 7, this indicated that on the average, the students rather strongly agreed with this statement.

The mode response was 6. More students rated the statement with a 6 than any other rating. A rating of 6 indicated a rather strong agreement to the statement.

b.   Since the mode is only slightly larger than the mean, there probably is no skewness present. If any skewness exists, the data would probably be skewed to the left since the mode is larger than the mean and because the mode is very close to the largest value possible.

2.51   a.   The first variable is gender. It has only two values which are not numerical, so it is qualitative. The next variable is group. There are three groups which are not numerical, so group is qualitative. The next variable is DIQ. This variable is measured on a numerical scale, so it is quantitative. The last variable is percent of pronoun errors. This variable is measured on the numerical scale, so it is quantitative.

b.   In order to compute numerical descriptive measures, the data must have numbers associated with them. Qualitative variables do not have meaningful numbers associated with them, so one cannot compute numerical measures.

c.   The mean of the DIQ scores for the SLI children is:

$$\bar{x} = \frac{\sum x}{n} = \frac{86 + 86 + 94 + \cdots + 95}{10} = \frac{936}{10} = 93.6$$

The median is the average of the middle two numbers after they have been arranged in order: 84, 86, 86, 87, 89, 94, 95, 98, 107, 110.

The median is $\frac{89 + 94}{2} = \frac{183}{2} = 91.5$

The mode is the value with the highest frequency. Since 86 occurred twice and no other value occurred more than once, the mode is 86.

d.   The mean of the DIQ scores for the YND children is:

$$\bar{x} = \frac{\sum x}{n} = \frac{110 + 92 + 92 + \cdots + 92}{10} = \frac{953}{10} = 95.3$$

The median is the average of the middle two numbers after they have been arranged in order: 86, 90, 90, 92, 92, 92, 96, 100, 105, 110.

The median is $\dfrac{92+92}{2} = \dfrac{184}{2} = 92$

The mode is the value with the highest frequency. Since 92 occurred three times and no other value occurred more than twice, the mode is 92.

e.   The mean of the DIQ scores for the OND children is:

$$\bar{x} = \frac{\sum x}{n} = \frac{110+113+113+\cdots+98}{10} = \frac{1019}{10} = 101.9$$

The median is the average of the middle two numbers after they have been arranged in order: 87, 92, 94, 95, 98, 108, 109, 110, 113, 113.

The median is $\dfrac{98+108}{2} = \dfrac{206}{2} = 103$

The mode is the value with the highest frequency. Since 113 occurred twice and no other value occurred more than once, the mode is 113.

f.   Of the three groups, the SLI group had the lowest mean DIQ score (93.6), the YND group had a slightly higher mean DIQ score (95.3), while the OND group had the highest mean DIQ score (101.9). Thus, the SLI and the YND groups appear to be fairly similar with regard to DIQ, while the OND group appears to be much higher.

Of the three groups, the SLI group had the lowest median DIQ score (91.5), the YND group had a slightly higher median DIQ score (92), while the OND group had the highest median DIQ score (103). Thus, again, the SLI and the YND groups appear to be fairly similar with regard to DIQ, while the OND group appears to be much higher.

Of the three groups, the SLI group had the lowest mode DIQ score (86), the YND group had a slightly higher mode DIQ score (92), while the OND group had the highest mode IDQ score (113). Thus, again, the SLI and the YND groups appear to be fairly similar with regard to DIQ, while the OND group appears to be much higher.

g.   For the SLI group, the mean percentage of pronoun errors is 30.17 and the median is 32.46. For the YND group, the mean percentage of pronoun errors is 46.88 and the median is 43.72. For the OND group, the mean percentage of pronoun errors is 0.00 and the median is 0.00.

The mean percentage of pronoun errors for the OND group (0) is much smaller than the other two groups (30.17 for the SLI group and 46.88 for the YND group). The median percentage of pronoun errors for the OND group (0) again is much smaller than that for the other two groups (32.46 for SLI and 43.72 for YND).

h. Since the mean percentage of pronoun errors is quite different for the three groups as is the median percentage of errors, it is not reasonable to use just a single number for the center of the distribution. Three "centers" should be calculated, one for each of the three groups.

2.53 a. Due to the "elite" superstars, the salary distribution is skewed to the right. Since this implies that the median is less than the mean, the players' association would want to use the median.

b. The owners, by the logic of part **a**, would want to use the mean.

2.55 a. $s^2 = \dfrac{\sum x^2 - \dfrac{\left(\sum x\right)^2}{n}}{n-1} = \dfrac{84 - \dfrac{20^2}{10}}{10-1} = 4.8889$ $\qquad s = \sqrt{4.8889} = 2.211$

b. $s^2 = \dfrac{\sum x^2 - \dfrac{\left(\sum x\right)^2}{n}}{n-1} = \dfrac{380 - \dfrac{100^2}{40}}{40-1} = 3.3333$ $\qquad s = \sqrt{3.3333} = 1.826$

c. $s^2 = \dfrac{\sum x^2 - \dfrac{\left(\sum x\right)^2}{n}}{n-1} = \dfrac{18 - \dfrac{17^2}{20}}{20-1} = .1868$ $\qquad s = \sqrt{.1868} = .432$

2.57 a. Range $= 42 - 37 = 5$

$s^2 = \dfrac{\sum x^2 - \dfrac{\left(\sum x\right)^2}{n}}{n-1} = \dfrac{7935 - \dfrac{199^2}{5}}{5-1} = 3.7$ $\qquad s = \sqrt{3.7} = 1.92$

b. Range $= 100 - 1 = 99$

$s^2 = \dfrac{\sum x^2 - \dfrac{\left(\sum x\right)^2}{n}}{n-1} = \dfrac{25,795 - \dfrac{303^2}{9}}{9-1} = 1,949.25$ $\qquad s = \sqrt{1,949.25} = 44.15$

c. Range $= 100 - 2 = 98$

$s^2 = \dfrac{\sum x^2 - \dfrac{\left(\sum x\right)^2}{n}}{n-1} = \dfrac{20,033 - \dfrac{295^2}{8}}{8-1} = 1,307.84$ $\qquad s = \sqrt{1,307.84} = 36.16$

2.59 This is one possibility for the two data sets.

Data Set 1: 1, 1, 2, 2, 3, 3, 4, 4, 5, 5
Data Set 2: 1, 1, 1, 1, 1, 5, 5, 5, 5, 5

$\bar{x}_1 = \dfrac{\sum x}{n} = \dfrac{1+1+2+2+3+3+4+4+5+5+}{10} = \dfrac{30}{10} = 3$

$$\bar{x}_2 = \frac{\sum x}{n} = \frac{1+1+1+1+1+5+5+5+5+5}{10} = \frac{30}{10} = 3$$

Therefore, the two data sets have the same mean. The variances for the two data sets are:

$$s_1^2 = \frac{\sum x^2 - \dfrac{\left(\sum x\right)^2}{n}}{n-1} = \frac{110 - \dfrac{30^2}{10}}{9} = \frac{20}{9} = 2.2222$$

$$s_2^2 = \frac{\sum x^2 - \dfrac{\left(\sum x\right)^2}{n}}{n-1} = \frac{130 - \dfrac{30^2}{10}}{9} = \frac{40}{9} = 4.4444$$

The dot diagram for the two data sets are shown below.

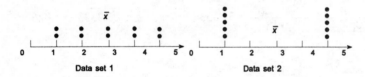

Data set 1   Data set 2

2.61   a.   Range = 3 − 0 = 3

$$s^2 = \frac{\sum x^2 - \dfrac{\left(\sum x\right)^2}{n}}{n-1} = \frac{15 - \dfrac{7^2}{5}}{5-1} = 1.3 \qquad s = \sqrt{1.3} = 1.1402$$

b.   After adding 3 to each of the data points,

Range = 6 − 3 = 3

$$s^2 = \frac{\sum x^2 - \dfrac{\left(\sum x\right)^2}{n}}{n-1} = \frac{102 - \dfrac{22^2}{5}}{5-1} = 1.3 \qquad s = \sqrt{1.3} = 1.1402$$

c.   After subtracting 4 from each of the data points,

Range = −1 − (−4) = 3

$$s^2 = \frac{\sum x^2 - \dfrac{\left(\sum x\right)^2}{n}}{n-1} = \frac{39 - \dfrac{(-13)^2}{5}}{5-1} = 1.3 \qquad s = \sqrt{1.3} = 1.1402$$

d.   The range, variance, and standard deviation remain the same when any number is added to or subtracted from each measurement in the data set.

2.63 a. Range = $1.55 - 1.37 = .18$

b. $s^2 = \dfrac{\sum x^2 - \dfrac{\left(\sum x\right)^2}{n}}{n-1} = \dfrac{17.3453 - \dfrac{11.77^2}{8}}{8-1} = .0041$

c. $s = \sqrt{.0041} = .064$

d. If the standard deviation of the daily ammonia levels during the morning drive-time is 1.45 ppm (compared to .064 ppm in the afternoon drive-time), then the morning drive-time has more variable ammonia levels.

2.65 a. The measures of variation on the printout are: standard deviation = SD = 6.0148 and variance = VARIANCE = 36.178

b. A value of the standard deviation that would make the age distribution more variable is any number greater than 6.0148.

c. A value of the standard deviation that would make the age distribution less variable is any number less than 6.0148.

d. If the largest age in the data set was omitted, then the standard deviation would decrease. The range of the data would be smaller, thus making the data less variable.

2.67 a. The unit of measurement of the variable of interest is dollars (the same as the mean and standard deviation). Based on this, the data are quantitative.

b. Since no information is given about the shape of the data set, we can only use Chebyshev's rule.

$900 is 2 standard deviations below the mean, and $2100 is 2 standard deviations above the mean. Using Chebyshev's rule, at least 3/4 of the measurements (or $3/4 \times 200 = 150$ measurements) will fall between $900 and $2100.

$600 is 3 standard deviations below the mean and $2400 is 3 standard deviations above the mean. Using Chebyshev's rule, at least 8/9 of the measurements (or $8/9 \times 200 \times 178$ measurements) will fall between $600 and $2400.

$1200 is 1 standard deviation below the mean and $1800 is 1 standard deviation above the mean. Using Chebyshev's rule, nothing can be said about the number of measurements that will fall between $1200 and $1800.

$1500 is equal to the mean and $2100 is 2 standard deviations above the mean. Using Chebyshev's rule, at least 3/4 of the measurements (or $3/4 \times 200 = 150$ measurements) will fall between $900 and $2100. It is possible that all of the 150 measurements will be between $900 and $1500. Thus, nothing can be said about the number of measurements between $1500 and $2100.

2.69   According to the Empirical Rule:

a.   Approximately 68% of the measurements will be contained in the interval $\bar{x} - s$ to $\bar{x} + s$.

b.   Approximately 95% of the measurements will be contained in the interval $\bar{x} - 2s$ to $\bar{x} + 2s$.

c.   Essentially all the measurements will be contained in the interval $\bar{x} - 3s$ to $\bar{x} + 3s$.

2.71   Using Chebyshev's rule, at least 8/9 of the measurements will fall within 3 standard deviations of the mean. Thus, the range of the data would be around 6 standard deviations. Using the Empirical Rule, approximately 95% of the observations are within 2 standard deviations of the mean. Thus, the range of the data would be around 4 standard deviations. We would expect the standard deviation to be somewhere between Range/6 and Range/4.

For our data, the range = 760 − 135 = 625.

The Range/6 = 625/6 = 104.17 and Range/4 = 625/4 = 156.25.

Therefore, I would estimate that the standard deviation of the data set is between 104.17 and 156.25.

It would not be feasible to have a standard deviation of 25. If the standard deviation were 25, the data would span 625/25 = 25 standard deviations. This would be extremely unlikely.

2.73   a.   More than half of the spillage amounts are less than or equal to 50 metric tons and almost all (44 out of 50) are below 104 metric tons. There appear to be three outliers, values which are much different than the others. These three values are larger than 216 metric tons.

b.   From the graph in part **a**, the data are not mound shaped. Thus, we must use Chebyshev's rule. This says that at least 8/9 of the measurements will fall within 3 standard deviations of the mean. Since most of the observations will be within 3 standard deviations of the mean, we could use this interval to predict the spillage amount of the next major oil spill. From the printout, the mean is 59.8 and the standard deviation is 53.36. The interval would be:

$$\bar{x} \pm 3s \Rightarrow 59.8 \pm 3(53.36) \Rightarrow 59.8 \pm 160.08 \Rightarrow (-100.28, 219.88)$$

Since an oil spillage amount cannot be negative, we would predict that the spillage amount of the next major oil spill will be between 0 and 219.88 metric tons.

2.75   a.   From the information given, we have $\bar{x} = 375$ and $s = 25$. From Chebyshev's rule, we know that at least three-fourths of the measurements are within the interval:

$$\bar{x} \pm 2s, \text{ or } (325, 425)$$

Thus, at most one-fourth of the measurements exceed 425. In other words, more than 425 vehicles used the intersection on at most 25% of the days.

b.   According to the Empirical Rule, approximately 95% of the measurements are within the interval:

$$\bar{x} \pm 2s, \text{ or } (325, 425)$$

This leaves approximately 5% of the measurements to lie outside the interval. Because of the symmetry of a mound-shaped distribution, approximately 2.5% of these will lie below 325, and the remaining 2.5% will lie above 425. Thus, on approximately 2.5% of the days, more than 425 vehicles used the intersection.

2.77　a.　Since no information is given about the distribution of the velocities of the Winchester bullets, we can only use Chebyshev's rule to describe the data. We know that at least 3/4 of the velocities will fall within the interval:

$$\bar{x} \pm 2s \Rightarrow 936 \pm 2(10) \Rightarrow 936 \pm 20 \Rightarrow (916, 956)$$

Also, at least 8/9 of the velocities will fall within the interval:

$$\bar{x} \pm 3s \Rightarrow 936 \pm 3(10) \Rightarrow 936 \pm 30 \Rightarrow (906, 966)$$

　　　b.　Since a velocity of 1,000 is much larger than the largest value in the second interval in part **a**, it is very unlikely that the bullet was manufactured by Winchester.

2.79　a.　For Adults:

		Interval	Percent	Frequency
$\bar{x} \pm s$	6.45 ± 2.89	$(3.56, 9.34)$	$\approx 68\%$	$46(.68) \approx 31.3$
$\bar{x} \pm 2s$	$6.45 \pm 2(2.89)$	$(0.67, 12.23)$	$\approx 95\%$	$46(.95) \approx 43.7$

For Adolescents:

		Interval	Percent	Frequency
$\bar{x} \pm s$	10.89 ± 2.48	$(8.41, 13.37)$	$\approx 68\%$	$19(.68) \approx 12.9$
$\bar{x} \pm 2s$	$10.89 \pm 2(2.48)$	$(5.93, 15.85)$	$\approx 95\%$	$19(.95) \approx 18.1$

　　　b.　See numbers in the above tables.

2.81　a.　Regardless of the shape of the distribution, most of the observations will fall within 3 standard deviations of the mean. Thus, for the SAT-Math scores, an interval likely to contain a student's change in score is:

$$\bar{x} \pm 3s \Rightarrow 19 \pm 3(65) \Rightarrow 19 \pm 195 \Rightarrow (-176, 214)$$

　　　b.　Regardless of the shape of the distribution, most of the observations will fall within 3 standard deviations of the mean. Thus, for the SAT-Verbal scores, an interval likely to contain a student's change in score is:

$$\bar{x} \pm 3s \Rightarrow 7 \pm 3(49) \Rightarrow 7 \pm 147 \Rightarrow (-140, 154)$$

　　　c.　For the SAT-Verbal, the maximum increase in scores is about 154 points. For the SAT-Math, the maximum increase in scores is approximately 214. Thus, a student is more likely to get a 140-point increase on the SAT-Math test.

2.83   Using the definition of a percentile:

	Percentile	Percentage Above	Percentage Below
a.	75th	25%	75%
b.	50th	50%	50%
c.	20th	80%	20%
d.	84th	16%	84%

2.85   We first compute z-scores for each x value.

a. $z = \dfrac{x - \mu}{\sigma} = \dfrac{100 - 50}{25} = 2$

b. $z = \dfrac{x - \mu}{\sigma} = \dfrac{1 - 4}{1} = -3$

c. $z = \dfrac{x - \mu}{\sigma} = \dfrac{0 - 200}{100} = -2$

d. $z = \dfrac{x - \mu}{\sigma} = \dfrac{10 - 5}{3} = 1.67$

The above z-scores indicate that the x value in part **a** lies the greatest distance above the mean and the x value of part **b** lies the greatest distance below the mean.

2.87   a.   If the distribution of scores was symmetric, the mean and median would be equal. The fact that the mean exceeds the median is an indication that the distribution of scores is skewed to the right.

b.   It means that 90% of the scores are below 660, and 10% are above 660. (This ignores the possibility of ties, i.e., other people obtaining a score of 660.)

c.   If you scored at the 94th percentile, 94% of the scores are below your score, while 6% exceed your score.

2.89   a.   The z-score for the Norwegian Star is: $z = \dfrac{x - \bar{x}}{s} = \dfrac{78 - 93.113}{5.184} = -2.92$
The score for the Norwegian Star is 2.92 standard deviations below the mean for all the cruise ships.

b.   The z-score for the Rotterdam is: $z = \dfrac{x - \bar{x}}{s} = \dfrac{97 - 93.113}{5.184} = .75$

The score for the Rotterdam is .75 standard deviations above the mean for all the cruise ships.

2.91 a. Since all the data are negative numbers, we must reverse the percentiles. Thus, the 10th percentile of the actual data will be the 90th percentile on the printout. From the printout, the 10th percentile is –0.2. Thus, 10% of the cylinder power measurements are below –0.2 and 90% are above –0.2.

b. From the printout, the 95th percentile is –0.06. Thus, 95% of the cylinder power measurements are below –.06 and 5% are above –0.06.

c. $$z = \frac{x - \bar{x}}{s} = \frac{-1.07 - (-0.1544)}{0.196767} = -4.65$$

A power measurement of –1.07 is 4.65 standard deviations below the mean.

Since the z-score computed in part c is so small (–4.65), it is an extremely unlikely value to observe. The cylinder value of –1.07 is an extreme value.

2.93 a. From the problem, $\mu = 2.7$ and $\sigma = .5$

$$z = \frac{x - \mu}{\sigma} \Rightarrow z\sigma = x - \mu \Rightarrow x = \mu + z\sigma$$

For $z = 2.0$, $x = 2.7 + 2.0(.5) = 3.7$

For $z = -1.0$, $x = 2.7 - 1.0(.5) = 2.2$

For $z = .5$, $x = 2.7 + .5(.5) = 2.95$

For $z = -2.5$, $x = 2.7 - 2.5(.5) = 1.45$

b. For $z = -1.6$, $x = 2.7 - 1.6(.5) = 1.9$

c. If we assume the distribution of GPAs is approximately mound-shaped, we can use the Empirical Rule.

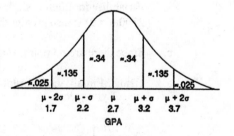

From the Empirical Rule, we know that –.025 or –2.5% of the students will have GPAs above 3.7 (with $z = 2$). Thus, the GPA corresponding to summa cum laude (top 2.5%) will be greater than 3.7 ($z > 2$).

We know that –.16 or 16% of the students will have GPAs above 3.2 ($z = 1$). Thus, the limit on GPAs for cum laude (top 16%) will be greater than 3.2 ($z > 1$).

We must assume the distribution is mound-shaped.

2.95   The 25th percentile, or lower quartile, is the measurement that has 25% of the measurements below it and 75% of the measurements above it. The 50th percentile, or median, is the measurement that has 50% of the measurements below it and 50% of the measurements above it. The 75th percentile, or upper quartile, is the measurement that has 75% of the measurements below it and 25% of the measurements above it.

2.97   a.   Median is approximately 4.

       b. Q_L is approximately 3 (Lower Quartile)

           Q_U is approximately 5.75 (Upper Quartile)

       c.   IQR $= Q_U - Q_L \approx 5.75 - 3 = 2.75$

       d.   The data set is skewed to the right since the right whisker is longer.

       e.   50% of the measurements are to the right of the median and 75% are to the left of the upper quartile.

       f.   There is one potential outlier, 14. There are 2 outliers, 18.5 and 19.5.

2.99   a.   The z-score is: $z = \dfrac{x - \bar{x}}{s} = \dfrac{3.3 - 7.3}{3.1849} = -1.26$

      b.   A PMI score of 3.3 would not be considered an outlier. The z-score is −1.26. A z-score this small is not considered unusual.

2.101   a.   The median average SAT score for 1990 (≈ 1034) is smaller than the median average SAT score for 2000 (≈ 1054).

       b.   The interquartile range for 2000 is slightly greater than that for 1990. Therefore, the variability for 2000 is slightly greater than that for 1990. The standard deviation for 2000 (65.9) is also greater than the standard deviation for 1990 (59.5).

       c.   Since there are no points outside the inner fences for either year, there are no outliers.

2.103   a.   Using Minitab, the box plot is:

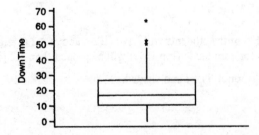

The median is about 18. The data appear to be skewed to the right since there are 4 suspect outliers to the right and none to the left. The variability of the data is fairly small because the IQR is fairly small, approximately $26 - 10 = 16$.

b. The customers associated with the suspected outliers are customers 238, 268, 269, and 264.

c. In order to find the z-scores, we must first find the mean and standard deviation.

$$\bar{x} = \frac{\sum x}{n} = \frac{815}{40} = 20.375$$

$$s^2 = \frac{\sum x^2 - \frac{\left(\sum x\right)^2}{n}}{n-1} = \frac{24129 - \frac{815^2}{40}}{40-1} = 192.90705$$

$$s = \sqrt{192.90705} = 13.89$$

The z-scores associated with the suspected outliers are:

Customer 238 $z = \dfrac{x - \bar{x}}{s} = \dfrac{47 - 20.375}{13.89} = 1.92$

Customer 268 $z = \dfrac{49 - 20.375}{13.89} = 2.06$

Customer 269 $z = \dfrac{50 - 20.375}{13.89} = 2.13$

Customer 264 $z = \dfrac{64 - 20.375}{13.89} = 3.14$

All but one of the z-scores is greater than 2. These are very unusual values.

2.105   For group 1, the median age is approximately 48. For group 2, the median age is approximately 47.5, while for group 3, the median age is approximately 46. These are all fairly close. The interquartile range for group 1 is from approximately 45 to 57 or 12. For group 2, the interquartile range is from approximately 46 to 58 or 12. For group 3, the interquartile range is from approximately 44 to 48 or 4. The interquartile range for group 3 is much smaller than for groups 1 and 2. This indicates that the variability in ages for group 3 is less than for the other two groups.

The inner fences for group 1 run from approximately 36 to 60. The inner fences for group 2 run from approximately 40 to 55, while the inner fences for group 3 run from approximately 42 to 50. This indicates that the variability for group1 is slightly larger than for group 2. Both group 1 and group 2 have greater variability in ages than group 3. All three groups have one suspect outlier. For group 1, the age of approximately 67 is a suspect outlier. For group 2, the age of approximately 64 is a suspect outlier. For group 3, the age of approximately 37 is a suspect outlier.

2.107   First, compute the z-score for 1.80: $z = \dfrac{x - \mu}{\sigma} = \dfrac{1.80 - 2.00}{.08} = -2.5$

If the data are actually mound-shaped, it would be very unusual (less than 2.5%) to observe a batch with 1.80% zinc phosphide if the true mean is 2.0%. Thus, if we did observe 1.8%, we would conclude that the mean percent of zinc phosphide in today's production is probably less than 2.0%.

2.109   Using Excel, the scatterplot is:

From the scatterplot, there does not appear to be much of a trend between variable 1 and variable 2. There is a slight positive linear trend - as variable 1 increases, variable 2 tends to increase. However, this relationship appears to be very weak.

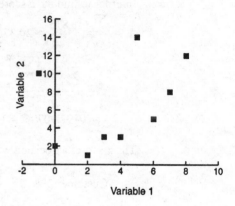

2.111   A scattergram of the data is:

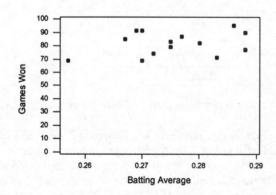

From the scattergram, it appears that there is a very slight trend. As the batting average increases, there is a slight increase in the number of games won.

2.113 a.   Using Excel, the scatterplot is:

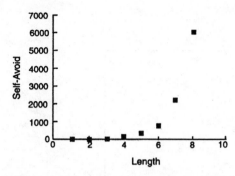

It appears that as length increases, the self-avoiding walks increase, but at an increasing rate.

b.   Using Excel, the scatterplot of the data is:

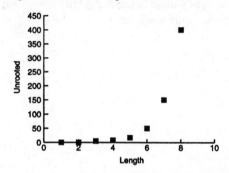

It appears that as the length increases, the unrooted walks also increase, but at an increasing rate.

2.115 a.   A scattergram of the data is:

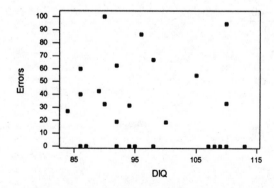

There does not appear to be much of a relationship between deviation intelligence quotient (DIQ) and the percent of pronoun errors. The points are scattered randomly.

b. A plot of the data for the SLI children only is:

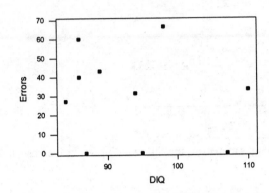

Again, there does not appear to be much of a trend between the DIQ scores and the proper use of pronouns. The data points are randomly scattered.

2.117 a. $z = \dfrac{x - \mu}{\sigma} = \dfrac{50 - 60}{10} = -1$

$z = \dfrac{70 - 60}{10} = 1$

$z = \dfrac{80 - 60}{10} = 2$

b. $z = \dfrac{x - \mu}{\sigma} = \dfrac{50 - 60}{5} = -2$

$z = \dfrac{70 - 60}{5} = 2$

$z = \dfrac{80 - 60}{5} = 4$

c. $z = \dfrac{x - \mu}{\sigma} = \dfrac{50 - 40}{10} = 1$

$z = \dfrac{70 - 40}{10} = 3$

$z = \dfrac{80 - 40}{10} = 4$

d. $z = \dfrac{x - \mu}{\sigma} = \dfrac{50 - 40}{100} = .1$

$z = \dfrac{70 - 40}{100} = .3$

$z = \dfrac{80 - 40}{100} = 4$

2.119 **a.** $s^2 = \dfrac{\sum x^2 - \dfrac{\left(\sum x\right)^2}{n}}{n-1} = \dfrac{246 - \dfrac{63^2}{22}}{22-1} = 3.1234$

b. $s^2 = \dfrac{\sum x^2 - \dfrac{\left(\sum x\right)^2}{n}}{n-1} = \dfrac{666 - \dfrac{106^2}{25}}{25-1} = 9.0233$

c. $s^2 = \dfrac{\sum x^2 - \dfrac{\left(\sum x\right)^2}{n}}{n-1} = \dfrac{76 - \dfrac{11^2}{7}}{7-1} = 9.7857$

2.121 **a.** $\sum x = 4 + 6 + 6 + 5 + 6 + 7 = 34$

$\sum x^2 = 4^2 + 6^2 + 6^2 + 5^2 + 6^2 + 7^2 = 198$

$\bar{x} = \dfrac{\sum x}{n} = \dfrac{34}{6} = 5.67$

$s^2 \dfrac{\sum x^2 - \dfrac{\left(\sum x\right)^2}{n}}{n-1} = \dfrac{198 - \dfrac{34^2}{6}}{6-1} = \dfrac{5.3333}{5} = 1.0667$

$s = \sqrt{1.0667} = 1.03$

b. $\sum x = -1 + 4 + (-3) + 0 + (-3) + (-6) = -9$

$\sum x^2 = (-1)^2 + 4^2 + (-3)^2 + 0^2 + (-3)^2 + (-6)^2 = 71$

$\bar{x} = \dfrac{\sum x}{n} = \dfrac{-9}{6} = -\$1.5$

$s^2 = \dfrac{\sum x^2 - \dfrac{\left(\sum x\right)^2}{n}}{n-1} = \dfrac{71 - \dfrac{(-9)^2}{6}}{6-1} = \dfrac{57.5}{5} = 11.5 \text{ dollars squared}$

$s = \sqrt{11.5} = \$3.39$

c. $\sum x = \dfrac{3}{5} + \dfrac{4}{5} + \dfrac{2}{5} + \dfrac{1}{5} + \dfrac{1}{16} = 2.0625$

$\sum x^2 = \left(\dfrac{3}{5}\right)^2 + \left(\dfrac{4}{5}\right)^2 + \left(\dfrac{2}{5}\right)^2 + \left(\dfrac{1}{5}\right)^2 + \left(\dfrac{1}{16}\right)^2 = 1.2039$

$\bar{x} = \dfrac{\sum x}{n} = \dfrac{2.0625}{5} = .4125\%$

$s^2 = \dfrac{\sum x^2 - \dfrac{\left(\sum x\right)^2}{n}}{n-1} = \dfrac{1.2039 - \dfrac{2.0625^2}{5}}{5-1} = \dfrac{.3531}{4} = .0883\% \text{ squared}$

$s = \sqrt{.0883} = .30\%$

d. (a) Range = 7 − 4 = 3

(b) Range = $4 − ($−6) = $10

(c) Range = $\dfrac{4}{5}\% - \dfrac{1}{16}\% = \dfrac{64}{80}\% - \dfrac{5}{80}\% = \dfrac{59}{80\%} = .7375\%$

2.123 Using Excel, the scatterplot is:

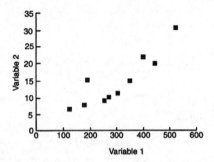

From the scatterplot, it appears that there is a trend. As variable 1 increases, variable 2 also tends to increase.

2.125 A pie chart of the data is:

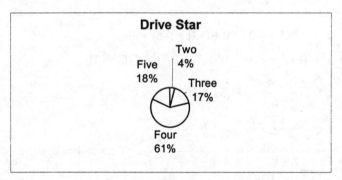

More than half of the cars received 4 star ratings (60.2%). A little less than a quarter of the cars tested received ratings of 3 stars or less.

2.127   a.   To display the status, we use a pie chart. From the pie chart, we see that 58% of the Beanie babies are retired and 42% are current.

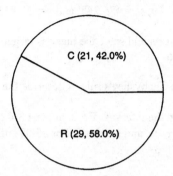

Pie Chart of Status

b.   Using Minitab, a histogram of the values is:

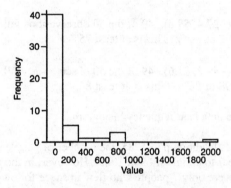

Most (40 of 50) Beanie babies have values less than $100. Of the remaining 10, 5 have values between $100 and $300, 1 has a value between $300 and $500, 1 has a value between $500 and $700, 2 have values between $700 and $900, and 1 has a value between $1900 and $2100.

2.129   a.   The mean COMPAS score for the male teachers is 26.4. This is just about in the middle of the possible range of from 0 to 50. This is the average score for the 68 male teachers.

b.   The mean COMPAS score for the female teachers is 24.5. This is just about in the middle of the possible range of from 0 to 50. This is the average score for the 48 female teachers.

c.   By the Empirical Rule, approximately 95% of the measurements will fall within 2 standard deviations of the mean. This interval for the male teachers would be:

$$\bar{x} \pm 2s \Rightarrow 26.4 \pm 2(10.6) \Rightarrow 26.4 \pm 21.2 \Rightarrow (5.2, 47.6)$$

d.   By the Empirical Rule, approximately 95% of the measurements will fall within 2 standard deviations of the mean. This interval for the female teachers would be:

$$\bar{x} \pm 2s \Rightarrow 24.5 \pm 2(11.2) \Rightarrow 24.5 \pm 22.4 \Rightarrow (2.1, 46.9)$$

Methods for Describing Sets of Data

2.131   a.   From the stem-and-leaf display, it appears that most states average less than 30 tornadoes a year (only 6 average more than 30). There is one observation which is much higher than any of the others: 137. Texas averages 137 tornadoes per year.

      b.   From the printout, $\bar{x}$ = 16.2000, median = 10, s^2 = 476.245, and s = 21.8230.

      c.   Since there is one very extreme observation and since the mean is larger than the median, the data are skewed to the right.

      d.   Since the data are skewed, we must use Chebyshev's rule to describe the data.

            At least 0 of the observations will fall in the interval $\bar{x} \pm s$, at least 3/4 or 75% of the observations will fall in the interval $\bar{x} \pm 2s$, and at least 8/9 or 88.9% of the observations will fall in the interval $\bar{x} \pm 3s$.

      e. $\bar{x} \pm s \Rightarrow 16.2 \pm 21.8 \Rightarrow (-5.6, 38.0)$. 47 of the 50 observations fall in this interval. This is the same as 47/50 = .94 or 94%. This is at least 0.

            $\bar{x} \pm 2s \Rightarrow 16.2 \pm 2(21.8) \Rightarrow (-27.4, 59.8)$. 49 of the 50 observations fall in this interval. This is the same as 49/50 = .98 or 98%. This is at least 75%.

            $\bar{x} \pm 3s \Rightarrow 16.2 \pm 3(21.8) \Rightarrow (-49.2, 81.6)$. 49 of the 50 observations fall in this interval. This is the same as 49/50 = .98 or 98%. This is at least 88.9%.

2.133   a.   The graph used to display the data is a frequency histogram.

      b.   Definitely! In the recent years leading up to 1985, the number of people going up in space was increasing each year to a high of 63 in 1985. However, in the year of the Challenger explosion, there were only 7 people who flew in space followed by only 6 in 1987. Then, in 1988, the number of people in space per year started to increase again.

2.135   a.   For each transect, three variables were measured. The number of seabirds found is quantitative. The length of the transect is also quantitative. Whether or not the transect was in an oiled area is qualitative.

      b.   The experimental unit is the transect.

      c.   A pie chart of the oiled and unoiled areas is:

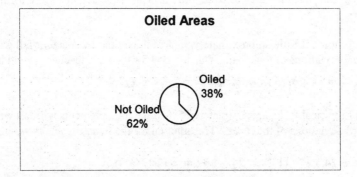

d.   Using MINITAB, a scattergram of the data is:

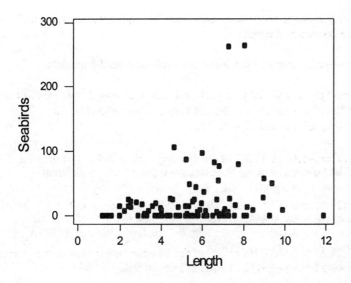

e.   From the histograms, the distributions of seabird densities are fairly similar for oiled and unoiled areas. The mean density for the unoiled area is 3.27, while the mean for the oiled area is 3.495. The median for the unoiled area is .89 and is .70 for the oiled area. These are both fairly similar.

f.   Using Chebyshev's Theorem, at least 75% of the observations will fall within 2 standard deviations of the mean. This interval for unoiled areas would be:

$$\bar{x} \pm 2s \Rightarrow 3.27 \pm 2(6.7) \Rightarrow 3.27 \pm 13.4 \Rightarrow (-10.13, 16.67)$$

g.   Using Chebyshev's Theorem, at least 75% of the observations will fall within 2 standard deviations of the mean. This interval for oiled areas would be:

$$\bar{x} \pm 2s \Rightarrow 3.495 \pm 2(5.968) \Rightarrow 3.495 \pm 11.936 \Rightarrow (-8.441, 15.431)$$

h.   From the above two intervals, we know that at least 75% of the observations for the unoiled area will fall between –10.31 and 16.67 and at most 25% of the observations will fall above 15.431 for the oiled areas. Thus, the unoiled areas would be more likely to have a seabird density of 16.

2.137 a.   Using Minitab, a stem-and-leaf display of the data is:

```
Stem-and-leaf of VELOCITY        N = 51
Leaf Unit = 100

  1     18  4
  3     18  79
 12     19  001112444
 18     19  566788
 20     20  12
 21     20  7
 21     21
 23     21  99
 (5)    22  11344
 23     22  5666777777889
 10     23  001222344
  1     23
  1     24
  1     24  9
```

b.   From this stem-and-leaf display, it is fairly obvious that there are two different distributions since there are two groups of data.

c.   Since there appears to be two distributions, we will compute two sets of numerical descriptive measures. We will call the group with the smaller velocities A1775A and the group with the larger velocities A1775B.

For A1775A:

$$\bar{x} = \frac{\sum x}{n} = \frac{408,707}{21} = 19,462.2$$

$$s^2 = \frac{\sum x^2 - \dfrac{\left(\sum x\right)^2}{n}}{n-1} = \frac{7,960,019,531 - \dfrac{408,707^2}{21}}{21-1} = 283,329.3$$

$$s = \sqrt{283,329.3} = 532.29$$

For A1775B:

$$\bar{x} = \frac{\sum x}{n} = \frac{685,154}{30} = 22,838.5$$

$$s^2 = \frac{\sum x^2 - \dfrac{\left(\sum x\right)^2}{n}}{n-1} = \frac{15,656,992,942 - \dfrac{685,154^2}{30}}{30-1} = 314,694.83$$

$$s = \sqrt{314,694.83} = 560.98$$

d.   To determine which of the two clusters this observation probably belongs to, we will compute z-scores for this observation for each of the two clusters.

For A1775A:
$$z = \frac{20,000 - 19,462.2}{532.29} = 1.01$$

Since this z-score is so small, it would not be unlikely that this observation came from this cluster.

For A1775B:

$$z = \frac{20,000 - 22,838.5}{560.98} = -5.06$$

Since this z-score is so large (in magnitude), it would be very unlikely that this observation came from this cluster.

Thus, this observation probably came from the cluster A1775A.

2.139   a.   The variable "Days in Jail Before Suicide" is measured on a numerical scale, so it is quantitative. The variables "Marital Status", "Race", "Murder/Manslaughter Charge", and "Time of Suicide" are not measured on a numerical scale, so they are all qualitative. The variable "Year" is measured on a numerical scale, so it is quantitative.

b.   Using MINITAB, the pie chart for the data is:

Pie Chart of Murder

No (23, 62.2%)

Yes (14, 37.8%)

Suicides are more likely to be committed by inmates charged with lesser crimes than by inmates charged with murder/manslaughter. Of the suicides reported, 62.2% are committed by those convicted of a lesser charge.

c. Using MINITAB, the pie chart for the data is:

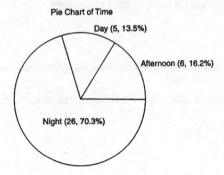

Pie Chart of Time

Day (5, 13.5%)

Afternoon (6, 16.2%)

Night (26, 70.3%)

Suicides are much more likely to be committed at night than any other time. Of the suicides reported, 70.3% were committed at night.

d. Using MINITAB, the descriptive statistics are:

Variable	N	Mean	Median	TrMean	StDev	SE Mean
Jail	37	41.4	15.0	30.3	66.7	11.0

Variable	Minimum	Maximum	Q1	Q3
Jail	1.0	309.0	4.0	41.5

The mean length of time an inmate spent in jail before committing suicide is 41.4 days. The median length of time an inmate spent in jail before committing suicide is 15 days. Since the mean is much larger than the median, the data are skewed to the right. Most of those committing suicide, commit it within 15 days of arriving in jail. However, there are a few inmates who spend many more days in jail before committing suicide.

e. First, compute the z-score associated with 200 days:

$$z = \frac{x - \bar{x}}{s} = \frac{200 - 41.4}{66.7} = 2.38$$

Using Chebyshev's rule, we know that at most $1/2^2 = 1/4$ of the observations will fall more than 2 standard deviations from the mean. Also, we know at most $1/3^2 = 1/9$ of the observations will fall more than 3 standard deviations from the mean. The score 200 is 2.38 standard deviations from the mean. Since we know that at most 1/4 of the observations will fall in this range, it looks like it would not be that unusual to see someone commit suicide after 200 days. However, if we look at the data, of the 37 observations, there are only 2 observations of 200 or larger. This proportion is 2/37 = .054. Using this information, it would be rather unusual for an inmate to commit suicide after 200 days.

f.   Using MINITAB, the stem-and-leaf plot of the data is:

```
Stem-and-leaf of Year      N = 37
Leaf Unit = 1.0
  ⊣
   1    196   7
   5    196   8889
  11    197   000001
  13    197   23
  16    197   455
  18    197   66
  (2)   197   99
  17    198   000111
  11    198   233
   8    198   55555
   3    198   77
   1    198   9
```

From the stem-and-leaf plot, it does not appear that the number of suicides have decreased over time.

Probability

3.1   a.   Since the probabilities must sum to 1,

$$P(E_3) = 1 - P(E_1) - P(E_2) - P(E_4) - P(E_5) = 1 - .1 - .2 - .1 - .1 = .5$$

     b. $P(E_3) = 1 - P(E_3) - P(E_2) - P(E_4) - P(E_5)$
             $\Rightarrow 2P(E_3) = 1 - .1 - .2 - .1 \Rightarrow 2P(E_3) = .6 \Rightarrow P(E_3) = .3$

     c. $P(E_3) = 1 - P(E_1) - P(E_2) - P(E_4) - P(E_5) = 1 - .1 - .1 - .1 - .1 = .6$

3.3 $P(A) = P(1) + P(2) + P(3) = .05 + .20 + .30 = .55$
      $P(B) = P(1) + P(3) + P(5) = .05 + .30 + .15 = .50$
      $P(C) = P(1) + P(2) + P(3) + P(5) = .05 + .20 + .30 + .15 = .70$

3.5   a. $P(A) = P(2) + P(3) + P(5) = 2/9 + 1/9 + 1/9 = 4/9$

     b. $P(B) = P(4) + P(6) = 2/9 + 1/9 = 3/9 = 1/3$

     c.   Since there are no events that occur in both A and B at the same time, $P(A$ and B simultaneously$) = 0$.

3.7   The simple events are:

        1, H 1, T
        2, H 2, T
        3, H 3, T
        4, H 4, T
        5, H 5, T
        6, H 6, T

Since each event is equally likely if the die and coin are fair, each event will have a probability of 1/12.

$P(A) = P(6, H) = 1/12$

$P(B) = P(2, T) + P(4, T) + P(6, T) = 1/12 + 1/12 + 1/12 = 3/12 = 1/4$

$P(C) = P(2, T) + P(4, T) + P(6, T) + P(2, H) + P(4, H) + P(6, H)$
      $= 1/12 + 1/12 + 1/12 + 1/12 + 1/12 + 1/12 = 6/12 = 1/2$

$P(D) = P(1, T) + P(2, T) + P(3, T) + P(4, T) + P(5, T) + P(6, T)$
      $= 1/12 + 1/12 + 1/12 + 1/12 + 1/12 + 1/12 = 6/12 = 1/2$

3.9   Each student will obtain slightly different proportions. However, the proportions should be close to $P(A) = 1/10$, $P(B) = 6/10$ and $P(C) = 3/10$.

3.11   Use the relative frequency as an estimate for the probability. Thus, P(physically assaulted) = 600 / 12,000 = .05.

3.13   a.   Let A = {chicken passes inspection with fecal contamination}. $P(A) = 1 / 100 = .01$.

       b.   Yes. The relative frequency of passing inspection with fecal contamination is $306 / 32,075 = .0095 \approx .01$.

3.15   a.   The sample points would correspond to the possible outcomes: 1, 2, 3, 4, 5, 6, 7, 8, and 9.

       b.   From the sample, the number of times each of these sample points occurred was different. We would expect the probability of each sample point to the close to the relative frequency of that outcome.

       c.   Reasonable probabilities would be the relative frequencies of the sample points:

First Digit	Frequency	Relative Frequency
1	109	109/743 = .147
2	75	75/743 = .101
3	77	77/743 = .104
4	99	99/743 = .133
5	72	72/743 = .097
6	117	117/743 = .157
7	89	89/743 = .120
8	62	62/743 = .083
9	43	43/743 = .058
Total	743	1.000

       d. P(first digit is 1 or 2) = P(first digit is 1) + P(first digit is 2) = $.147 + .101 = .248$

       e. P(first digit > 5) = P(first digit is 6) + P(first digit is 7) + P(first digit is 8) + P(first digit is 9) = $.157 + .120 + .083 + .058 = .418$

3.17   a.   The sample points would be the possible answers to the question. Thus, the sample points would be: Total population, Agricultural change, Presence of industry, Growth, and Population concentration.

       b.   Reasonable probabilities would correspond to the proportions given in the pie chart. These would be:

            P(Total population) = .18, P(Agricultural change) = .05, P(Presence of industry) = .27, P(growth) = .05, and P(Population concentration) = .45.

       c.   Sample points related to population are Total population and Population concentration. P(population-related) = P(Total population) + P(Population concentration) = $.18 + .45$ = .63.

3.19   a. P(Most salient role is spouse) $= 424 / 1{,}102 = .385$

        b. P(Most salient role is parent or grandparent)
              $= 269 / 1{,}102 + 148 / 1{,}102 = .244 + .134 = .378$

        c. P(Most salient role does not involve a spouse or relative)
              $= 73 / 1{,}102 + 59 / 1{,}102 + 34 / 1{,}102 + 36 / 1{,}102 = .066 + .054 + .031 + .033 = .184$

3.21   If each of 3 players uses a fair coin, the sample space is:

        HHH   *HTT*
        HHT   *THT*
        HTH   *TTH*
        THH   *TTT*

        Since each event is equally likely, each event will have probability 1/8.

        The probability of odd man out on the first roll is:

        $P(HHT) + P(HTH) + P(THH) + P(HTT) + P(THT) + P(TTH)$
        $= 1/8 + 1/8 + 1/8 + 1/8 + 1/8 + 1/8 = 6/8 = 3/4$

        If one player uses a two-headed coin, the sample space will be (assume 3rd player has the two-headed coin):

        HHH   *THH*
        HTH   *TTH*

        Since each event is equally likely, each will have probability 1/4.

        The probability the 3rd player is odd man out is $P(TTH) = 1/4$.

        From the first part, the probability the 3rd player is odd man out is $P(HHT) + P(TTH) = 1/8 + 1/8 = 1/4$. Thus, the two probabilities are the same.

3.23   a.   The odds in favor of Snow Chief are $\dfrac{1}{3}$ to $\left(1 - \dfrac{1}{3}\right)$ or $\dfrac{1}{3}$ to $\dfrac{2}{3}$ or 1 to 2

        b.   If the odds are 1 to 1, P(Snow Chief will win) $= \dfrac{1}{1+1} = \dfrac{1}{2} = .5$

        c.   If the odds against Snow Chief winning are 3 to 2, the odds for Snow Chief winning are 2 to 3. The probability Snow Chief will win is $\dfrac{2}{2+3} = \dfrac{2}{5} = .4$

3.25   a. A: {*HHH, HHT, HTH, THH, TTH, THT, HTT*}
              B: {*HHH, TTH, THT, HTT*}
              $A \cup B$: {*HHH, HHT, HTH, THH, TTH, THT, HTT*}
              A^c: {*TTT*}
              $A \cap B$: {*HHH, TTH, THT, HTT*}

b. If the coin is fair, then each of the 8 possible outcomes are equally likely, with probability 1/8.

$$P(A) = \frac{7}{8} \qquad P(B) = \frac{4}{8} = \frac{1}{2} \qquad P(A \cup B) = \frac{7}{8}$$

$$P(A^c) = \frac{1}{8} \qquad P(A \cap B) = \frac{4}{8} = \frac{1}{2}$$

c. $P(A \cup B) = P(A) + P(B) - P(A \cap B) = \frac{7}{8} + \frac{1}{2} - \frac{1}{2} = \frac{7}{8}$

d. No. $P(A \cap B) = \frac{1}{2}$ which is not 0.

3.27   a. $P(A) = P(E_1) + P(E_2) + P(E_3) + P(E_5) + P(E_6) = \frac{1}{5} + \frac{1}{5} + \frac{1}{5} + \frac{1}{20} + \frac{1}{10} = \frac{15}{20} = \frac{3}{4}$

b. $P(B) = P(E_2) + P(E_3) + P(E_4) + P(E_7) = \frac{1}{5} + \frac{1}{5} + \frac{1}{20} + \frac{1}{5} = \frac{13}{20}$

c. $P(A \cup B) = P(E_1) + P(E_2) + P(E_3) + P(E_4) + P(E_5) + P(E_6) + P(E_7)$

$$= \frac{1}{5} + \frac{1}{5} + \frac{1}{5} + \frac{1}{20} + \frac{1}{20} + \frac{1}{10} + \frac{1}{5} = 1$$

d. $P(A \cap B) = P(E_2) + P(E_3) = \frac{1}{5} + \frac{1}{5} = \frac{2}{5}$

e. $P(A^c) = 1 - P(A) = 1 - \frac{3}{4} = \frac{1}{4}$

f. $P(B^c) = 1 - P(B) = 1 - \frac{13}{20} = \frac{7}{20}$

g. $P(A \cup A^c) = P(E_1) + P(E_2) + P(E_3) + P(E_4) + P(E_5) + P(E_6) + P(E_7)$

$$= \frac{1}{5} + \frac{1}{5} + \frac{1}{5} + \frac{1}{20} + \frac{1}{20} + \frac{1}{10} + \frac{1}{5} = 1$$

h. $P(A^c \cap B) = P(E_4) + P(E_7) = \frac{1}{20} + \frac{1}{5} = \frac{5}{20} = \frac{1}{4}$

3.29 a. $P(A) = .50 + .10 = .05 = .65$

b. $P(B) = .10 + .07 + .50 + .05 = .72$

c. $P(C) = .25$

d. $P(A^c) = .25 + .07 + .03 = .35$ (Note: $P(A^c) = 1 - P(A) = 1 - .65 = .35$)

e. $P(A \cup B) = P(B) = .10 + .07 + .50 + .05 = .72$

f. $P(A \cap B) = P(A) = .50 + .10 + .05 = .65$

g. Two events are mutually exclusive if they have no sample points in common or if the probability of their intersection is 0.

$P(A \cap B) = .50 + .10 + .05 = .65$. Since this is not 0, A and B are not mutually exclusive.

$P(A \cap C) = 0$. Since this is 0, A and C are mutually exclusive.

$P(B \cap C) = 0$. Since this is 0, B and C are mutually exclusive.

3.31 Suppose we define the following event:

$A = $ {eighth-grader scores above 653 on mathematics assessment test}

Then the probability that a randomly selected eighth-grader has a score of 653 or below on the mathematics assessment test is:

$P(A^c) = 1 - P(A) = 1 - .05 = .95$.

3.33 a. The event $A \cap B$ is the event the outcome is black and odd. The event is $A \cap B$: {11, 13, 15, 17, 29, 31, 33, 35}

b. The event $A \cup B$ is the event the outcome is black or odd or both. The event $A \cup B$ is {2, 4, 6, 8, 10, 11, 13, 15, 17, 20, 22, 24, 26, 28, 29, 31, 33, 35, 1, 3, 5, 7, 9, 19, 21, 23, 25, 27}

c. Assuming all events are equally likely, each has a probability of 1/38.

$$P(A) = 18\left(\frac{1}{38}\right) = \frac{18}{38} = \frac{9}{19}$$

$$P(B) = 18\left(\frac{1}{38}\right) = \frac{18}{38} = \frac{9}{19}$$

$$P(A \cap B) = 8\left(\frac{1}{38}\right) = \frac{8}{38} = \frac{4}{19}$$

$$P(A \cup B) = 28\left(\frac{1}{38}\right) = \frac{28}{38} = \frac{14}{19}$$

$$P(C) = 18\left(\frac{1}{38}\right) = \frac{18}{38} = \frac{9}{19}$$

d. The event $A \cap B \cap C$ is the event the outcome is odd and black and low. The event $A \cap B \cap C$ is {11, 13, 15, 17}.

e. $P(A \cup B) = P(A) + P(B) - P(A \cap B) = \dfrac{9}{19} + \dfrac{9}{19} - \dfrac{4}{19} = \dfrac{14}{19}$

f. $P(A \cap B \cap C) = 4\left(\dfrac{1}{38}\right) = \dfrac{4}{38} = \dfrac{2}{19}$

g. The event $A \cup B \cup C$ is the event the outcome is odd or black or low. The event $A \cup B \cup C$ is:

    {1, 2, 3, ... , 29, 31, 33, 35}

    or

    {All simple events except 00, 0, 30, 32, 34, 36}

h. $P(A \cup B \cup C) = 32\left(\dfrac{1}{38}\right) = \dfrac{32}{38} = \dfrac{16}{19}$

3.35 a. The event $A \cap G$ is the event the elderly hearing impaired subject has less than 1 year of experience and uses the hearing aids from 8–16 hours per day.

    $P(A \cap G) = .15$

    These two events are not mutually exclusive because the probability of their intersection is not 0.

    b. The event $C \cup E$ is the event the elderly hearing impaired subject has more than 10 years of experience or uses the hearing aids from 1–4 hours per day.

    $P(C \cup E) = P(C) + P(E) - P(C \cap E) = .21 + .18 - .03 = .36$

    These two events are not mutually exclusive because the probability of their intersection is not 0.

c.   The event $C \cap D$ is the event the elderly hearing impaired subject has more than 10 years of experience and uses the hearing aids less than 1 hour per day.

$P(C \cap D) = .00$

These two events are mutually exclusive because the probability of their intersection is 0.

d.   The event $A \cup G$ is the event the elderly hearing impaired subject has less than 1 year of experience or uses the hearing aids from 8–16 hours per day.

$P(A \cup G) = P(A) + P(G) - P(A \cap G) = .42 + .55 - .15 = .82$

These two events are not mutually exclusive because the probability of their intersection is not 0.

e.   The event $A \cup D$ is the event the elderly hearing impaired subject has less than 1 year of experience or uses the hearing aids less than 1 hour per day.

$P(A \cup D) = P(A) + P(D) - P(A \cap D) = .42 + .08 - .04 = .46$

These two events are not mutually exclusive because the probability of their intersection is not 0.

3.37   First, we will define the following events:

A: {Student is Anglo-American}
B: {Student has had little or no contact with mentally ill people}
C: {Student has had close contact with mentally ill people}
D: {Student is an Asian Immigrant}

a. $P(A \cap B) = 30/139 = .216$
b. $P(C) = (16 + 0 + 1 + 4)/139 = .151$
c. $P(A^c) = 1 - P(A) = 1 - (30 + 17 + 16)/139 = 1 - .453 = .547$
d. $P(C \cup D) = (16 + 0 + 1 + 4 + 17 + 5 + 20 + 3)/139 = 66/139 = .475$

3.39   a.   First, we will set up the sample space. The possible ways to select 2 firms from the 4 are:

$(G_1, G_2) \ (G_1, P_1) \ (G_1, P_2) \ (G_2, P_1) \ (G_2, P_2) \ (P_1, P_2)$

We will assume that all first have an equal chance of being selected. Thus, the probability of any sample point being chosen is 1/6.

$P(P_1 \text{ is selected}) = P[(G_1, P_1)] + P[(G_2, P_1)] + P[(P_1, P_2)]$
$= 1/6 + 1/6 + 1/6 = 3/6 = 1/2.$

b. P(at least one firm is in good financial condition)

$= P[(G_1, G_2)] + P[(G_1, P_1)] + P[(G_1, P_2)] + P[(G_2, P_1)] + P[(G_2, P_2)]$
$= 1/6 + 1/6 + 1/6 + 1/6 + 1/6 = 5/6$

3.41 $P(A \mid B) = \dfrac{P(A \cap B)}{P(B)} = \dfrac{.15}{.60} = .25$ $\qquad$ $P(B \mid A) = \dfrac{P(A \cap B)}{P(A)} = \dfrac{.15}{.30} = .5$

3.43 a.  Each simple event has a probability of $\dfrac{1}{4}$.

b. $P(A) = P(HH) + P(HT) + P(TH) = 3\left(\dfrac{1}{4}\right) = \dfrac{3}{4}$

$\qquad P(B) = P(HT) + P(TH) = 2\left(\dfrac{1}{4}\right) = \dfrac{1}{2}$

$\qquad P(A \cap B) = P(B) = \dfrac{1}{2}$

c. $P(A \mid B) = \dfrac{P(A \cap B)}{P(B)} = \dfrac{1/2}{1/2} = 1$ $\qquad$ $P(B \mid A) = \dfrac{P(A \cap B)}{P(A)} = \dfrac{1/2}{3/4} = \dfrac{2}{3}$

3.45 Let A = {Executive cheated at golf} and B = {Executive lied in business}. From the problem, $P(A) = .55$ and $P(A \cap B) = .20$.

$\qquad P(B \mid A) = P(A \cap B) / P(A) = .20 / .55 = .364$

3.47 Define the following events:

A: {Winner is from National League}

B: {Winner is from American League}

C: {Winner is from Eastern Division}

D: {Winner is from Central Division}

E: {Winner is from Western Division}

a. $P(C \mid B) = \dfrac{P(C \cap B)}{P(B)} = \dfrac{6/10}{7/10} = \dfrac{6}{7} = .857$

b. $P(A \mid D) = \dfrac{P(A \cap D)}{P(D)} = \dfrac{1/10}{2/10} = \dfrac{1}{2} = .5$

c. $P(C^c \mid A) = \dfrac{P(C^c \cap A)}{P(A)} = \dfrac{1/10}{3/10} = \dfrac{1}{3} = .333$

3.49   Define the following event:

A: {Highest degree is Bachelors}

a.   To find the probability that the highest degree obtained by the first CEO out of the 5 chosen is a bachelor's degree is the same as finding the probability that the highest degree obtained is a bachelor's degree when choosing only 1 CEO. Thus,

$P(A) = 7 / 25 = .28$

b.   The highest degree obtained by each of the first 4 CEOs selected is a bachelor's degree. Thus, for the fifth selection, it is the same as selecting 1 CEO from the remaining 21, where 4 of those having bachelor's degrees have been removed. Thus,

$P(A \mid$ first 4 have bachelors$) = 3 / 21 = .125$.

3.51   Define the following events:

A: {Player is white}

B: {Player is black}

C: {Player is a guard}

D: {Player is a forward}

E: {Player is a center}

a. $P(E \mid A) = \dfrac{P(E \cap A)}{P(A)} = \dfrac{28/368}{84/368} = \dfrac{28}{84} = .333$

b. $P(E \mid B) = \dfrac{P(E \cap B)}{P(B)} = \dfrac{34/368}{284/368} = \dfrac{34}{284} = .120$

c.   Events A and E are independent if $P(E \mid A) = P(E)$. From part a, $P(E \mid A) = .333$.

$P(E) = 62 / 368 = .168$

Since $P(E \mid A) \neq P(E)$, the events are not independent.

d.   Since the events {White player} and {Center} are not independent, this supports the theory of "stacking". Also, from part a, we found that P(Center $\mid$ White) = .333 and from part b, P(Center $\mid$ Black) = .120. Thus, the probability of playing Center depends on race. Again, this supports the theory of "stacking".

3.53   Define the following events:

        A: {Crime involved white attacker and white victim}
        B: {Crime involved serious injuries}
        C: {Crime involved fatalities}
        D: {Crime involved no injury}
        E: {Crime involved non-white attacker and white victim}

a. $P(A) = \dfrac{6{,}521}{11{,}717} = .557$

b. $P(B) = \dfrac{881}{11{,}717} = .075$

c. $P(C \mid A) = \dfrac{P(A \cap C)}{P(A)} = \dfrac{183/11{,}717}{6{,}521/11{,}717} = \dfrac{183}{6{,}521} = .028$

d. $P(E \mid D) = \dfrac{P(E \cap D)}{P(D)} = \dfrac{1{,}801/11{,}717}{3{,}573/11{,}717} = \dfrac{1{,}801}{3{,}573} = .504$

3.55   a.   Define the following events:

        W: {Player wins the game Go}
        F: {Player plays first (black stones)}

    $P(W \cap F) = 319/577 = .553$

b. $P(W \cap F \mid CA) = 34/34 = 1$
       $P(W \cap F \mid CB) = 69/79 = .873$
       $P(W \cap F \mid CC) = 66/118 = .559$
       $P(W \cap F \mid BA) = 40/54 = .741$
       $P(W \cap F \mid BB) = 52/95 = .547$
       $P(W \cap F \mid BC) = 27/79 = .342$
       $P(W \cap F \mid AA) = 15/28 = .536$
       $P(W \cap F \mid AB) = 11/51 = .216$
       $P(W \cap F \mid AC) = 3/39 = .077$

c.   There are three combinations where the player with the black stones (first) is ranked higher than the player with the white stones: CA, CB, and BA.

    $P(W \cap F \mid CA \cup CB \cup BA) = (34 + 69 + 40)/(34 + 79 + 54) = 143/167 = .856$

d.   There are three combinations where the players are of the same level: CC, BB, and AA.

    $P(W \cap F \mid CC \cup BB \cup AA) = (66 + 52 + 15)/(118 + 95 + 28) = 133/241 = .552$

3.57 $S = \{HHH, HHT, HTH, THH, TTH, THT, HTT, TTT\}$.

These simple events would be equally probable, implying each would occur with a probability of 1/8.

a. $P(A) = \dfrac{7}{8}$ $P(B) = \dfrac{3}{8}$ $P(C) = \dfrac{3}{8}$ $P(D) = \dfrac{4}{8} = \dfrac{1}{2}$

$P(A \cap B) = P(B) = \dfrac{3}{8}$ $P(A \cap D) = P(\text{One H}) = \dfrac{3}{8}$

$P(B \cap C) = 0$ $P(B \cap D) = 0$

b. $P(B \mid A) = \dfrac{P(B \cap A)}{P(A)} = \dfrac{3/8}{7/8} = \dfrac{3}{7}$

$P(A \mid D) = \dfrac{P(A \cap D)}{P(D)} = \dfrac{3/8}{4/8} = \dfrac{3}{4}$

$P(C \mid B) = \dfrac{P(C \cap B)}{P(B)} = \dfrac{0}{3/8} = 0$

c. $P(A) \cdot P(B) = (7/8)(3/8) = 21/64 \neq 3/8 = P(A \cap B)$; therefore, A and B are not independent.

$P(A) \cdot P(C) = (7/8)(3/8) = 21/64) \neq 3/8 = P(A \cap C)$; therefore, A and C are not independent.

$P(A) \cdot P(D) = (7/8)(1/2) = 7/16 \neq 3/8 = P(A \cap D)$; therefore, A and D are not independent.

$P(B) \cdot P(C) = (3/8)(3/8) = 9/64 \neq 0 = P(B \cap C)$; therefore, B and C are not independent.

$P(B) \cdot P(D) = (3/8)(1/2) = 3/16 \neq 0 = P(B \cap D)$; therefore, B and D are not independent.

$P(C) \cdot P(D) = (3/8)(1/2) = 3/16 \neq 3/8 = P(C \cap D)$; therefore, C and D are not independent.

3.59   The 36 possible outcomes obtained when tossing two dice are listed below:

(1, 1) (1, 2) (1, 3) (1, 4) (1, 5) (1, 6)
(2, 1) (2, 2) (2, 3) (2, 4) (2, 5) (2, 6)
(3, 1) (3, 2) (3, 3) (3, 4) (3, 5) (3, 6)
(4, 1) (4, 2) (4, 3) (4, 4) (4, 5) (4, 6)
(5, 1) (5, 2) (5, 3) (5, 4) (5, 5) (5, 6)
(6, 1) (6, 2) (6, 3) (6, 4) (6, 5) (6, 6)

A: {(1, 2), (1, 4), (1, 6), (2, 1), (2, 3), (2, 5), (3, 2), (3, 4), (3, 6), (4, 1), (4, 3), (4, 5), (5, 2), (5, 4), (5, 6), (6, 1), (6, 3), (6, 5)}

B: {(3, 6), (4, 5), (5, 4), (5, 6), (6, 3), (6, 5), (6, 6)}

$A \cap B$: {(3, 6), (4, 5), (5, 4), (5, 6), (6, 3), (6, 5)}

If A and B are independent, then $P(A)P(B) = P(A \cap B)$.

$$P(A) = \frac{18}{36} = \frac{1}{2} \quad P(B) = \frac{7}{36} \quad P(A \cap B) = \frac{6}{36} = \frac{1}{6}$$

$$P(A)P(B) = \frac{1}{2} \cdot \frac{7}{36} = \frac{7}{72} \neq \frac{1}{6} = P(A \cap B). \text{ Thus, } A \text{ and } B \text{ are not independent.}$$

3.61   a.   Dependent events are not always mutually exclusive. From Exercise 3.48, events A and B are dependent, but they are not mutually exclusive because $P(A \cap B) = 1/6 \neq 0$.

       b.   Mutually exclusive events are always dependent. Let A and B be mutually exclusive events where $P(A) \neq 0$ and $P(B) \neq 0$. Then $P(A \cap B) = 0$ (because they are mutually exclusive), but $P(A)P(B) \neq 0$. Therefore, A and B are dependent.

       c.   Independent events are never mutually exclusive. Let A and B be independent events where $P(A) \neq 0$ and $P(B) \neq 0$. Then $P(A)P(B) = P(A \cap B)$ (because A and B are independent), but $P(A \cap B) \neq 0$. Therefore, A and B are not mutually exclusive.

3.63   Define the following events:

A: {School has inadequate plumbing}

B: {Schools plans to make repairs}

From the problem, we know that $P(A) = .25$ and $P(B \mid A) = .38$

Thus, $P(A \cap B) = P(B \mid A) \, P(A) = .25 \, (.38) = .095$

3.65   a.   Define the following event:

         A: {Psychic picks box with crystal}
         If psychic is just guessing, $P(A) = 1/10 = .1$

       b. P(Psychic guesses correct at least once in 7 trials)
           $= 1 - P$(Psychic does not guess correct in 7 trials)

           $= 1 - P(A^c \cap A^c \cap A^c \cap A^c \cap A^c \cap A^c \cap A^c) = 1 - .9(.9)(.9)(.9)(.9)(.9)(.9)$
           $= 1 - .478 = .522$ (This assumes that the trials are independent)

       c.   If the psychic is just guessing, then P(Psychic does not guess correct in 7 trials) = .478.

           Thus, if the person was not a psychic (and was merely guessing) we would expect the person to guess wrong in all seven trials about half the time. This would not be a rare event. Thus, this outcome would support the notion that the person is guessing.

3.67   Let us define the following events:

         S: {School is a subscriber}
         N: {School never uses the CCN broadcast}
         F: {School uses the CCN broadcast more than 5 times per week}

From the problem, $P(S) = .40$, $P(N|S) = .05$, and $P(F|S) = .20$.

a. $P(S \cap N) = P(N|S)\ P(S) = .05(.40) = .02$

b. $P(S \cap F) = P(F|S)\ P(S) = .20(.40) = .08$

3.69   First, define the following events:

A: {Driver was African American}
B: {Driver exceeded the speed limit by at least 5 mph}
C: {Driver was stopped for speeding}

a. $P(A) = .14$
     $P(B) = .98$
     $P(A|B) = .15$
     $P(A|C) = .35$

b. $P(A \cap B) = P(A|B)\ P(B) = .15(.98) = .147$

c. $P(A^c|B) = 1 - P(A|B) = 1 - .15 = .85$

d. $P(A^c|C) = 1 - P(A|C) = 1 - .35 = .65$

e.   If events A and B are independent, then $P(A|B) = P(A)$. $P(A|B) = .15$ and $P(A) = .14$.
     For this example, $P(A|B) \neq P(A)$, but they are very close to each other. To a close
     approximation, A and B are independent.

f.   If African Americans are not stopped more often for speeding than expected, then
     events A and C should be independent. We already know from part e that events A and
     B are essentially independent - i.e. the proportion of the speeders who are African
     American is equal to the proportion of African Americans.

     Are events A and C independent? If they are, then $P(A|C) = P(A)$. From the above,
     $P(A|C) = .35$ and $P(A) = .14$. Thus, events A and C are not independent. This means
     that the proportion of those stopped for speeding who are African American is not equal
     to the proportion of African Americans in the population. Since the $P(A|C) > P(A)$,
     this implies that on this stretch of highway, African Americans are stopped for speeding
     at a higher rate than would be expected.

3.71   Define the following events:

A: {Seed carries single spikelets}
B: {Seed carries paired spikelets}
C: {Seed produces ears with single spikelets}
D: {Seed produces ears with paired spikelets}

From the problem, $P(A) = .4$, $P(B) = .6$, $P(C|A) = .29$, $P(D|A) = .71$, $P(C|B) = .26$, and
$P(D|B) = .74$.

a. $P(A \cap C) = P(C|A)P(A) = .29(.4) = .116$

b. $P(D) = P(A \cap D) + P(B \cap D) = P(D \mid A)P(A) + P(D \mid B)P(B) = .71(.4) + .74(.6)$
$= .284 + .444 = .728$

3.73 Define the following event:

A: {Antigens match}

From the problem, $P(A) = .25$ and $P(A') = 1 - P(A) = 1 - .25 = .75$

a. The probability that one sibling has a match is $P(A) = .25$

b. The probability that all three will match is:

$P(AAA) = P(A \cap A \cap A) = P(A)P(A)P(A) = .25^3 = .0156$

c. The probability that none of the three match is:

$P(A^c A^c A^c) = P(A^c \cap A^c \cap A^c) = P(A^c)P(A^c)P(A^c) = .75^3 = .4219$

d. For this part, $P(A) = .001$ and $P(A^c) = 1 - P(A) = 1 - .001 = .999$

$P(A) = .001$

$P(AAA) = P(A \cap A \cap A) = P(A)P(A)P(A) = .001^3 = .000000001$

$P(A^c A^c A^c) = P(A^c \cap A^c \cap A^c) = P(A^c)P(A^c)P(A^c) = .999^3 = .9970$

3.75 a. If the coin is balanced, then $P(H) = P(T) = .5$. To find the probability of any sequence, you find the probability of the intersection of the simple events. Thus,

$P(H\,H\,H\,H\,H\,H\,H\,H\,H\,H) = P(H \cap H \cap H \cap H \cap H \cap H \cap H \cap H \cap H)$

$= P(H)P(H)P(H)P(H)P(H)P(H)P(H)P(H)P(H)P(H) = .5(.5)(.5)(.5)(.5)(.5)(.5)(.5)(.5)$

$= .00098$

$P(H\,H\,T\,T\,H\,T\,T\,H\,H\,H) = P(H \cap H \cap T \cap T \cap H \cap T \cap T \cap H \cap H \cap H)$

$= P(H)P(H)P(T)P(T)P(H)P(T)P(T)P(H)P(H)P(H) = .5(.5)(.5)(.5)(.5)(.5)(.5)(.5)(.5)$

$= .00098$

$P(T\,T\,T\,T\,T\,T\,T\,T\,T\,T) = P(T \cap T \cap T \cap T \cap T \cap T \cap T \cap T \cap T \cap T)$

$= P(T)P(T)P(T)P(T)P(T)P(T)P(T)P(T)P(T)P(T) = .5(.5)(.5)(.5)(.5)(.5)(.5)(.5)(.5)$

$= .00098$

b. P(All heads or all tails) = P(All heads $\cup$ All tails)

$= P$(All heads) $+ P$(All tails) $- P$(All heads $\cap$ All tails) $= .00098 + .00098 - 0$

$= .00196$

c. P(10 coin tosses result in a mix of heads and tails) $= 1 - P$(All heads or all tails)

$= 1 - .00196 = .99804$

d. From part b, we know that the probability of getting all heads or all tails is .00196.

From part c, we know that the probability of getting a mix of heads and tails is .99804.

If we know that one of these sequences actually occurred, then we would we conclude that it was probably the one with the mix of heads and tails because the probability of a mix is almost 1.

3.77 a. $\dbinom{600}{3} = \dfrac{600!}{(3!)(597!)} = \dfrac{(600)(599)(598)}{(1)(2)(3)} = 35,820,200$

b. All sets are equally likely in random sampling, so that the probability of any particular sample is:

$$\dfrac{1}{35,820,200}$$

c. Repeated samples would be highly unlikely.

3.79 a. Decide on a starting point on the random number table. Then take the first n numbers reading down, and this would be the sample. Group the digits on the random number table into groups of 7 (for part **b**) or groups of 4 (for part **c**). Eliminate any duplicates and numbers that begin with zero since they are not valid telephone numbers.

b. Starting in Row 6, column 5, take the first 10 seven-digit numbers reading down. The telephone numbers are:

277-5653
988-7231
188-7620
174-5318
530-6059
709-9779
496-2669
889-7433
482-3752
772-3313

c. Starting in Row 10, column 7, take the first 5 four-digit numbers reading down. The 5 telephone numbers are:

> 373-3886
> 373-5686
> 373-1866
> 373-3632
> 373-6768

3.81 a. The probability that account 3,241 is chosen is 1 out of 5,382 or 1/5,382 = .000186

b. Using MINITAB, click on **calc** in the menu, then **Random data**, and **Uniform**. In the screen that appears, enter **10** in the **Generate rows of data** blank, and enter **C1** in the **Store in column** blank. Enter **1** in the **Lower endpoint** blank and **5382** in the **Upper endpoint** blank. Then click **OK**. The ten randomly generated numbers between 1 and 5,382 will appear in column 1. One example of the data generated is:

> 904
> 4780
> 5095
> 4693
> 2077
> 508
> 3002
> 2928
> 3584
> 832

c. No. If we are looking at the exact numbers chosen, both samples are equally likely. The total number of ways one can draw a sample of size 10 from a population of 5382 is a combination of 5382 items taken 10 at a time or:

$$\binom{5280}{10} = 1.036 \times 10^{27}$$

The probability of getting either of these two exact samples is $\dfrac{1}{1.036} \times 10^{27}$.

3.83 Since the number of possible outcomes from tossing a coin is always two on each flip, the multiplicative rule would yield a product of a series of 2's.

a. $(2)(2) = 2^2 = 4$

b. $(2)(2)(2) = 2^3 = 8$

c. $(2)(2)(2)(2)(2) = 2^5 = 32$

d. 2^n

3.85 a. $\dbinom{7}{3} = \dfrac{7!}{(3!)(4!)} = \dfrac{7 \cdot 6 \cdot 5}{3 \cdot 2 \cdot 1} = 35$

b. $\dbinom{6}{2} = \dfrac{6!}{(2!)(4!)} = \dfrac{6 \cdot 5}{2 \cdot 1} = 15$

c. $\dbinom{30}{2} = \dfrac{30!}{(2!)(28!)} = \dfrac{30 \cdot 29}{2 \cdot 1} = 435$

d. $\dbinom{10}{8} = \dfrac{10!}{(8!)(2!)} = \dfrac{10 \cdot 9}{2 \cdot 1} = 45$

e. $\dbinom{q}{r} = \dfrac{q!}{r!(q-r!)}$

3.87 To solve each of these parts, we think of the teams being formed in sequence. First we determine how many ways we can select employees for the first team. Then, from the remaining employees, how many ways can we select from them for the second team. Then the third team would consist of the remaining, unselected employees. The multiplicative rule would then be used to determine the total number of ways these sets of three teams can be selected.

a. $\dbinom{10}{3}\dbinom{7}{3}\dbinom{4}{4} = \left(\dfrac{10!}{3!7!}\right)\left(\dfrac{7!}{3!4!}\right)\left(\dfrac{4!}{4!0!}\right) = \left(\dfrac{10 \cdot 9 \cdot 8}{3 \cdot 2 \cdot 1}\right)\left(\dfrac{7 \cdot 6 \cdot 5}{3 \cdot 2 \cdot 1}\right)1 = 4200$

b. $\dbinom{10}{2}\dbinom{8}{3}\dbinom{5}{5} = \left(\dfrac{10!}{2!8!}\right)\left(\dfrac{8!}{3!5!}\right)\left(\dfrac{5!}{5!0!}\right) = \left(\dfrac{10 \cdot 9}{2 \cdot 1}\right)\left(\dfrac{8 \cdot 7 \cdot 6}{3 \cdot 2 \cdot 1}\right)1 = 2520$

c. $\dbinom{10}{1}\dbinom{9}{4}\dbinom{5}{5} = \left(\dfrac{10!}{1!9!}\right)\left(\dfrac{9!}{4!5!}\right)\left(\dfrac{5!}{5!0!}\right) = \dfrac{10}{1}\left(\dfrac{9 \cdot 8 \cdot 7 \cdot 6}{4 \cdot 3 \cdot 2 \cdot 1}\right)1 = 1260$

d. $\dbinom{10}{2}\dbinom{8}{4}\dbinom{4}{4} = \left(\dfrac{10!}{2!8!}\right)\left(\dfrac{8!}{4!4!}\right)\left(\dfrac{4!}{4!0!}\right) = \left(\dfrac{10 \cdot 9}{2 \cdot 1}\right)\left(\dfrac{8 \cdot 7 \cdot 6 \cdot 5}{4 \cdot 3 \cdot 2 \cdot 1}\right)1 = 3150$

3.89 a. Since order is not important, the number of choices is:

$$\dbinom{8}{5} = \dfrac{8!}{5!3!} = 56$$

b. Order is important because it is necessary to distinguish between the positions (guard, forward, and center). However, the two players chosen as guards and the two forwards are indistinguishable once they are chosen. That is, choosing player #1 as the first guard and player #3 as the second guard results in the same team as choosing player #3 for the first guard and player #1 for the second guard. Denote the five positions to be filled as G_1, G_2, F_1, F_2, and C. The number of distinct teams, assuming each position was distinct is:

$$P_5^8 = \dfrac{8!}{3!} = 8(7)(6)(5)(4) = 6720$$

However, since G_1 and G_2 are indistinguishable, this number must be divided by 2, or $6720/2 = 3360$ possible teams. Similarly, F_1 and F_2 are indistinguishable, so that there are in fact only $3360/2 = 1680$ possible teams.

c. In this situation, all five positions are distinct. Thus, there are $P_5^8 = 6720$ possible teams.

3.91 a. To visit all 4 cities and then return home, the order of the visits is important. The number of different orders is a permutation of 4 things taken 4 at a time or

$$P_4^4 = \frac{4!}{(4-4)!} = \frac{4 \cdot 3 \cdot 2 \cdot 1}{1} = 24$$

b. If the salesperson visits B first, then he only has 2 choices for the next city (D or E), since B and C are not connected. Once in D or E, he has 2 choices for the next city (if in D, can choose C or E; if in E, can choose D or C). Once in the third city, there is only one choice for the fourth city. The possible trips with B first are $2 \times 2 \times 1 = 4$: BDEC, BDCE, BEDC, BECD.

Similarly, if city C is the first city, there are 2 choices for the second city (D or E), 2 choices for the third, and one for the fourth. Thus, there are $2 \times 2 \times 1 = 4$ choices: CDEB, CDBE, CEDB, CEBD.

If city D is the first chosen, then to visit all four cities exactly once, city B or C must be chosen next. The third city must be E, and the fourth city the only one left. Thus, there are $2 \times 1 \times 1 = 2$ choices: DBEC, DCEB.

Similarly, if city E is the first chosen, then B or C must be the second city, D must be the third, and the fourth city must be the one remaining. Thus, there are $2 \times 1 \times 1 = 2$ choices: EBDC, ECDB.

The total number of routes are $4 + 4 + 2 + 2 = 12$.

3.93 a. The total number of distributors that make up Elaine's group is $6(5) = 30$.

b. The total number of distributors that make up Elaine's group now is $6(5)(7)(5) = 1,050$.

3.95 a. From the problem, we have three students and three hospitals. Suppose we consider the three hospitals as three positions. Using the permutations rule, there are

$$P = P_3^3 = \frac{3!}{(3-3)!} = \frac{3 \cdot 2 \cdot 1}{1} = 6 \text{ different assignments possible.}$$

These assignments are:

$$[S_1 H_A \quad S_2 H_B \quad S_3 H_C] \quad [S_1 H_A \quad S_3 H_B \quad S_2 H_C]$$
$$[S_2 H_A \quad S_1 H_B \quad S_3 H_C] \quad [S_2 H_A \quad S_3 H_B \quad S_1 H_C]$$
$$[S_3 H_A \quad S_1 H_B \quad S_2 H_C] \quad [S_3 H_A \quad S_2 H_B \quad S_1 H_C]$$

b. Assuming that all of the above assignments are equally likely, the probability of any one of them is 1/6. The probability that student #1 is assigned to hospital B is $2/6 = 1/3$.

3.97 a. The number of ways to select 5 members from 15 is a combination of 15 things taken 5 at a time, or

$$\binom{15}{5} = \frac{15!}{5!(15-5)!} = \frac{15 \cdot 14 \cdot 13 \cdots 1}{5 \cdot 4 \cdot 3 \cdot 2 \cdot 1 \cdot 10 \cdot 9 \cdots 1} = 3003$$

b. To get no democrats, all 5 members chosen must be Republicans. The number of ways to select 5 members from 8 is a combination of 8 things taken 5 at a time or

$$\binom{8}{5} = \frac{8!}{5!(8-5)!} = \frac{8 \cdot 7 \cdot 6 \cdot 5 \cdots 1}{5 \cdot 4 \cdot 3 \cdot 2 \cdot 1 \cdot 3 \cdot 2 \cdot 1} = 56$$

The probability that no Democrat is appointed is the number of ways to select no Democrats divided by the total number of ways to select 5 members, or

$$P(\text{no Democrats}) = \frac{56}{3003} = .0186$$

Since this probability is so small, we have either seen a rare event or the appointments were not made at random.

c. To get a majority of Republicans, 3, 4, or 5 members must be Republicans. We computed the number of ways to get 5 Republicans in part **b**. The number of ways to select 4 Republicans from 8 and 1 Democrat from 7 is:

$$\binom{8}{4}\binom{7}{1} = \frac{8!}{4!(8-4)!} \cdot \frac{7!}{1!(7-1)!}$$

$$= \frac{8 \cdot 7 \cdots 1}{4 \cdot 3 \cdot 2 \cdot 1 \cdot 4 \cdot 3 \cdot 2 \cdot 1} \cdot \frac{7 \cdot 6 \cdots 1}{1 \cdot 6 \cdot 5 \cdots 1} = 490$$

The number of ways to select 3 Republicans from 8 and 2 Democrats from 7 is:

$$\binom{8}{3}\binom{7}{2} = \frac{8!}{3!(8-3)!} \cdot \frac{7!}{2!(7-2)!}$$

$$= \frac{8 \cdot 7 \cdots 1}{3 \cdot 2 \cdot 1 \cdot 5 \cdot 4 \cdot 3 \cdot 2 \cdot 1} \cdot \frac{7 \cdot 6 \cdots 1}{2 \cdot 1 \cdot 5 \cdot 4 \cdot 3 \cdot 2 \cdot 1}$$

The probability that the majority of committee members are Republican is the number of ways to get 3, 4, or 5 Republicans divided by the total number of ways to get 5 members, or

$$P(\text{Majority Republican}) = \frac{56 + 490 + 1176}{3003} = \frac{1722}{3003} = .5734$$

This is not an unusual event because the probability is so high. There would be no reason to doubt the governor made the selections randomly.

3.99   The professor will choose 5 of 10 questions. Of these 10 questions, the student has prepared answers for 7, and has not prepared the other three. The 5 questions on the exam can be chosen in

$$\binom{10}{5} = \frac{10!}{5!5!} = 252 \text{ ways}$$

a.   If the student has prepared all 5 exam questions, the professor has chosen 5 from the 7 prepared questions. This can be done in

$$\binom{7}{5} = 21 \text{ ways}$$

Hence, the desired probability is 21/252.

b.   Define the following events.

   A: {The student is prepared for no questions}
   B: {The student is prepared for 1 question}
   C: {The student is prepared for 2 questions}

A will occur if the professor picks 0 questions from the 7 prepared and 5 from the 3 unprepared. This is impossible, as is B. C will occur if the professor picks 2 questions from the 7 prepared and 3 questions from the 3 unprepared. This can happen in

$$\binom{7}{2}\binom{3}{3} = 21(1) = 21 \text{ ways}$$

Hence, P(prepared for less than 3) = $P(A) + P(B) + P(C) = 0 + 0 + 21/252 = 21/252$

c.   The professor must pick 4 from the 7 prepared questions and 1 from the remaining 3 questions. This can happen in

$$\binom{7}{4}\binom{3}{1} = 35(3) = 105 \text{ ways}$$

Hence, P(prepared for exactly 4) = 105/252

3.101 a.   The number of ways you can choose 6 numbers from 53 is a combination of 53 items taken 6 at a time:

$$\binom{53}{6} = 22,957,480$$

If you purchase a single ticket, your probability of winning would be 1/22,957,480 = .000000043.

b.   The number of ways you can choose 6 numbers from 30 is a combination of 30 items taken 6 at a time:

$$\binom{30}{6} = 593,775$$

If you purchase a single ticket, your probability of winning would be 1/593,775 = .000001684.

c. The number of ways you can choose 6 numbers from 40 is a combination of 40 items taken 6 at a time:

$$\binom{40}{6} = 3{,}838{,}380$$

If you purchase a single ticket, your probability of winning would be $1/3{,}838{,}380 = .00000026$.

d. One would prefer to play Delaware's game, because the chances of winning are the highest among the three.

e. The total number of ways you can select 6 numbers from the 7 is a combination of 7 numbers taken 6 at a time:

$$\binom{7}{6} = 7$$

The 7 sets of 6 numbers are: (2, 7, 18, 23, 30, 32), (2, 7, 18, 23, 30, 39), (2, 7, 18, 23, 32, 39), (2, 7, 18, 30, 32, 39), (2, 7, 23, 30, 32, 39), (2, 18, 23, 30, 32, 39), and (7, 18, 23, 30, 32, 39).

f. If you wheel seven numbers, the odds of winning is $7/3{,}838{,}380 = .000001823$. This is seven times the probability of winning with a single ticket. However, you also have to pay 7 times more to play the seven combinations.

3.103 a.

(1, H)	(2, 1)	(3, H)	(4, 1)	(5, H)	(6, 1)
(1, T)	(2, 2)	(3, T)	(4, 2)	(5, T)	(6, 2)
	(2, 3)		(4, 3)		(6, 3)
	(2, 4)		(4, 4)		(6, 4)
	(2, 5)		(4, 5)		(6, 5)
	(2, 6)		(4, 6)		(6, 6)

b. Each simple event is an intersection of two independent events. Each simple event whose first element is 1, 3, or 5 has probability

$$\left(\frac{1}{6}\right)\left(\frac{1}{2}\right) = \frac{1}{12}$$

while each simple event whose first element is 2, 4, or 6 has probability

$$\left(\frac{1}{6}\right)\left(\frac{1}{6}\right) = \frac{1}{36}$$

c. $P(A) = P\{(1, H), (3, H), (5, H)\} = \dfrac{1}{12} + \dfrac{1}{12} + \dfrac{1}{12} = \dfrac{3}{12} = \dfrac{1}{4}$

$P(B) = P\{(1, H), (1, T), (3, H), (3, T), (5, H), (5, T)\} = \dfrac{6}{12} = \dfrac{1}{2}$

d. A^c: {all except $(1, H)$, $(3, H)$, and $(5, H)$}

     B^c: {$(2, 1), (2, 2), (2, 3), (2, 4), (2, 5), (2, 6), (4, 1), (4, 2), (4, 3), (4, 4), (4, 5), (4, 6),$
           $(6, 1), (6, 2), (6, 3), (6, 4), (6, 5), (6, 6)$}

     $A \cap B$: {$(1, H), (3, H), (5, H)$}

     $A \cup B$: {$(1, H), (1, T), (3, H), (3, T), (5, H), (5, T)$}

e. $P(A^c) = 1 - P(A) = 1 - \dfrac{1}{4} = \dfrac{3}{4}$

     $P(B^c) = 1 - P(B) = 1 - \dfrac{1}{2} = \dfrac{1}{2}$

     $P(A \cap B) = P(A) = \dfrac{1}{4}$

     $P(A \cup B) = P(A) + P(B) - P(A \cap B) = \dfrac{1}{4} + \dfrac{1}{2} - \dfrac{1}{4} = \dfrac{1}{2}$

     $P(A \mid B) = \dfrac{P(A \cap B)}{P(B)} = \dfrac{1/4}{1/2} = \dfrac{1}{2}$

     $P(B \mid A) = \dfrac{P(A \cap B)}{P(A)} = \dfrac{1/4}{1/4} = 1$

f. $P(A \cap B) \neq 0$, so that A and B are not mutually exclusive.

     $P(A \mid B) \neq P(A)$, so that A and B are not independent.

3.105   a.   Two events are mutually exclusive if neither could happen at the same time. It is possible that the St. Louis Cardinals could win the World Series and Mark McGuire could hit 70 home runs at the same time. Therefore, these events would not be mutually exclusive.

      b.   A person is buying one computer. The computer purchased could not be an IBM and an Apple computer at the same time. These events are mutually exclusive.

      c.   It is possible that student A responds within 5 seconds (less than 5 seconds) and has the fastest time of 2.3 seconds at the same time. Therefore, these events are not mutually exclusive.

3.107   a.   Because events A and B are independent, we have:

         $P(A \cap B) = P(A)P(B) = (.3)(.1) = .03$

     Thus, $P(A \cap B) \neq 0$, and the two events cannot be mutually exclusive.

      b. $P(A \mid B) = \dfrac{P(A \cap B)}{P(B)} = \dfrac{.03}{.1} = .3$ $P(B \mid A) = \dfrac{P(A \cap B)}{P(A)} = \dfrac{.03}{.1} = .1$

      c. $P(A \cup B) = P(A) + P(B) - P(A \cap B) = .3 + .1 - .03 = .37$

3.109   From the problem, to find probabilities, we must convert the percents to proportions by dividing by 100. Thus, $P(1 \text{ star}) = 0$, $P(2 \text{ stars}) = .0408$, $P(3 \text{ stars}) = .1735$, $P(4 \text{ stars}) = .6020$, and $P(5 \text{ stars}) = .1837$.

Probability

a. False. The probability of an event cannot be 4. The probability of an event must be between 0 and 1.

b. True. $P(4 \text{ or } 5 \text{ stars}) = P(4 \text{ stars}) + P(5 \text{ stars}) = .6020 + .1837 = .7857$.

c. True. No cars have a rating of 1 star, thus $P(1 \text{ star}) = 0$.

d. False. $P(2 \text{ stars}) = .0408$ and $P(5 \text{ stars}) = .1837$. Since .0408 us smaller than .1837, the car has a better chance of having a 5-star rating than a 2-star rating.

3.111 a. From the problem, it states the 25% of American adults smoke cigarettes. Thus, $P(A) = .25$.

b. Again, from the problem, it says that of the smokers, 13% attempted to quit smoking. Thus, $P(B \mid A) = .13$.

c. $P(A^c) = 1\ P(A) = 1 - .25 = .75$. The probability that an American adult does not smoke is .75.

d. $P(A \cap B) = P(B \mid A)\ P(A) = .13(.25) = .0325$.

3.113 a. The sample points would correspond to the possible outcomes: Violence to others, sympathetic, Harm to self, Comic images, and Criticism of definitions.

b. From the sample, the number of times each of these sample points occurred was different. We would expect the probability of each sample point to be close to the relative frequency of that outcome.

c. Reasonable probabilities would be the relative frequencies of the sample points:

Media Coverage	Frequency	Relative Frequency
Violence to others	373	373/562 = .6637
Sympathetic	102	102/562 = .1815
Harm to self	71	71/562 = .1263
Comic images	12	12/562 = .0214
Criticism of definitions	4	4/562 = .0071
Total	562	1.0000

d. This probability would be $.1815 + .0214 = .2029$.

3.115 a. Let S denote the event that an insect travels toward the pheromone and F the event that an insect travels toward the control. The sample space associated with the experiment of releasing five insects has 32 simple events:

SSSSS	FSSFS	SSFFF	FFSFS
SSSSF	FSSSF	SFSFF	FFFSS
SSSFS	SFFSS	SFFSF	FFFFS
SSFSS	SFSFS	SFFFS	FFFSF
SFSSS	SFSSF	FSSFF	FFSFF
FSSSS	SSFFS	FSFSF	FSFFF
FFSSS	SSFSF	FSFFS	SFFFF
FSFSS	SSSFF	FFSSF	FFFFF

If the pheromone under study has no effect, then each simple event is equally likely and occurs with probability 1/32.

The probability that all five insects travel toward the pheromone is $P(SSSSS) = 1/32 = .03125$.

b. The probability that exactly four of the five insects travel toward the pheromone is:

$$P(SSSSF) + P(SSSFS) + P(SSFSS) + P(SFSSS) + P(FSSSS) = 5/32 = .15625$$

c. Since the probability that exactly four of the five insects travel toward the pheromone is not small (.15625), this is not an unusual event. (This probability was computed assuming the pheromone has no effect.)

3.117 a. Define the following event:

F_i: {Player makes a foul shot on ith attempt} $P(F_i) = .8$

$P(F_i') = 1 - P(F_i) = 1 - .8 = .2$

The event "the player scores on both shots" is $F_1 \cap F_2$. If the throws are independent, then:

$$P(F_1 \cap F_2) = P(F_1)P(F_2) = .8(.8) = .64$$

The event "the player scores on exactly one shot" is:

$$(F_1 \cap F_2^c) \cup (F_1^c \cap F_2) = P(F_1 \cap F_2^c) + P(F_1^c \cap F_2)$$
$$= P(F_1)P(F_2^c) + P(F_1^c)P(F_2)$$
$$= .8(.2) + .2(.8) = .16 + .16 = .32$$

The event "the player scores on neither shot" is $F_1^c \cap F_2^c$.

$$P(F_1^c \cap F_2^c) = P(F_1^c)P(F_2^c) = .2(.2) = .04$$

b.   We know $P(F_1) = .8$, $P(F_2 | F_1) = .9$, and $P(F_2 | F_1^c) = .7$

The probability the player scores on both shots is:

$$P(F_1 \cap F_2) = P(F_2 | F_1)P(F_1) = .9(.8) = .72$$

The probability the player scores on exactly one shot is:

$$
\begin{aligned}
P(F_1 \cap F_2^c) + P(F_1^c \cap F_2) &= P(F_2^c | F_1)P(F_1) + P(F_2 | F_1^c)P(F_1^c) \\
&= [1 - P(F_2 | F_1)]P(F_1) + P(F_2 | F_1^c)P(F_1^c) \\
&= (1 - .9)(.8) + .7(.2) = .08 + .14 = .22
\end{aligned}
$$

The probability the player scores on neither shot is:

$$P(F_1^c \cap F_2^c) = P(F_2^c | F_1^c)P(F_1^c) = [1 - P(F_2 | F_1^c)]P(F_1^c) = (1 - .7)(.2) = .06$$

c.   Two consecutive foul shots are probably dependent. The outcome of the second shot probably depends on the outcome of the first.

3.119   Define the following events:

> S: {Patient receiving cyclosporine survives through the first year}
> F: {Patient receiving cyclosporine does not survive through the first year}

From the problem, $P(S) = .80$. Since $F = S^c$, $P(F) = P(S^c) = 1 - .80 = .20$

If four patients receive the transplant and cyclosporine, the possible outcomes are:

> *SSSS, SSSF, SSFS, SFSS, FSSS, SSFF, SFSF, SFFS, FSSF, FSFS, FFSS, SFFF, FSFF, FFSF, FFFS, FFFF*

a. $P(SSSS) = P(S \cap S \cap S \cap S) = P(S)P(S)P(S)P(S) = .8(.8)(.8)(.8) = .4096$ (assuming the events are independent)

b. $P(FFFF) = P(F \cap F \cap F \cap F) = P(F)P(F)P(F)P(F) = .2(.2)(.2)(.2) = .0016$

c. $P(\text{at least one patient is alive}) = 1 - P(FFFF) = 1 - .0016 = .9984$

3.121   a. Define the following events:

> A_1: {Component 1 works properly}
> A_2: {Component 2 works properly}
> B_3: {Component 3 works properly}
> B_4: {Component 4 works properly}
> A: {Subsystem A works properly}
> B: {Subsystem B works properly}

The probability a component fails is .1, so the probability a component works properly is $1 - .1 = .9$.

Subsystem *A* works properly if both components 1 and 2 work properly.

$$P(A) = P(A_1 \cap A_2) = P(A_1)P(A_2) = .9(.9) = .81$$
(since the components operate independently)

Similarly, $P(B) = P(B_1 \cap B_2) = P(B_1)P(B_2) = .9(.9) = .81$

The system operates properly if either subsystem *A* or *B* operates properly.

The probability the system operates properly is:

$$P(A \cup B) = P(A) + P(B) - P(A \cap B) = P(A) + P(B) - P(A)P(B)$$
$$= .81 + .81 - .81(.81) = .9639$$

b.   The probability exactly 1 subsystem fails is:

$$P(A \cap B^c) + P(A^c \cap B) = P(A)P(B^c) + P(A^c)P(B)$$
$$= .81(1 - .81) + (1 - .81).81 = .1539 + .1539 = .3078$$

c.   The probability the system fails is the probability that both subsystems fail or:

$$P(A^c \cap B^c) = P(A^c)P(B^c) = (1 - .81)(1 - .81) = .0361$$

d.   The system operates correctly 99% of the time means it fails 1% of the time. The probability 1 subsystem fails is .19. The probability *n* subsystems fail is $.19^n$. Thus, we must find *n* such that

$.19^n \le .01$

Thus, $n = 3$.

3.123   Define the following events:

A: {Crime involved a non-white attacker}
B: {Crime involved no injury}

a. $P(A) = \dfrac{18}{11,717} + \dfrac{111}{11,717} + \dfrac{594}{11,717} + \dfrac{214}{11,717} + \dfrac{18}{11,717} + \dfrac{141}{11,717} + \dfrac{1,656}{11,717} + \dfrac{1,801}{11,717}$

$= \dfrac{4,553}{11,717} = .3886$

b. $P(B) = \dfrac{1,422}{11,717} + \dfrac{136}{11,717} + \dfrac{214}{11,717} + \dfrac{1,801}{11,717} = \dfrac{3,573}{11,717} = .3049$

c. $P(A \cap B) = \dfrac{214}{11,717} + \dfrac{1,801}{11,717} = \dfrac{2,105}{11,717} = .1720$

d.   No, race of attacker and degree of injury are not independent.

$$P(A)P(B) = .3886(.3049) = .1185 \ne P(A \cap B) = .1720$$

3.125 We will let H_i represent high-anxiety person "i" being selected, and a similar definition of L_i for low-anxiety person "i". Then:

a. S:
$$\begin{bmatrix}
(H_1,H_2,H_3),(H_1,H_2,L_1),(H_1,H_2,L_2),(H_1,H_2,L_3) \\
(H_1,H_3,L_1),(H_1,H_3,L_2),(H_1,H_3,L_3),(H_1,L_1,L_2) \\
(H_1,L_1,L_3),(H_1,L_2,L_3),(H_2,H_3,L_1),(H_2,H_3,L_2) \\
(H_2,H_3,L_3,)(H_2,L_1,L_2)(H_2,L_1,L_3),(H_2,L_2,L_3) \\
(H_3,L_1,L_2,)(H_3,,L_1,L_3)(H_3,L_2,L_3),(L_1,L_2,L_3)
\end{bmatrix}$$

b. If the psychologist is guessing, each of the simple events would be equally likely, implying each would have a probability of 1/20.

c. Only one simple event, (H_1, H_2, H_3), indicates all guesses are correct, so the probability is 1/20.

d. Ten of the twenty simple events have at least two correct guesses, so the probability is 1/2.

3.127 a. The company is to choose $n = 5$ persons from the group of $N = 30$ finalists. The number of simple events is given by the combinations rule:

$$\binom{30}{5} = \frac{30}{5!25!} = 142,506$$

b. In order to choose no minority candidates, the company must choose 0 of the 7 minority candidates and 5 of the 23 nonminority candidates. This can be done in

$$\binom{7}{0}\binom{23}{5} = \left(\frac{7!}{0!7!}\right)\left(\frac{23!}{5!18!}\right) = 1(33,649) = 33,649$$

ways; thus, the probability that none of the minority candidates is hired is:

$$\frac{33,649}{142,506} \approx .236$$

c. The probability that no more than one minority candidate is hired is:

$$P(\text{No more than } 1) = P(0 \text{ hired}) + P(1 \text{ hired}) = \frac{33,649}{142,506} + P(1 \text{ hired})$$

In order to hire exactly one minority candidate, the company must select 1 of the 7 minority candidates and 4 of the 23 nonminority candidates. This selection can be made in

$$\binom{7}{1}\binom{23}{4} = \left(\frac{7!}{1!6!}\right)\left(\frac{23!}{4!19!}\right) = 7(8,855) = 61,985$$

ways. Therefore,

$$P(1 \text{ minority hired}) = \frac{61,985}{142,506} \text{ and}$$

$$P(\text{No more than 1 minority hired}) = \frac{33,649}{142,506} + \frac{61,985}{142,506} = \frac{95,634}{142,506} \approx .671$$

3.129 There are fifteen {(1, 4), (1, 5), (2, 3), (2, 5), (3, 2), (3, 5), (4, 1), (4, 5), (5, 1), (5, 2), (5, 3), (5, 4), (5, 5), (5, 6), (6, 5)} dice throws out of a possible thirty-six which would allow a player to begin the game. Since these are equally likely simple events, the probability of rolling any of these is 15/36. The sample space for this experiment would have the following form:

$$S = \{F, NF, NNF, NNNF, ...\}$$

where $F \Rightarrow$ a five is rolled
and $N \Rightarrow$ a five is not rolled

For example, NNF represents the event that the first and second rolls result in non-fives with the third roll resulting in a five, allowing the player to begin. These dice rolls are independent. If we let the event that the player begins the game on the ith roll be represented by E_i, then

$$P(E_1) = P(F) = \frac{15}{36} = .4167$$

$$P(E_2) = P(N \cap F) = P(N) \cdot P(F) \quad \text{(by independence)}$$

$$= \frac{21}{36} \cdot \frac{15}{36} = \frac{315}{1296} = .2431$$

$$P(E_3) = P(N \cap N \cap F) = P(N) \cdot P(N) \cdot P(F) = \left(\frac{21}{36}\right)^2 \cdot \left(\frac{15}{36}\right) = \frac{6615}{46656} = .1418$$

$$P(E_n) = P(\underbrace{N \cap N \cap \cdots \cap N}_{(n-1)N\text{'s}} \cap F) = \left(\frac{21}{36}\right)^{n-1}\left(\frac{15}{36}\right)$$

3.131 We will work with the following events:

A: {Woman is pregnant}
B: {Pregnancy test is positive}

The given probabilities can be written as:

$$P(A) = .75, \ P(B|A) = .99, \ P(B|A^c) = .02$$

Then, $P(A|B) = \dfrac{P(A \cap B)}{P(B)} = \dfrac{P(B|A) \cdot P(A)}{P(A \cap B) + P(A^c \cap B)}$

$$= \dfrac{P(B|A) \cdot P(A)}{P(B|A) \cdot P(A) + P(B|A^c) \cdot P(A^c)}$$

$$= \dfrac{(.99)(.75)}{(.99)(.75) + (.02)(.25)} = \dfrac{.7425}{.7475} \approx .993$$

3.133 We first determine the number of simple events for this experiment. According to the partitions rule, the number of ways of dealing 52 cards so that each of the four players receives 13 cards is:

$$\dfrac{52!}{13!13!13!13!} = 5.3644738 \times 10^{28}$$

If the cards are well shuffled, then the probability of each simple event is:

$$\dfrac{1}{5.3644738} \text{ times } 10^{-28}$$

We now need to count the number of simple events that result in one player receiving all the diamonds, another all the hearts, another all the spades, and another all the clubs. To do this, first consider a particular *ordering* that yields the desired result:

Player #1 receives 13 diamonds
Player #2 receives 13 hearts
Player #3 receives 13 spades
Player #4 receives 13 clubs

This particular *ordered* event can happen in

$$\binom{13}{13}\binom{13}{13}\binom{13}{13}\binom{13}{13} = 1$$

way. However, the *ordering* of the players is not relevant to us.

Since there are $P_4^4 = \dfrac{4!}{(4-4)!} = 24$ possible orderings of the players, we conclude that the probability of the rare event observed in Dubuque is:

$$\dfrac{24}{5.3644738 \times 10^{28}} \approx 4.4739 \times 10^{-28}$$

3.135   a.   Refer to the simple events listed in Exercise 3.22.

$$P(\text{win on first roll}) = P(7 \text{ or } 11) = \frac{8}{36} = \frac{2}{9}$$

b. $P(\text{lose on first roll}) = P(\text{sum of } 2 \text{ or } 3) = \frac{3}{36} = \frac{1}{12}$

c.   If the player rolls a 4 on the first roll, the game will end on the next roll if:

     1)   the player rolls a 4
     2)   the player rolls a 7 or an 11
     3)   the player rolls a 2 or 3

This probability is the union of three mutually exclusive events.

$$P(\text{game ends on a second roll}) = \frac{3}{36} + \frac{8}{36} + \frac{3}{36} = \frac{14}{36} = \frac{7}{18}$$

Discrete Random Variables

4.1   A random variable is a rule that assigns one and only one value to each simple event of an experiment.

4.3   a.   Since we can count the number of words spelled correctly, the random variable is discrete.

b.   Since we can assume values in an interval, the amount of water flowing through the Hoover Dam in a day is continuous.

c.   Since time is measured on an interval, the random variable is continuous.

d.   Since we can count the number of bacteria per cubic centimeter in drinking water, this random variable is discrete.

e.   Since we cannot count the amount of carbon monoxide produced per gallon of unleaded gas, this random variable is continuous.

f.   Since weight is measured on an interval, weight is a continuous random variable.

4.5   a.   The reaction time difference is continuous because it lies within an interval.

b.   Since we can count the number of violent crimes, this random variable is discrete.

c.   Since we can count the number of near misses in a month, this variable is discrete.

d.   Since we can count the number of winners each week, this variable is discrete.

e.   Since we can count the number of free throws made per game by a basketball team, this variable is discrete.

f.   Since distance traveled by a school bus lies in some interval, this is a continuous random variable.

4.7   a.   We know $\sum p(x) = 1$. Thus, $p(2) + p(3) + p(5) + p(8) + p(10) = 1$

$$\Rightarrow p(5) = 1 - p(2) - p(3) - p(8) - p(10) = 1 - .15 - .10 - .25 - .25 = .25$$

b. $P(x = 2 \text{ or } x = 10) = P(x = 2) + P(x = 10) = .15 + .25 = .40$

c. $P(x \le 8) = P(x = 2) + P(x = 3) + P(x = 5) + P(x = 8) = .15 + .10 + .25 + .25 = .75$

4.9   a. $P(x \le 12) = P(x = 10) + P(x = 11) + P(x = 12) = .2 + .3 + .2 = .7$

b. $P(x > 12) = 1 - P(x \le 12) = 1 - .7 = .3$

c. $P(x \le 14) = P(x = 10) + P(x = 11) + P(x = 12) + P(x = 13) + P(x = 14)$
$$= .2 + .3 + .2 + .1 + .2 = 1$$

d. $P(x = 14) = .2$

e. $P(x \le 11 \text{ or } x > 12) = P(x \le 11) + P(x > 12)$
$$= P(x = 10) + P(x = 11) + P(x = 13) + P(x = 14)$$
$$= .2 + .3 + .1 + .2 = .8$$

4.11   a.   The simple events are (where H = head, T = tail):

	HHH	HHT	HTH	THH	HTT	THT	TTH	TTT
x = # heads	3	2	2	2	1	1	1	0

b.   If each event is equally likely, then $P(\text{simple event}) = \dfrac{1}{n} = \dfrac{1}{8}$

$$p(3) = \frac{1}{8}, \; p(2) = \frac{1}{8} + \frac{1}{8} + \frac{1}{8} = \frac{3}{8}, \; p(1) = \frac{1}{8} + \frac{1}{8} + \frac{1}{8} = \frac{3}{8}, \text{ and } p(0) = \frac{1}{8}$$

c.

d. $P(x = 2 \text{ or } x = 3) = p(2) + p(3) = \dfrac{3}{8} + \dfrac{1}{8} = \dfrac{4}{8} = \dfrac{1}{2}$

4.13   a. $p(0) + p(1) + p(2) + p(3) + p(4) = .09 + .30 + .37 + .20 + .04 = 1.00$

b. $P(x = 3 \text{ or } 4) = p(3) + p(4) = .20 + .04 = .24$

c. $P(x < 2) = p(0) + p(1) = .09 + .30 = .39$

4.15   Suppose we define the following events:

P: {Student scores perfect 1600}
N: {Student does not score perfect 1600}

From the text, $P(P) = 5/10,000 = .0005$. Thus, $P(N) = 1 - P(P) = 1 - .0005 = .9995$

a.   If three students are randomly selected, the possible outcomes for the 3 students are as follows:
$PPP \;\; PPN \;\; PNP \;\; NPP \;\; PNN \;\; NPN \;\; NNP \;\; NNN$

Each of the sample points listed is really the intersection of independent events.

Thus, $P(PPP) = P(P \cap P \cap P) = P(P)P(P)P(P) = .0005(.0005)(.0005) = .0000$

$P(PPN) = P(P \cap P \cap N) = P(P)P(P)P(N) = .0005(.0005)(.9995) = .0000$
$\qquad = P(PNP) = P(NPP)$

$P(PNN) = P(P \cap N \cap N) = P(P)P(N)P(N) = .0005(.9995)(.9995) = .0005$
$\qquad = P(NPN) = P(NNP)$

$P(NNN) = P(N \cap N \cap N) = P(N)P(N)P(N) = .9995(.9995)(.9995) = .9985$

Thus, $P(x = 0) = P(NNN) = .9985$
$\qquad P(x = 1) = P(PNN) + P(NPN) + P(NNP) = .0005(3) = .0015$
$\qquad P(x = 2) = P(PPN) + P(PNP) + P(NPP) = .0000(3) = .0000$
$\qquad P(x = 3) = P(PPP) = .0000$

b.   The graph of $p(x)$ is:

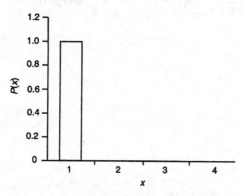

c. $P(x \le 1) = P(x = 0) + P(x = 1) = .9985 + .0015 = 1.0000$

4.17   a. $P(x \ge 10) = p(10) + p(15) + p(30) + p(50) + p(99) = .05 + .085 + .01 + .004 + .001 = .150$

b. $P(x < 0) = p(-4) + p(-2) + p(-1) = .02 + .06 + .07 = .15$

4.19   a.   The probability that the system maintains a capacity of more than 1 through the $a_1 - a_2$ path is:

$P(x > 1 \mid a_1)P(x > 1 \mid a_2) = P(x = 2 \text{ or } x = 3 \mid a_1)P(x = 2 \text{ or } x = 3 \mid a_2)$
$= [P(x = 2 \mid a_1) + P(x = 3 \mid a_1)] [P(x = 2 \mid a_2) + P(x = 3 \mid a_2)]$
$= [.25 + .60] [.60 + 0] = .85(.60) = .51$

b.   The probability that the system maintains a capacity of 1 through the $a_1 - a_3 - a_6$ path is:

$P(x = 1 \mid a_1)P(x = 1 \mid a_3)P(x = 1 \mid a_6)$
$= (.10)(.90)(.25) = .0225$

4.21   The sample space of the experiment would be:

S: {BBB, BBG, BGB, GBB, GGB, GBG, BGG, GGG}

where B and G represent a boy and girl respectively. Since the female parent always donates an X chromosome, the gender of any child is determined by the chromosome donated by the father; an X chromosome donated by the father will produce a girl, a Y, a boy. Each child therefore has a .5 probability of being either gender (as might be expected). From this, it follows that each of the above simple events has a probability of 1/8. If we let z represent the number of male offspring, the probability distribution of z is:

z	0	1	2	3
$p(z)$	1/8	3/8	3/8	1/8

Then,

P(At least one boy) $= P(z \geq 1) = 1 - P(z = 0) = 1 - 1/8 = 7/8$

4.23   a. $\mu = E(x) = \sum xp(x) = 1(.2) + 2(.4) + 4(.2) + 10(.2) = .2 + .8 + .8 + 2 = 3.8$

b. $\sigma^2 = E[(x - \mu)^2] = \sum (x - \mu)^2 p(x)$
$$= (1 - 3.8)^2(.2) + (2 - 3.8)^2(.4) + (4 - 3.8)^2(.2) + (10 - 3.8)^2(.2)$$
$$= 1.568 + 1.296 + .008 + 7.688 = 10.56$$

c. $\sigma = \sqrt{10.59} = 3.2496$

d.   The average value of x over many trials is 3.8.

e.   No. The random variable can only take on values 1, 2, 4, or 10.

f.   Yes. It is possible that μ can be equal to an actual value of x.

4.25   a.   It would seem that the mean of both would be 1 since they both are symmetric distributions centered at 1.

b. $P(x)$ seems more variable since there appears to be greater probability for the two extreme values of 0 and 2 than there is in the distribution of y.

c.   For x: $\mu = E(x) = \sum xp(x) = 0(.3) + 1(.4) + 2(.3) = 0 + .4 + .6 = 1$
$$\sigma^2 = E[(x - \mu)^2] = \sum (x - \mu)^2 p(x)$$
$$= (0 - 1)^2(.3) + (1 - 1)^2(.4) + (2 - 1)^2(.3) = .3 + 0 + .3 = .6$$

For y: $\mu = E(y) = \sum yp(y) = 0(.1) + 1(.8) + 2(.1) = 0 + .8 + .2 = 1$
$$\sigma^2 = E[(y - \mu)^2] = \sum (y - \mu)^2 p(y)$$
$$= (0 - 1)^2(.1) + (1 - 1)^2(.8) + (2 - 1)^2(.1) = .1 + 0 + .1 = .2$$

The variance for x is larger than that for y.

4.27 a. $E(x) = \sum xp(x) = 0(.09) + 1(.30) + 2(.37) + 3(.20) + 4(.04) = 0 + .30 + .74 + .60 + .16 = 1.8$

In a random samples of 4 homes, the average number of homes with high dust mite levels is 1.8.

b. $\sigma^2 = E[(x - \mu)^2] = \sum (x - \mu)^2 p(x)$
$= (0 - 1.8)^2 (.09) + (1 - 1.8)^2(.30) + (2 - 1.8)^2(.37) + (3 - 1.8)^2(.20) + (4 - 1.8)^2(.04)$

$= .2916 + .1920 + .0148 + .2880 + .1936 = .98$

$\sigma = \sqrt{.98} = .9899$

c. $\mu \pm 2\sigma \Rightarrow 1.8 \pm 2(.9899) \Rightarrow 1.8 \pm 1.9798 \Rightarrow (-.1798, 3.7798)$

$P(-.1798 < x < 3.7798) = p(0) + p(1) + p(2) + p(3) = .09 + .30 + .37 + .20 = .96$

Chebyshev's Theorem says that the interval $\mu \pm 2\sigma$ will contain at least .75 of the data. The Empirical Rule says that approximately .95 of the data will be contained in the interval. Both Chebyshev's Theorem and the Empirical Rule fit the distribution.

4.29 a.   Let $x =$ damage incurred by business. The probability distribution of x is:

x	$p(x)$
$0	.70
$300,000	.30

b. $E(x) = \sum xp(x) = 0 (.70) + 300,000(.30) = 0 + 90,000 = \$90,000.$

For similar situations, on average, the amount of damage incurred by this business is $90,000.

4.31 Let $x =$ winnings in the Florida lottery. The probability distribution for x is:

x	$p(x)$
$-\$1$	13,999,999/14,000,000
$\$6,999,999$	1/14,000,000

The expected net winnings would be:

$\mu = E(x) = (-1)(13,999,999/14,000,000) + 6,999,999(1/14,000,000) = -\$.50$

The average winnings of all those who play the lottery is $-\$.50.$

4.33 a.   Let $x =$ number of training units necessary to master the complex computer software program.

$$\mu = E(x) = \sum xp(x) = 1(.1) + 2(.25) + 3(.4) + 4(.15) + 5(.1)$$
$$= .1 + .5 + 1.2 + .6 + .5 = 2.9$$

This is the average number of units necessary to master the complex software program.

Median = 3 (first observation where the cumulative probability is $\geq .5$)

At least half of the observations are less than or equal to 3 and at least half of the observations are greater than or equal to 3.

b. $P(x \leq k) \geq .75 \Rightarrow k = 3$
$P(x \leq k) \geq .90 \Rightarrow k = 4$

c. $\mu = E(x) = \sum xp(x) = 1(.25) + 2(.35) + 3(.40) = .25 + .70 + 1.2 = 2.15$

This is smaller than the answer in part **a**. Again, this is the average number of units necessary to master the complex software program.

Median = 2 (first observation where the cumulative probability is $\geq .5$)

$P(x \leq k) \geq .75 \Rightarrow k = 3$
$P(x \leq k) \geq .90 \Rightarrow k = 3$

4.35 a. $\dfrac{6!}{2!(6-2)!} = \dfrac{6!}{2!4!} = \dfrac{6 \cdot 5 \cdot 4 \cdot 3 \cdot 2 \cdot 1}{(2 \cdot 1)(4 \cdot 3 \cdot 2 \cdot 1)} = 15$

b. $\dbinom{5}{2} = \dfrac{5!}{2!(5-2)!} = \dfrac{5!}{2!3!} = \dfrac{5 \cdot 4 \cdot 3 \cdot 2 \cdot 1}{(2 \cdot 1)(3 \cdot 2 \cdot 1)} = 10$

c. $\dbinom{7}{0} = \dfrac{7!}{0!(7-0)!} = \dfrac{7!}{0!7!} = \dfrac{7 \cdot 6 \cdot 5 \cdot 4 \cdot 3 \cdot 2 \cdot 1}{(1)(7 \cdot 6 \cdot 5 \cdot 4 \cdot 3 \cdot 2 \cdot 1)} = 1$

(Note: $0! = 1$)

d. $\dbinom{6}{6} = \dfrac{6!}{6!(6-6)!} = \dfrac{6!}{6!0!} = \dfrac{6 \cdot 5 \cdot 4 \cdot 3 \cdot 2 \cdot 1}{(6 \cdot 5 \cdot 4 \cdot 3 \cdot 2 \cdot 1)(1)} = 1$

e. $\dbinom{4}{3} = \dfrac{4!}{3!(4-3)!} = \dfrac{4!}{3!1!} = \dfrac{4 \cdot 3 \cdot 2 \cdot 1}{(3 \cdot 2 \cdot 1)(1)} = 4$

4.37 a. $P(x = 1) = \dfrac{5!}{1!4!}(.2)^1(.8)^4 = \dfrac{5 \cdot 4 \cdot 3 \cdot 2 \cdot 1}{(1)(4 \cdot 3 \cdot 2 \cdot 1)}(.2)^1(.8)^4 = 5(.2)^1(.8)^4 = .4096$

b. $P(x = 2) = \dfrac{4!}{2!2!}(.6)^2(.4)^2 = \dfrac{4 \cdot 3 \cdot 2 \cdot 1}{(2 \cdot 1)(2 \cdot 1)}(.6)^2(.4)^2 = 6(.6)^2(.4)^2 = .3456$

c. $P(x = 0) = \dfrac{3!}{0!3!}(.7)^0(.3)^3 = \dfrac{3 \cdot 2 \cdot 1}{(1)(3 \cdot 2 \cdot 1)}(.7)^0(.3)^3 = 1(.7)^0(.3)^3 = .027$

d. $P(x = 3) = \dfrac{5!}{3!2!}(.1)^3(.9)^2 = \dfrac{5\cdot4\cdot3\cdot2\cdot1}{(3\cdot2\cdot1)(2\cdot1)}(.1)^3(.9)^2 = 10(.1)^3(.9)^2 = .0081$

e. $P(x = 2) = \dfrac{4!}{2!2!}(.4)^2(.6)^2 = \dfrac{4\cdot3\cdot2\cdot1}{(2\cdot1)(2\cdot1)}(.4)^2(.6)^2 = 6(.4)^2(.6)^2 = .3456$

f. $P(x = 1) = \dfrac{3!}{1!2!}(.9)^1(.1)^2 = \dfrac{3\cdot2\cdot1}{(1)(2\cdot1)}(.9)^1(.1)^2 = 3(.9)^1(.1)^2 = .027$

4.39   a. $\mu = np = 25(.5) = 12.5$

$\sigma^2 = np(1 - p) = 25(.5)(.5) = 6.25$
$\sigma = \sqrt{\sigma^2} = \sqrt{6.25} = 2.5$

b. $\mu = np = 80(.2) = 16$

$\sigma^2 = np(1 - p) = 80(.2)(.8) = 12.8$
$\sigma = \sqrt{\sigma^2} = \sqrt{12.8} = 3.578$

c. $\mu = np = 100(.6) = 60$

$\sigma^2 = np(1 - p) = 100(.6)(.4) = 24$
$\sigma = \sqrt{\sigma^2} = \sqrt{24} = 4.899$

d. $\mu = np = 70(.9) = 63$

$\sigma^2 = np(1 - p) = 70(.9)(.1) = 6.3$
$\sigma = \sqrt{\sigma^2} = \sqrt{6.3} = 2.510$

e. $\mu = np = 60(.8) = 48$

$\sigma^2 = np(1 - p) = 60(.8)(.2) = 9.6$
$\sigma = \sqrt{\sigma^2} = \sqrt{9.6} = 3.098$

f. $\mu = np = 1,000(.04) = 40$

$\sigma^2 = np(1 - p) = 1,000(.04)(.96) = 38.4$
$\sigma = \sqrt{\sigma^2} = \sqrt{38.4} = 6.197$

4.41   a.   The simple events listed below are all equally likely, implying a probability of 1/32 for each. The list is in a regular pattern such that the first simple event would yield $x = 0$, the next five yield $x = 1$, the next ten yield $x = 2$, the next ten also yield $x = 3$, the next five yield $x = 4$, and the final one yields $x = 5$. The resulting probability distribution is given below the simple events.

$$\begin{bmatrix} \text{FFFFF, FFFFS, FFFSF, FFSFF, FSFFF, SFFFF, FFFSS, FFSFS} \\ \text{FSFFS, SFFFS, FFSSF, FSFSF, SFFSF, FSSFF, SFSFF, SSFFF} \\ \text{FFSSS, FSFSS, SFFSS, FSSFS, SFSFS, SSFFS, FSSSF, SFSSF} \\ \text{SSFSF, SSSFF, FSSSS, SFSSS, SSFSS, SSSFS, SSSSF, SSSSS} \end{bmatrix}$$

x	0	1	2	3	4	5
$p(x)$	1/32	5/32	10/32	10/32	5/32	1/32

b. $P(x = 0) = \dfrac{5!}{0!5!}\left(\dfrac{1}{2}\right)^0\left(\dfrac{1}{2}\right)^5 = \dfrac{5\cdot4\cdot3\cdot2\cdot1}{(1)(5\cdot4\cdot3\cdot2\cdot1)}\left(\dfrac{1}{2}\right)^0\left(\dfrac{1}{2}\right)^5$

$\qquad\qquad = 1\left(\dfrac{1}{2}\right)^0\left(\dfrac{1}{2}\right)^5 = \dfrac{1}{32} = .03125$

$\qquad P(x = 1) = \dfrac{5!}{1!4!}\left(\dfrac{1}{2}\right)^1\left(\dfrac{1}{2}\right)^4 = \dfrac{5\cdot4\cdot3\cdot2\cdot1}{(1)(4\cdot3\cdot2\cdot1)}\left(\dfrac{1}{2}\right)^1\left(\dfrac{1}{2}\right)^4$

$\qquad\qquad = 5\left(\dfrac{1}{2}\right)^1\left(\dfrac{1}{2}\right)^4 = \dfrac{5}{32} = .15625$

$\qquad P(x = 2) = \dfrac{5!}{2!3!}\left(\dfrac{1}{2}\right)^2\left(\dfrac{1}{2}\right)^3 = \dfrac{5\cdot4\cdot3\cdot2\cdot1}{(2\cdot1)(3\cdot2\cdot1)}\left(\dfrac{1}{2}\right)^2\left(\dfrac{1}{2}\right)^3$

$\qquad\qquad = 10\left(\dfrac{1}{2}\right)^2\left(\dfrac{1}{2}\right)^3 = \dfrac{10}{32} = .3125$

$\qquad P(x = 3) = \dfrac{5!}{3!2!}\left(\dfrac{1}{2}\right)^3\left(\dfrac{1}{2}\right)^2 = \dfrac{5\cdot4\cdot3\cdot2\cdot1}{(3\cdot2\cdot1)(2\cdot1)}\left(\dfrac{1}{2}\right)^3\left(\dfrac{1}{2}\right)^2$

$\qquad\qquad = 10\left(\dfrac{1}{2}\right)^3\left(\dfrac{1}{2}\right)^2 = \dfrac{10}{32} = .3125$

$\qquad P(x = 4) = \dfrac{5!}{4!1!}\left(\dfrac{1}{2}\right)^4\left(\dfrac{1}{2}\right)^1 = \dfrac{5\cdot4\cdot3\cdot2\cdot1}{(4\cdot3\cdot2\cdot1)(1)}\left(\dfrac{1}{2}\right)^4\left(\dfrac{1}{2}\right)^1$

$\qquad\qquad = 5\left(\dfrac{1}{2}\right)^4\left(\dfrac{1}{2}\right)^1 = \dfrac{5}{32} = .15625$

$$P(x = 5) = \frac{5!}{5!0!}\left(\frac{1}{2}\right)^5\left(\frac{1}{2}\right)^0 = \frac{5 \cdot 4 \cdot 3 \cdot 2 \cdot 1}{(5 \cdot 4 \cdot 3 \cdot 2 \cdot 1)(1)}\left(\frac{1}{2}\right)^5\left(\frac{1}{2}\right)^0$$

$$= 1\left(\frac{1}{2}\right)^5\left(\frac{1}{2}\right)^0 = \frac{1}{32} = .15625$$

    c.   From Table II, $n = 5$, $p = .5$:

$$P(x = 0) = .031$$
$$P(x = 1) = .188 - .031 = .157$$
$$P(x = 2) = .500 - .188 = .312$$
$$P(x = 3) = .812 - .500 = .312$$
$$P(x = 4) = .969 - .812 = .157$$
$$P(x = 5) = 1 - .969 = .031$$

4.43   a.   For this experiment, there are $n = 200$ smokers (n identical trials). For each smoker, there are 2 possible outcomes: $S =$ smoker enters treatment program and $F =$ smoker does not enter treatment program. The probability of S (smoker enters treatment program) is the same from trial to trial. This probability is $P(S) = p = .05$. $P(F) = 1 - P(S) = 1 - .05 = .95$. The trials are independent and $x =$ number of smokers entering treatment program in 200 trials. Thus, x is a binomial random variable.

    b. $p = P(S) = .05$. Of all the smokers, only .05 or 5% enter treatment programs.

    c. $E(x) = np = 200(.05) = 10$. For all samples of 200 smokers, the average number of smokers who enter a treatment program will be 10.

4.45   a.   From the problem, x is a binomial random variable with $n = 10$ and $p = .8$.

        $P(x = 3) = P(x \leq 3) - P(x \leq 2) = .001 - .000 = .001$ (from Table II, Appendix A)

    b. $P(x \leq 7) = .322$

    c. $P(x > 4) = 1 - P(x \leq 4) = 1 - .006 = .994$

4.47   a.   In order to be a binomial random variable, the five characteristics must hold.

        1.   For this problem, there are 5 items scanned. We will assume that these 5 trials are identical.

        2.   For each item scanned, there are 2 possible outcomes: priced incorrectly (S) or priced correctly (F).

        3.   The probability of being priced incorrectly remains constant from trial to trial. For this problem, we will assume that the probability of being priced incorrectly is $P(S) = 1/30$ for each trial.

        4.   We will assume that whether one item is priced incorrectly is independent of any other.

        5.   The random variable x is the number of items priced incorrectly in 5 trials.

        Thus, x is a binomial random variable.

b. The estimate of p, the probability of an item being priced incorrectly is 1/30.

c. $P(x = 1) = \binom{5}{1}(1/30)^1(29/30)^4 = .1455$

d. $P(x \geq 1) = 1 - P(x = 0) = 1 - \binom{5}{0}(1/30)^0(29/30)^5 = 1 - .8441 = .1559$

4.49 a. For this problem, let x = the number of students who have nasal allergies in 25 trials. The variable x is a binomial random variable with $n = 25$ and $p = .8$.

$P(x < 20) = P(x \leq 19) = .383$ (Using Table II, Appendix A)

b. $P(x > 15) = 1 - P(x \leq 15) = 1 - .017 = .983$

c. $E(x) = np = 25(.8) = 20$. On the average, out of 25 students, we would expect to find 20 with nasal allergies.

4.51 a. Let x = number of women out of 15 that have been abused. Then x is a binomial random variable with $n = 15$, S = woman has been abused, F = woman has not been abused, $P(S) = p = 1/3$ and $q = 1$ $p = 1 - 1/3 = 2/3$.

$P(x \geq 4) = 1 - P(x < 4) = 1 - P(x = 0) - P(x = 1) - P(x = 2) - P(x = 3)$

$$= 1 - \frac{15!}{0!\,15!}\left(\frac{1}{3}\right)^0\left(\frac{2}{3}\right)^{15} - \frac{15!}{1!\,14!}\left(\frac{1}{3}\right)^1\left(\frac{2}{3}\right)^{14} - \frac{15!}{2!\,13!}\left(\frac{1}{3}\right)^2\left(\frac{2}{3}\right)^{13}$$

$$- \frac{15!}{3!\,12!}\left(\frac{1}{3}\right)^3\left(\frac{2}{3}\right)^{12}$$

$$= 1 - .0023 - .0171 - .0599 - .1299 = .7908$$

b. For $p = .1$,
$P(x \geq 4) = 1 - P(x < 4) = 1 - P(x = 0) - P(x = 1) - P(x = 2) - P(x = 3)$

$$= 1 - \frac{15!}{0!\,5!}.1^0.9^{15} - \frac{15!}{1!\,4!}.1^1.9^{14} - \frac{15!}{2!\,3!}.1^2.9^{13} - \frac{15!}{3!\,2!}.1^3.9^{12}$$

$$= 1 - .2059 - .3432 - .2669 - .1285 = .0555$$

c. We sampled 15 women and actually found that 4 had been abused. If $p = 1/3$, the probability of observing 4 or more abused women is .7908. If $p = .1$, the probability of observing 4 or more abused women is only .0555. If $p = .1$, we would have seen a very unusual event because the probability is so small (.0555). If $p = 1/3$, we would have seen an event that was very common because the probability was very large (.7908). Since we normally do not see rare events, the true probability of abuse is probably close to 1/3.

4.53 a. We must assume that the probability that a specific type of ball meets the requirements is always the same from trial to trial and the trials are independent. To use the binomial probability distribution, we need to know the probability that a specific type of golf ball meets the requirements.

b. For a binomial distribution,

$$\mu = np$$
$$\sigma = \sqrt{npq}$$

In this example, n = two dozen = $2 \cdot 12 = 24$.

$p = .10$ (Success here means the golf ball *does not* meet standards.)
$q = .90$
$\mu = np = 24(.10) = 2.4$
$\sigma = \sqrt{npq} = \sqrt{24(.10)(.90)} = 1.47$

c. In this situation,

p = Probability of success
   = Probability golf ball *does* meet standards
   = .90
$q = 1 - .90 = .10$
$n = 24$
$E(x) = \mu = np = 24(.90) = 21.60$
$\sigma = \sqrt{npq} = \sqrt{24(.10)(.90)} = 1.47$ (Note that this is the same as in part **b**.)

4.55 Assuming the supplier's claim is true,

$$\mu = np = 500(.001) = .5$$
$$\sigma = \sqrt{npq} = \sqrt{500(.001)(.999)} = \sqrt{.4995} = .707$$

If the supplier's claim is true, we would only expect to find .5 defective switches in a sample of size 500. Therefore, it is not likely we would find 4.

Based on the sample, the guarantee is probably inaccurate.

Note: $z = \dfrac{x - \mu}{\sigma} = \dfrac{4 - 5}{.707} = 4.95$

This is an unusually large z-score.

4.57 a. The random variable x is discrete since it can assume a countable number of values (0, 1, 2, ...).

b. This is a Poisson probability distribution with $\lambda = 3$.

c. In order to graph the probability distribution, we need to know the probabilities for the possible values of x. Using Table III of Appendix A with $\lambda = 3$:

$p(0) = .050$
$p(1) = P(x \le 1) - P(x = 0) = .199 - .050 = .149$
$p(2) = P(x \le 2) - P(x \le 1) = .423 - .199 = .224$
$p(3) = P(x \le 3) - P(x \le 2) = .647 - .423 = .224$
$p(4) = P(x \le 4) - P(x \le 3) = .815 - .647 = .168$
$p(5) = P(x \le 5) - P(x \le 4) = .916 - .815 = .101$
$p(6) = P(x \le 6) - P(x \le 5) = .966 - .916 = .050$
$p(7) = P(x \le 7) - P(x \le 6) = .988 - .966 = .022$
$p(8) = P(x \le 8) - P(x \le 7) = .996 - .988 = .008$
$p(9) = P(x \le 9) - P(x \le 8) = .999 - .996 = .003$
$p(10) \approx .001$

The probability distribution of x in graphical form is:

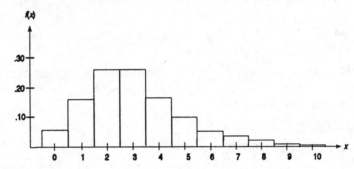

d. $\mu = \lambda = 3$
$\sigma^2 = \lambda = 3$
$\sigma = \sqrt{3} = 1.7321$

e. The mean of x is the same as the mean of the probability distribution, $\mu = \lambda = 3$.

The standard deviation of x is the same as the standard deviation of the probability distribution, $\sigma = 1.7321$.

4.59 $\mu = \lambda = 1.5$

Using Table III of Appendix A:

a. $P(x \le 3) = .934$

b. $P(x \ge 3) = 1 - P(x \le 2) = 1 - .809 = .191$

c. $P(x = 3) = P(x \le 3) - P(x \le 2) = .934 - .809 = .125$

d. $P(x = 0) = .223$

e. $P(x > 0) = 1 - P(x = 0) = 1 - .223 = .777$

f. $P(x > 6) = 1 - P(x \le 6) = 1 - .999 = .001$

4.61　a.　To graph the Poisson probability distribution with $\lambda = 3$, we need to calculate $p(x)$ for $x = 0$ to 10. Using Table III, Appendix A,

$$p(0) = .050$$
$$p(1) = P(x \le 1) - P(x = 0) = .199 - .050 = .149$$
$$p(2) = P(x \le 2) - P(x \le 1) = .423 - .199 = .224$$
$$p(3) = P(x \le 3) - P(x \le 2) = .647 - .423 = .224$$
$$p(4) = P(x \le 4) - P(x \le 3) = .815 - .647 = .168$$
$$p(5) = P(x \le 5) - P(x \le 4) = .916 - .815 = .101$$
$$p(6) = P(x \le 6) - P(x \le 5) = .966 - .916 = .050$$
$$p(7) = P(x \le 7) - P(x \le 6) = .988 - .966 = .022$$
$$p(8) = P(x \le 8) - P(x \le 7) = .996 - .988 = .008$$
$$p(9) = P(x \le 9) - P(x \le 8) = .999 - .996 = .003$$
$$p(10) = P(x \le 10) - P(x \le 9) \approx 1 - .999 = .001$$

The graph is shown here.

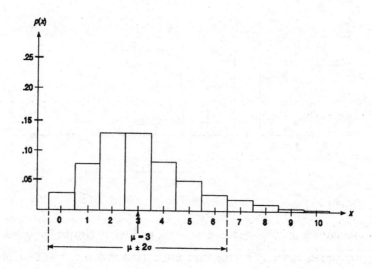

b.　$\mu = \lambda = 3$

$\sigma = \sqrt{\lambda} = \sqrt{3} = 1.7321$

$\mu \pm 2\sigma \Rightarrow 3 \pm 2(1.7321) \Rightarrow (-.4642, 6.4642)$

c.　$P(\mu - 2\sigma < x < \mu + 2\sigma) = P(-.4642 < x < 6.4642) = P(0 \le x \le 6)$
$$= P(x \le 6) = .966$$

4.63　a.　Using Table III, Appendix A, with $\lambda = 1$, $P(x = 0) = .368$.

b.　Using Table III, Appendix A, with $\lambda = 1$, $P(x > 1) = 1 - P(x \le 1) = 1 - .736 = .264$.

c.　Using Table III, Appendix A, with $\lambda = 1$, $P(x \le 2) = .920$.

4.65　a.　$P(x = 0) = \dfrac{\lambda^0 e^{-\lambda}}{0!} = \dfrac{3.8^0 e^{-3.8}}{0!}$ .0224

　　　b.　$P(x = 1) = \dfrac{\lambda^1 e^{-\lambda}}{1!} = \dfrac{3.8^1 e^{-3.8}}{1!} = .0850$

　　　c.　$E(x) = \mu = \lambda = 3.8$

　　　　　$\sigma^2 = \lambda = 3.8$
　　　　　$\sigma = \sqrt{3.8} = 1.9494$

4.67　a.　$\sigma = \sqrt{\lambda} = \sqrt{4} = 2$

　　　b.　$P(x > 10) = 1 - P(x \le 10) = 1 - .997 = .003$ from Table III, Appendix A, with $\lambda = 4$. Since the probability is so small (.003), it would be very unlikely that the plant would yield a value that would exceed the EPA limit.

4.69　a.　Using Table III and $\lambda = 6.2$, $P(x = 2) = P(x \le 2) - P(x \le 1) = .054 - .015 = .039$
　　　　　$P(x = 6) = P(x \le 6) - P(x \le 5) = .574 - .414 = .160$
　　　　　$P(x = 10) = P(x \le 10) - P(x \le 9) = .949 - .902 = .047$

　　　b.　The plot of the distribution is:

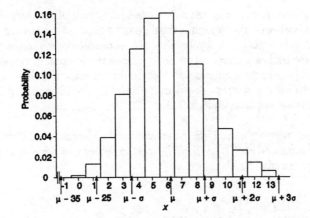

　　　c.　$\mu = \lambda = 6.2$, $\sigma = \sqrt{\lambda} = \sqrt{6.2} = 2.490$
　　　　　$\mu \pm \sigma \Rightarrow 6.2 \pm 2.49 \Rightarrow (3.71, 8.69)$
　　　　　$\mu \pm 2\sigma \Rightarrow 6.2 \pm 2(2.49) \Rightarrow 6.2 \pm 4.98 \Rightarrow (1.22, 11.18)$
　　　　　$\mu \pm 3\sigma \Rightarrow 6.2 \pm 3(2.49) \Rightarrow 6.2 \pm 7.47 \Rightarrow (-1.27, 13.67)$

　　　　　See the plot in part **b**.

d. First, we need to find the mean number of customers per hour. If the mean number of customers per 10 minutes is 6.2, then the mean number of customers per hour is $6.2(6) = 37.2 = \lambda$.

$\mu = \lambda = 37.2$ and $\sigma = \sqrt{\lambda} = \sqrt{37.2} = 6.099$

$\mu \pm 3\sigma \Rightarrow 37.2 \pm 3(6.099) \Rightarrow 37.2 \pm 18.297 \Rightarrow (18.903, 55.498)$

Using Chebyshev's Rule, we know at least 8/9 or 88.9% of the observations will fall within 3 standard deviations of the mean. The number 75 is way beyond the 3 standard deviation limit. Thus, it would be very unlikely that more than 75 customers entered the store per hour on Saturdays.

4.71 Let x = number of cars that will arrive within the next 30 minutes. If the average number of cars that arrive in 60 minutes is 10, the average number of cars that arrive in 30 minutes is $\lambda = 5$.

In order for no cars to be in line at closing time, no more than 1 car can arrive in the next 30 minutes.

$P(x \leq 1) = .040$ from Table III, Appendix A, with $\lambda = 5$.

Since this probability is so small, it is very likely that at least one car will be in line at closing time. The probability that at least one car will be in line at closing is $1 - .040 = .960$.

4.73 A binomial random variable is characterized by n trials where the trials are identical. In order for the trials to be identical, we have to sample from an infinite population or sample from a finite population with replacement. The probability of success on any trial remains constant from trial to trial and the trials are independent. A hypergeometric random variable is characterized by n trials where the elements are selected from a finite population without replacement. The probability of success on any trial depends on what happened on previous trials, and thus the trials are not independent.

Both a binomial random variable and a hypergeometric random variable are characterized by n trials, 2 possible outcomes on each trial (S or F), and the random variable represents the number of successes in n trials.

4.75 a. $P(x = 1) = \dfrac{\binom{r}{x}\binom{N-r}{n-x}}{\binom{N}{n}} = \dfrac{\binom{3}{1}\binom{5-3}{3-1}}{\binom{5}{3}} = \dfrac{\frac{3!}{1!2!}\frac{2!}{2!0!}}{\frac{5!}{3!2!}} = \dfrac{3(1)}{10} = .3$

b. $P(x = 3) = \dfrac{\binom{r}{x}\binom{N-r}{n-x}}{\binom{N}{n}} = \dfrac{\binom{3}{3}\binom{9-3}{5-3}}{\binom{9}{5}} = \dfrac{\frac{3!}{3!0!}\frac{6!}{2!4!}}{\frac{9!}{5!4!}} = \dfrac{1(15)}{126} = .119$

c. $P(x=2) = \dfrac{\binom{r}{x}\binom{N-r}{n-x}}{\binom{N}{n}} = \dfrac{\binom{2}{2}\binom{4-2}{2-2}}{\binom{4}{2}} = \dfrac{\frac{2!}{2!0!}\frac{2!}{0!2!}}{\frac{4!}{2!2!}} = \dfrac{1(1)}{6} = .167$

d. $P(x=0) = \dfrac{\binom{r}{x}\binom{N-r}{n-x}}{\binom{N}{n}} = \dfrac{\binom{2}{0}\binom{4-2}{2-0}}{\binom{4}{2}} = \dfrac{\frac{2!}{0!2!}\frac{2!}{2!0!}}{\frac{4!}{2!2!}} = \dfrac{1(1)}{6} = .167$

4.77   With $N = 12$, $n = 8$ and $r = 6$, x can take on values 2, 3, 4, 5, or 6.

a. $P(x=2) = \dfrac{\binom{r}{x}\binom{N-r}{n-x}}{\binom{N}{n}} = \dfrac{\binom{6}{2}\binom{12-6}{8-2}}{\binom{12}{8}} = \dfrac{\frac{6!}{2!4!}\frac{6!}{6!0!}}{\frac{12!}{8!4!}} = \dfrac{15(1)}{495} = .030$

$P(x=3) = \dfrac{\binom{r}{x}\binom{N-r}{n-x}}{\binom{N}{n}} = \dfrac{\binom{6}{3}\binom{12-6}{8-3}}{\binom{12}{8}} = \dfrac{\frac{6!}{3!3!}\frac{6!}{5!1!}}{\frac{12!}{8!4!}} = \dfrac{20(6)}{495} = .242$

$P(x=4) = \dfrac{\binom{r}{x}\binom{N-r}{n-x}}{\binom{N}{n}} = \dfrac{\binom{6}{4}\binom{12-6}{8-4}}{\binom{12}{8}} = \dfrac{\frac{6!}{4!2!}\frac{6!}{4!2!}}{\frac{12!}{8!4!}} = \dfrac{15(15)}{495} = .455$

$P(x=5) = \dfrac{\binom{r}{x}\binom{N-r}{n-x}}{\binom{N}{n}} = \dfrac{\binom{6}{5}\binom{12-6}{8-5}}{\binom{12}{8}} = \dfrac{\frac{6!}{5!1!}\frac{6!}{3!3!}}{\frac{12!}{8!4!}} = \dfrac{6(20)}{495} = .242$

$P(x=6) = \dfrac{\binom{r}{x}\binom{N-r}{n-x}}{\binom{N}{n}} = \dfrac{\binom{6}{6}\binom{12-6}{8-6}}{\binom{12}{8}} = \dfrac{\frac{6!}{6!0!}\frac{6!}{2!4!}}{\frac{12!}{8!4!}} = \dfrac{1(15)}{495} = .030$

The probability distribution of x in tabular form is:

x	$p(x)$
2	.030
3	.242
4	.455
5	.242
6	.030

b. $\mu = \dfrac{nr}{N} = \dfrac{8(6)}{12} = 4$

$\sigma^2 = \dfrac{r(N-r)n(N-n)}{N^2(N-1)} = \dfrac{6(12-6)8(12-8)}{12^2(12-1)} = \dfrac{1,152}{1,584} = .7273$

$\sigma = \sqrt{.7273} = .853$

c. $\mu \pm 2\sigma \Rightarrow 4 \pm 2(.853) \Rightarrow 4 \pm 1.706 \Rightarrow (2.294, 5.706)$

The graph of the distribution is:

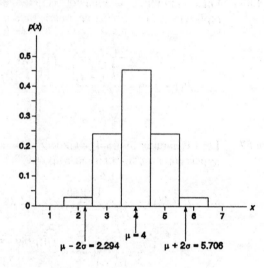

d. $P(2.294 < x < 5.706) = P(3 \le x \le 5) = .242 + .455 + .242 = .939$

4.79 For this problem, $N = 100$, $n = 10$, and $x = 4$.

a. If the sample is drawn without replacement, the hypergeometric distribution should be used. The hypergeometric distribution requires that sampling be done without replacement.

b. If the sample is drawn with replacement, the binomial distribution should be used. The binomial distribution requires that sampling be done with replacement.

4.81 a. Let x = number of facilities chosen that treat hazardous waste on-site in 10 trials. For this problem, N = 209, r = 8, and n = 10.

$$E(x) = \mu = \frac{nr}{N} = \frac{10(8)}{209} = .383$$

b. $$P(x = 4) = \frac{\binom{r}{x}\binom{N-r}{n-x}}{\binom{N}{n}} = \frac{\binom{8}{4}\binom{201}{6}}{\binom{209}{10}} = \frac{\frac{8!}{4!4!}\frac{201!}{6!195!}}{\frac{209!}{10!199!}} = .0002$$

4.83 Let x = number of "clean" cartridges selected in 5 trials. For this problem, $N = 158$, $r = 122$, and $n = 5$.

$$P(x = 5) = \frac{\binom{r}{x}\binom{N-r}{n-x}}{\binom{N}{n}} = \frac{\binom{122}{5}\binom{36}{0}}{\binom{158}{5}} = \frac{\frac{122!}{5!117!}\frac{36!}{0!36!}}{\frac{158!}{5!153!}} = .2693$$

4.85 Let x = number of spoiled bottles in the sample of 3. Since the sampling will be done without replacement, x is a hypergeometric random variable with $N = 12$, $n = 3$, and $r = 1$.

$$P(x = 1) = \frac{\binom{r}{x}\binom{N-r}{n-x}}{\binom{N}{n}} = \frac{\binom{1}{1}\binom{12-1}{3-1}}{\binom{12}{3}} = \frac{\frac{1!}{1!0!}\frac{11!}{2!9!}}{\frac{12!}{3!9!}} = \frac{55}{220} = .25$$

4.87 Let x = number of grants awarded to the north side in 140 trials. The random variable x has a hypergeometric distribution with $N = 743$, $n = 140$, and $r = 601$.

a. $$\mu = E(x) = \frac{nr}{N} = \frac{140(601)}{743} = 113.24$$

$$\sigma^2 = \frac{r(N-r)n(N-n)}{N^2(N-1)} = \frac{601(743-601)140(743-140)}{743^2(743-1)} = 17.5884$$

$$\sigma = \sqrt{17.5884} = 4.194$$

b. If the grants were awarded at random, we would expect approximately 113 to be awarded to the north side. We observed 140. The z-score associated with 140 is:

$$z = \frac{x-\mu}{\sigma} = \frac{140-113.24}{4.194} = 6.38$$

Because this z-score is so large, it would be extremely unlikely to observe all 140 grants to the north side if they are randomly selected. Thus, we would conclude that the grants were not randomly selected.

4.89   a.   The length of time that an exercise physiologist's program takes to elevate her client's heart rate to 140 beats per minute is measured on an interval and thus, is continuous.

       b.   The number of crimes committed on a college campus per year is a whole number such as 0, 1, 2, etc. This variable is discrete.

       c.   The number of square feet of vacant office space in a large city is a measurement of area and is measured on an interval. Thus, this variable is continuous.

       d.   The number of voters who favor a new tax proposal is a whole number such as 0, 1, 2, etc. This variable is discrete.

4.91   a.   Poisson

       b.   Binomial

       c.   Binomial

4.93   a. $P(x = 1) = \binom{3}{1}(.1)^1(.9)^{3-1} = \dfrac{3!}{1!2!}(.1)^1(.9)^2 = 3(.1)(.81) = .243$

       b. $P(x = 4) = P(x \le 4) - P(x \le 3) = .238 - .107 = .131$ from Table II, Appendix A.

       c. $P(x = 0) = \binom{2}{0}(.4)^0(.6)^{2-0} = \dfrac{2!}{0!2!}(1)(.6)^2 = .36$

       d. $P(x = 4) = P(x \le 4) - P(x \le 3) = .969 - .812 = .157$ from Table II, Appendix A.

       e. $P(x = 12) = P(x \le 12) - P(x \le 11) = .184 - .056 = .128$ from Table II, Appendix A.

       f. $P(x = 8) = P(x \le 8) - P(x \le 7) = .954 - .833 = .121$ from Table II, Appendix A.

4.95   Using Table III, Appendix A,

       a.   When $\lambda = 2$, $p(3) = P(x \le 3) - P(x \le 2) = .857 - .677 = .180$

       b.   When $\lambda = 1$, $p(4) = P(x \le 4) - P(x \le 3) = .996 - .981 = .015$

       c.   When $\lambda = .5$, $p(2) = P(x \le 2) - P(x \le 1) = .986 - .910 = .076$

4.97   The random variable x would have a binomial distribution:

       1. n identical trials. A sample of married women was taken and would be considered small compared to the entire population of married women, so the trials would be very close to identical.

       2.   There are only two possible outcomes for each trial. In this experiment, there are only two possible outcomes: the woman would marry the same man (S) or she would not (F).

3. The probability of Success stays the same from trial to trial. Here, $P(S) = .80$ and $P(F) = .20$. In reality, these probabilities would not be exactly the same from trial to trial, but rounded off to 4 decimal places, they would be the same.

4. The trials are independent. In this experiment, the trials would not be exactly independent because we would be sampling without replacement from a finite population. However, if the sample size is fairly small compared to the population size, the trials will be essentially independent.

5. The binomial random variable x would be the number of successes in n trials. For this experiment, x = the number of women out of those sampled who would marry the same man again.

Thus, x would possess (approximately) a binomial distribution.

4.99   a.   In order for this to be a valid probability distribution, $\sum p(x) = 1$ and $0 \leq p(x) \leq 1$ for all values of x.

In this distribution, $0 \leq p(x) \leq 1$ for all values of x and $\sum p(x) = .051 + .099 + .093 + .635 + .122 = 1.000$. Thus, this is a valid probability distribution.

b. $P(x = 1) = .051$

c. $P(x \geq 4) = P(x = 4) + P(x = 5) = .635 + .122 = .757$

d. $P(x = 2 \text{ or } x = 3) = P(x = 2) + P(x = 3) = .099 + .093 = .192$

e. $E(x) = \mu = \sum x p(x) = 1(.051) + 2(.099) + 3(.093) + 4(.635) + 4(.122) = 3.678$

f.   The average rating of all books rated is 3.678.

4.101   a. $\mu = E(x) = np = 800(.65) = 520$
$\sigma = \sqrt{npq} = \sqrt{800(.65)(.35)} = 13.491$

b.   Half of the 800 food items is 400. A value of $x = 400$ would have a z-score of:

$$z = \frac{x - \mu}{\sigma} = \frac{400 - 520}{13.491} = -8.895$$

Since the z-score associated with 400 items is so small (-8.895), it would be virtually impossible to observe less than half without any traces of pesticides if the 65% value was correct.

4.103   Let x = number of parents who yell at their child before, during, or after a meet in 20 trials. Then x has a binomial distribution with $n = 20$ and $p = .05$.

$P(x \geq 1) = 1 - P(x = 0) = 1 - .358 = .642$ (Using Table II, Appendix A)

4.105 Let x = the number of calls that are a hoax; then x is a binomial random variable with $n = 5$, $p = \dfrac{1}{6}$ and $q = \dfrac{5}{6}$.

$$p(x) = \binom{5}{x}\left(\frac{1}{6}\right)^{x}\left(\frac{5}{6}\right)^{5-x} \qquad x = 0, 1, 2, 3, 4, 5$$

a. The probability that none of the calls was a hoax is:

$$P(x = 0) = p(0) = \binom{5}{0}\left(\frac{1}{6}\right)^{0}\left(\frac{5}{6}\right)^{5-0} = \frac{5!}{0!5!}\left(\frac{1}{6}\right)^{0}\left(\frac{5}{6}\right)^{5} = 1(1)(.4019) = .4019$$

b. The probability that three of the callers really needed assistance is the probability that two of the calls were hoaxes:

$$P(x = 2) = p(2) = \binom{5}{2}\left(\frac{1}{6}\right)^{2}\left(\frac{5}{6}\right)^{5-2} = \frac{5!}{2!3!}\left(\frac{1}{6}\right)^{2}\left(\frac{5}{6}\right)^{3} = 10(.02778)(.5787)$$
$$= .1608$$

c. We must assume that the characteristics of a binomial experiment are satisfied. That is, the five calls received are independent with the same probability of a hoax on each call.

d. Since approximately 1/6 of the 10,000 calls costing $30.00 apiece are hoaxes, the approximate cost would be:

$$\$30 \cdot \frac{1}{6} \cdot 10{,}000 = \$50{,}000$$

4.107 Using Table II with $n = 25$ and $p = .8$:

a. $P(x < 15) = P(x \le 14) = .006$

b. Since the probability of such an event is so small when $p = .8$, if less than 15 insects die we would conclude that the insecticide is not as effective as claimed.

4.109 Let x = number of people responding to a questionnaire in 20 trials. Then x is a binomial random variable with $n = 20$ and $p = .4$.

a. $P(x > 12) = 1 - P(x \le 12) = 1 - .979 = .021$ (from Table II, Appendix A)

b. We know from the Empirical Rule that almost all the observations are larger than $\mu - 2\sigma$. ($\approx 95\%$ are between $\mu - 2\sigma$ and $\mu + 2\sigma$). Thus $\mu - 2\sigma > 100$.

For the binomial, $\mu = np = n(.4)$ and $\sigma = \sqrt{npq} = \sqrt{n(.4)(.6)} = \sqrt{.24n}$

$\mu - 2\sigma > 100 \Rightarrow .4n - 2\sqrt{.24n} > 100 \Rightarrow .4n - .98\sqrt{n} - 100 > 0$

Solving for $\sqrt{n}$, we get:

$$\sqrt{n} = \frac{.98 \pm \sqrt{.98^2 - 4(.4)(-100)}}{2(.4)} = \frac{.98 \pm 12.687}{.8}$$

$$\Rightarrow \sqrt{n} = 17.084 \Rightarrow n = 17.084^2 = 291.9 \approx 292$$

4.111   Define x as the number of components that operate successfully. The random variable x is a binomial random variable (the components operate independently and there are only two possible outcomes) with $n = 4$ and $p = .85$.

$$P(\text{system fails}) = P(x = 0) = \binom{4}{0}.85^0(.15)^{4-0} = \frac{4!}{0!4!}.85^0(.15)^4 = .15^4 = .0005$$

Continuous Random Variables

5.1   a.   For a uniform random variable,

$$f(x) = \begin{cases} \dfrac{1}{d-c} = \dfrac{1}{30-10} = \dfrac{1}{20} & 10 \le x \le 30 \\ 0 & \text{otherwise} \end{cases}$$

b. $\mu = \dfrac{c+d}{2} = \dfrac{10+30}{2} = 20$

$\sigma = \dfrac{d-c}{\sqrt{12}} = \dfrac{30-10}{\sqrt{12}} = 5.774$

c.

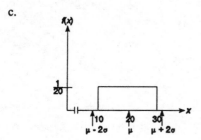

$\mu \pm 2\sigma \Rightarrow 20 \pm 2(5.774) \Rightarrow 20 \pm 11.584 \Rightarrow (8.452, 31.548)$

5.3   a.   For a uniform random variable,

$$f(x) = \begin{cases} \dfrac{1}{d-c} = \dfrac{1}{4-2} = \dfrac{1}{2} & 2 \le x \le 4 \\ 0 & \text{otherwise} \end{cases}$$

b. $\mu = \dfrac{c+d}{2} = \dfrac{2+4}{2} = 3$

$\sigma = \dfrac{d-c}{\sqrt{12}} = \dfrac{4-2}{\sqrt{12}} = .577$

c. $\mu \pm \sigma \Rightarrow 3 \pm .577 \Rightarrow (2.423, 3.577)$

$P(2.423 \le x \le 3.577) = (3.577 - 2.423)\dfrac{1}{2} = .577$

d. $P(x > 2.78) = (4 - 2.78)\dfrac{1}{2} = .61$

e. $P(2.4 \le x \le 3.7) = (3.7 - 2.4)\dfrac{1}{2} = .65$

f. $P(x < 2) = 0$

5.5   For the uniform random variable,

$$f(x) = \begin{cases} \dfrac{1}{d-c} = \dfrac{1}{200-100} = \dfrac{1}{100} & 100 \le x \le 200 \\ 0 & \text{otherwise} \end{cases}$$

a. $\mu = \dfrac{c+d}{2} = \dfrac{100+200}{2} = 150$ $\sigma = \dfrac{d-c}{\sqrt{12}} = \dfrac{200-100}{\sqrt{12}} = 28.87$

        $\mu \pm 2\sigma \Rightarrow 150 \pm 2(28.87) \Rightarrow 150 \pm 57.74 \Rightarrow (92.26, 207.74)$

        $P(x < 92.26) + P(x > 207.74) = 0 + 0 = 0$

b. $\mu \pm 3\sigma \Rightarrow 150 \pm 3(28.87) \Rightarrow 150 \pm 86.61 \Rightarrow (63.39, 236.61)$

        $P(63.39 < x < 236.61) = P(100 \le x \le 200) = 1$

c. $P(92.26 < x < 207.74) = P(100 \le x \le 200) = 1$

5.7   For the uniform distribution,

$$f(x) = \begin{cases} \dfrac{1}{d-c} = \dfrac{1}{90-75} = \dfrac{1}{15} & 75 \le x \le 90 \\ 0 & \text{otherwise} \end{cases}$$

a. $P(x > 80) = (90 - 80)\left(\dfrac{1}{15}\right) = .667$

b. $P(80 < x < 85) = (85 - 80)\left(\dfrac{1}{15}\right) = .333$

c. $E(x) = \mu = \dfrac{c+d}{2} = \dfrac{75+90}{2} = 82.5°F$

5.9   To construct a relative frequency histogram for the data, we can use 7 measurement classes.

$$\text{Interval width} = \frac{\text{Largest number} - \text{smallest number}}{\text{Number of classes}} = \frac{98.0716 - .7434}{7} = 13.9$$

We will use an interval width of 14 and a starting value of .74335.

The measurement classes, frequencies, and relative frequencies are given in the table below.

Class	Measurement Class	Class Frequency	Class Relative Frequency
1	.74335 – 14.74335	6	6/40 = .15
2	14.74335 – 28.74335	4	.10
3	28.74335 – 42.74335	6	.15
4	42.74335 – 56.74335	6	.15
5	56.74335 – 70.74335	5	.125
6	70.74335 – 84.74335	4	.10
7	84.74335 – 98.74335	9	.225
		40	1.000

The histogram looks like the data could be from a uniform distribution. The last class (84.74335 – 98.74335) has a few more observations in it than we would expect. However, we cannot expect a perfect graph from a sample of only 40 observations.

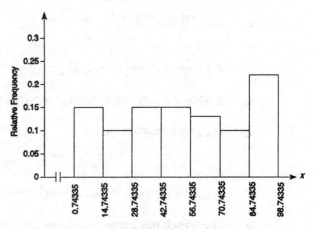

5.11 a. The amount dispensed by the beverage machine is a continuous random variable since it can take on any value between 6.5 and 7.5 ounces.

b. Since the amount dispensed is random between 6.5 and 7.5 ounces, x is a uniform random variable.

$$f(x) = \frac{1}{d-c} \quad (c \le x \le d)$$

$$\frac{1}{d-c} = \frac{1}{7.5-6.5} = \frac{1}{1} = 1$$

Therefore, $f(x) = \begin{cases} 1 & (6.5 \le x \le 7.5) \\ 0 & \text{otherwise} \end{cases}$

The graph is as follows:

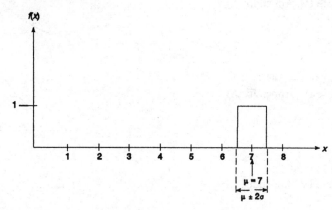

c. $\mu = \dfrac{c+d}{2} = \dfrac{6.5+7.5}{2} = \dfrac{14}{2} = 7$

$\sigma = \dfrac{d-c}{\sqrt{12}} = \dfrac{7.5-6.5}{\sqrt{12}} = .2887$

$\mu \pm 2\sigma \Rightarrow 7 \pm 2(.2887) \Rightarrow 7 \pm .5774 \Rightarrow (6.422, 7.577)$

d. $P(x \geq 7) = (7.5 - 7)(1) = .5$

e. $P(x < 6) = 0$

f. $P(6.5 \leq x \leq 7.25) = (7.25 - 6.5)(1) = .75$

g. The probability that the next bottle filled will contain more than 7.25 ounces is:

$P(x > 7.25) = (7.5 - 7.25)(1) = .25$

The probability that the next 6 bottles filled will contain more than 7.25 ounces is:

$P[(x > 7.25) \cap (x > 7.25) \cap (x > 7.25) \cap (x > 7.25) \cap (x > 7.25) \cap (x > 7.25)]$
$= [P(x > 7.25)]^6 = .25^6 = .0002$

5.13   Let x = number of inches a gouge is from one end of the spindle. Then x has a uniform distribution with $f(x)$ as follows:

$$f(x) = \begin{cases} \dfrac{1}{d-c} = \dfrac{1}{18-0} = \dfrac{1}{18} & 0 \leq x \leq 18 \\ 0 & \text{otherwise} \end{cases}$$

In order to get at least 14 consecutive inches without a gouge, the gouge must be within 4 inches of either end. Thus, we must find:

$P(x < 4) + P(x > 14) = (4 - 0)(1/18) + (18 - 14)(1/18)$
$= 4/18 + 4/18 = 8/18 = .4444$

5.15 a. $P(0 < z < 2.00) = .4772$
(from Table IV, Appendix A)

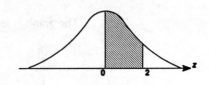

b. $P(0 < z < 1.00) = .3413$
(from Table IV, Appendix A)

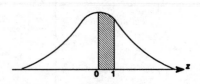

c. $P(0 < z < 3) = .4987$
(from Table IV, Appendix A)

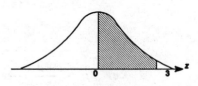

d. $P(0 < z < .58) = .2190$
(from Table IV, Appendix A)

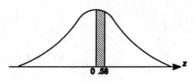

e. $P(-2.00 < z < 0) = .4772$
(from Table IV, Appendix A)

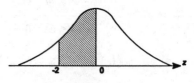

f. $P(-1.00 < z < 0) = .3413$
(from Table IV, Appendix A)

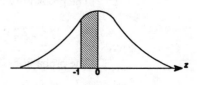

g. $P(-1.69 < z < 0) = .4545$
(from Table IV, Appendix A)

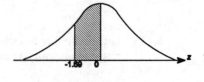

h. $P(-.58 < z < 0) = .2190$
(from Table IV, Appendix A)

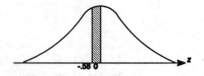

5.17 a. $P(z = 1) = 0$, since a single point does not have an area.

b. $P(z \leq 1) = P(z \leq 0) + P(0 < z \leq 1)$
$\quad\quad\quad\quad = A_1 + A_2$
$\quad\quad\quad\quad = .5 + .3413$
$\quad\quad\quad\quad = .8413$
$\quad\quad\quad\quad$(Table IV, Appendix A)

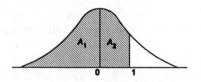

c. $P(z < 1) = P(z \leq 1) = .8413$ (Refer to part **b**.)

d. $P(z > 1) = 1 - P(z \leq 1) = 1 - .8413 = .1587$ (Refer to part **b**.)

e. $P(-1 \leq z \leq 1) = A_1 + A_2$
$\quad\quad\quad\quad\quad = .3413 + .3413$
$\quad\quad\quad\quad\quad = .6826$

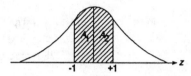

f. $P(-2 \leq z \leq 2) = A_1 + A_2$
$\quad\quad\quad\quad\quad = .4772 + .4772$
$\quad\quad\quad\quad\quad = .9544$

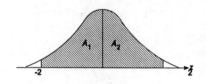

g. $P(-2.16 < z \leq 0.55) = A_1 + A_2$
$\quad\quad\quad\quad\quad\quad = .4846 + .2088$
$\quad\quad\quad\quad\quad\quad = .6934$

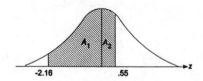

h. $P(-.42 < z < 1.96)$
$\quad\quad = P(-.42 \leq z \leq 0) + P(0 \leq z \leq 1.96)$
$\quad\quad = A_1 + A_2$
$\quad\quad = .1628 + .4750$
$\quad\quad = .6378$

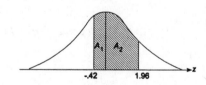

5.19 Using Table IV, Appendix A:

a. $P(z \geq z_0) = .05$
$A_1 = .5 - .05 = .4500$
Looking up the area .4500 in Table IV gives
$z_0 = 1.645$.

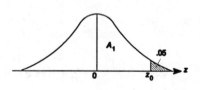

b. $P(z \geq z_0) = .025$
$A_1 = .5 - .025 = .4750$
Looking up the area .4750 in Table IV
gives $z_0 = 1.96$.

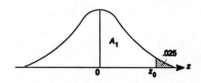

c. $P(z \le z_0) = .025$
$A_1 = .5 - .025 = .4750$
Looking up the area .4750 in Table IV gives
$z = 1.96$. Since z_0 is to the left of 0, $z_0 = -1.96$.

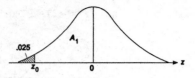

d. $P(z \ge z_0) = .10$
$A_1 = .5 - .1 = .4$
Looking up the area .4000 in Table IV
gives $z_0 = 1.28$.

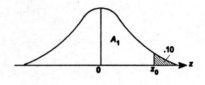

e. $P(z > z_0) = .10$
$A_1 = .5 - .1 = .4$
$z_0 = 1.28$ (same as in **d**)

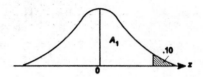

5.21   a. $z = 1$

b. $z = -1$

c. $z = 0$

d. $z = -2.5$

e. $z = 3$

5.23   a. $P(x < \mu - 2\sigma) + P(x > \mu + 2\sigma) = P(z < -2) + P(z > 2)$
$$= (.5 - .4772) + (.5 - .4772)$$
$$= 2(.5 - .4772) = .0456$$
(from Table IV, Appendix A)

$P(x < \mu - 3\sigma) + P(x > \mu + 3\sigma) = P(z < -3) + P(z > 3)$
$$= (.5 - .4987) + (.5 - .4987)$$
$$= 2(.5 - .4987) = .0026$$
(from Table IV, Appendix A)

b. $P(\mu - \sigma < x < \mu + \sigma) = P(-1 < z < 1)$
$$= P(-1 < z < 0) + P(0 < z < 1)$$
$$= .3413 + .3413$$
$$= 2(.3413) = .6826$$
(from Table IV, Appendix A)

$P(\mu - 2\sigma < z < \mu + 2\sigma) = P(-2 < z < 2)$
$$= P(-2 < z < 0) + P(0 < z < 2)$$
$$= .4772 + .4772$$
$$= 2(.4772) = .9544$$
(from Table IV, Appendix A)

c. $P(x \leq x_0) = .80$. Find x_0.

$$P(x \leq x_0) = P\left(z \leq \frac{x_0 - 300}{30}\right) = P(z \leq z_0) = .80$$

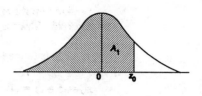

$A_1 = .80 - .5 = .3000$

Looking up area .3000 in Table IV, $z_0 = .84$.

$$z_0 = \frac{x_0 - 300}{30} \Rightarrow .84 = \frac{x_0 - 300}{30} \Rightarrow x_0 = 325.2$$

$P(x \leq x_0) = .10$. Find x_0.

$$P(x \leq x_0) = P\left(z \leq \frac{x_0 - 300}{30}\right) = P(z \leq z_0) = .10$$

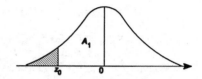

$A_1 = .50 - .10 = .4000$

Looking up area .4000 in Table IV, $z_0 = -1.28$.

$$z_0 = \frac{x_0 - 300}{30} \Rightarrow -1.28 = \frac{x_0 - 300}{30} \Rightarrow x_0 = 261.6$$

5.25 a. $P(x \geq x_0) = .5 \Rightarrow P\left(z \geq \frac{x_0 - 30}{8}\right)$

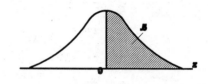

$$= P(x \geq z_0) = .5$$
$$\Rightarrow z_0 = 0 = \frac{x_0 - 30}{8}$$
$$\Rightarrow x_0 = 8(0) + 30 = 30$$

b. $P(x < x_0) = .025 \Rightarrow P\left(z < \frac{x_0 - 30}{8}\right)$

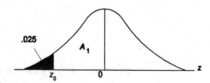

$$= P(z < z_0) = .025$$

$A_1 = .5 - .025 = .4750$

Looking up the area .4750 in Table IV gives $z_0 = 1.96$. Since z_0 is to the left of 0, $z_0 = -1.96$.

$$z_0 = -1.96 = \frac{x_0 - 30}{8} \Rightarrow x_0 = 8(-1.96) + 30 = 14.32$$

c. $P(x > x_0) = .10 \Rightarrow P\left(z > \frac{x_0 - 30}{8}\right)$

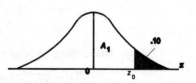

$$= P(z > z_0) = .10$$

$A_1 = .5 - .10 = .4000$

Looking up the area .4000 in Table IV gives $z_0 = 1.28$.

$$z_0 = 1.28 = \frac{x_0 - 30}{8} \Rightarrow x_0 = 8(1.28) + 30 = 40.24$$

d. $P(x > x_0) = .95 \Rightarrow P\left(z > \dfrac{x_0 - 30}{8}\right)$

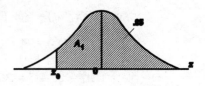

$= P(z > z_0) = .95$

$A_1 = .95 - .50 = .4500$

Looking up the area .4500 in Table IV gives $z_0 = 1.645$.

Since z_0 is to the left of 0, $z_0 = -1.645$.

$z_0 = -1.645 = \dfrac{x_0 - 30}{8} \Rightarrow x_0 = 8(-1.645) + 30 = 16.84$

e. $P(x < x_0) = .10 \Rightarrow P\left(z < \dfrac{x_0 - 30}{8}\right)$

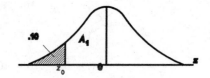

$= P(z < z_0) = .10$

$A_1 = .5 - .10 = .4000$

Looking up the area .4000 in Table IV gives $z_0 = 1.28$.

Since z_0 is to the left of 0, $z_0 = -1.28$.

$z_0 = -1.28 = \dfrac{x_0 - 30}{8} \Rightarrow x_0 = 8(-1.28) + 30 = 19.76$

f. $P(x < x_0) = .80 \Rightarrow P\left(z < \dfrac{x_0 - 30}{8}\right)$

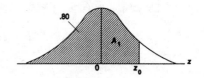

$= P(z < z_0) = .80$

$A_1 = .80 - .5 = .3000$

Looking up the area .3000 in Table IV gives $z_0 = 0.84$.

$z_0 = 0.84 = \dfrac{x_0 - 30}{8} \Rightarrow x_0 = 8(0.84) + 30 = 36.72$

g. $P(x > x_0) = .01 \Rightarrow P\left(z > \dfrac{x_0 - 30}{8}\right)$

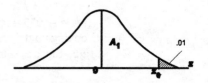

$= P(z > z_0) = .01$

$A_1 = .5 - .01 = .4900$

Looking up the area .4900 in Table IV gives $z_0 = 2.33$.

$z_0 = 2.33 = \dfrac{x_0 - 30}{8} \Rightarrow x_0 = 8(2.33) + 30 = 48.64$

5.27 The random variable x has a normal distribution with $\sigma = 25$.

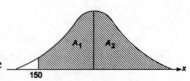

We know $P(x > 150) = .90$. So, $A_1 + A_2 = .90$. Since $A_2 = .50$, $A_1 = .90 - .50 = .40$. Look up the area .40 in the body of Table IV; (take the closest value) $z_0 = -1.28$.

To find μ, substitute all the values into the z-score formula:

$$z = \frac{x - \mu}{\sigma}$$

$$\Rightarrow -1.28 = \frac{150 - \mu}{25}$$

$$\Rightarrow \quad \mu = 150 + 25(1.28) = 182$$

5.29   a.   Let x = alkalinity level of water specimens collected from the Han River.

Using Table IV, Appendix A,

$$P(x > 45) = P\left(z > \frac{45 - 50}{3.2} \right) = P(z > -1.56) = .5 + .4406 = .9406.$$

b.   Using Table IV, Appendix A,

$$P(x < 55) = P\left(z < \frac{55 - 50}{3.2} \right) = P(z < 1.56) = .5 + .4406 = .9406.$$

c.   Using Table IV, Appendix A,

$$P(51 < x < 52) = P\left(\frac{51 - 50}{3.2} < z < \frac{52 - 50}{3.2} \right) = P(.31 < z < .63) = .2357 - .1217 = .1140.$$

5.31   a.   Let x = batting average of an American League player. Using Table IV, Appendix A,

$$P(x \geq .300) = P\left(z \geq \frac{.300 - .268}{.031} \right) = P(z \geq 1.03) = .5 - .3485 = .1515$$

b.   Let x = batting average of a National League player. Using Table IV, Appendix A,

$$P(x \geq .300) = P\left(z \geq \frac{.300 - .262}{.039} \right) = P(z \geq .97) = .5 - .3340 = .1660$$

5.33   a.   Using Table IV, Appendix A, with $\mu = 20.2$ and $\sigma = .65$,

$$P(20 < x < 21) = P\left(\frac{20 - 20.2}{.65} < z < \frac{21 - 20.2}{.65} \right) = P(-.31 < z < 1.23)$$
$$= P(-.31 < z < 0) + P(0 \leq z < 1.23) = .1217 + .3907 = .5124$$

b. $P(x < 19.84) = P\left(z < \frac{1984 - 20.2}{.65} \right) = P(z < -.55) = .5 - .2088 = .2912$

c. $P(x > 22.01) = P\left(z > \frac{22.01 - 20.2}{.65} \right) = P(z > 2.78) = .5 - .4973 = .0027$

5.35 a. Using Table IV, Appendix A, with $\mu = 9.06$ and $\sigma = 2.11$,

$$P(x < 6) = \left(z < \frac{6 - 9.06}{2.11} \right) = P(z < -1.45) = .5 - P(-1.45 \le z \le 0)$$
$$= .5 - .4265 = .0735$$

b. $P(8 < x\ 10) = P\left(\frac{8 - 9.06}{2.11} < z < \frac{10 - 9.06}{2.11} \right) = P(-.50 < z < .45)$

$$= P(-.50 < z < 0) + P(0 < z < .45) = .1915 + .1736 = .3651$$

c. $P(x < x_0) = .2000$.

If $P(x < x_0) = .2000$, then $P\left(z < \frac{x_0 - 9.06}{2.11} \right) = P(z < z_0) = .2000$

If $P(z < z_0)$, then $P(z_0 < z < 0) = .3000$. Looking up .3000 in Table IV, the z-score is .84. Since $z_0 < 0$, $z_0 = -.84$. Now, we must convert z_0 back to an x score.

$$z_0 = \frac{x_0 - 9.06}{2.11} \Rightarrow -.84 = \frac{x_0 - 9.06}{2.11} \Rightarrow x_0 = 2.11(-.84) + 9.06 = 7.29$$

5.37 Using Table IV, Appendix A, with $\mu = 50$ and $\sigma = 12$,

$$P(x > d) = .30 \Rightarrow P\left(z > \frac{d - 50}{12} \right) = P(z > z_0) = .30$$

$A_1 = .5 - .30 = .2000$
Looking up the area .2000 in Table IV gives $z_0 = .52$.

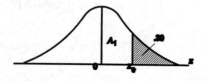

$$z_0 = .52 = \frac{d - 50}{12} \Rightarrow d = 12(.52) + 50 = 56.24 \text{ cm}$$

5.39 a. Using Table IV, Appendix A, with $\mu = 208$ and $\sigma = 25$,

$$P(x < 200) = P\left(z < \frac{200 - 208}{25} \right) = P(z < -.32) = .5 - .1255 = .3745$$

b. Let y = number of patients out of three who have cholesterol levels below 200. Assuming that the patients are independent, then y is a binomial random variable with $n = 3$ and $p = .3745$.

$$P(y \ge 1) = 1 - P(y = 0) = 1 - \binom{3}{0}.3745^0.6255^3 = 1 - .2447 = .7553$$

5.41   a.   If z is a standard normal random variable,

$Q_L = z_L$ is the value of the standard normal distribution which has 25% of the data to the left and 75% to the right.

Find z_L such that $P(z < z_L) = .25$

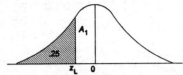

$A_1 = .50 - .25 = .25$.

Look up the area $A_1 = .25$ in the body of Table IV of Appendix A; $z_L = -.67$ (taking the closest value). If interpolation is used, $-.675$ would be obtained.

$Q_U = z_U$ is the value of the standard normal distribution which has 75% of the data to the left and 25% to the right.

Find z_U such that $P(z < z_U) = .75$

$$A_1 + A_2 = P(z \le 0) + P(0 \le z \le z_U)$$
$$= .5 + P(0 \le z \le z_U)$$
$$= .75$$

Therefore, $P(0 \le z \le z_U) = .25$.

Look up the area .25 in the body of Table IV of Appendix A; $z_U = .67$ (taking the closest value).

b.   Recall that the inner fences of a box plot are located $1.5(Q_U - Q_L)$ outside the hinges (Q_L and Q_U).

To find the lower inner fence,

$$Q_L - 1.5(Q_U - Q_L) = -.67 - 1.5(.67 - (-.67))$$
$$= -.67 - 1.5(1.34)$$
$$= -2.68 \ (-2.70 \text{ if } z_L = -.675 \text{ and } z_U = +.675)$$

The upper inner fence is:

$$Q_U + 1.5(Q_U - Q_L) = .67 + 1.5(.67 - (-.67))$$
$$= .67 + 1.5(1.34)$$
$$= 2.68 \ (+2.70 \text{ if } z_L = -.675 \text{ and } z_U = +.675)$$

c.   Recall that the outer fences of a box plot are located $3(Q_U - Q_L)$ outside the hinges (Q_L and Q_U).

To find the lower outer fence,

$$Q_L - 3(Q_U - Q_L) = -.67 - 3(.67 - (.67))$$
$$= -.67 - 3(1.34)$$
$$= -4.69 \ (-4.725 \text{ if } z_L = -.675 \text{ and } z_U = +.675)$$

The upper outer fence is:

$$Q_U + 3(Q_U - Q_L) = .67 + 3(.67 - (-.67))$$
$$= .67 + 3(1.34)$$
$$= 4.69 \ (4.725 \text{ if } z_L = -,675 \text{ and } z_U = +.675)$$

d. $P(z < -2.68) + P(z > 2.68)$
$= 2P(z > 2.68)$
$= 2(.5000 - .4963)$
 (Table IV, Appendix A)
$= 2(.0037) = .0074$

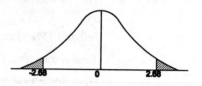

(or $2(.5000 - .4965) = .0070$ if -2.70 and 2.70 are used)

$P(z < -4.69) + P(z > 4.69)$
$= 2P(z > 4.69)$
$\approx 2(.5000 - .5000) \approx 0$

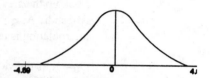

e. In a normal probability distribution, the probability of an observation being beyond the inner fences is only .0074 and the probability of an observation being beyond the outer fences is approximately zero. Since the probability is so small, there should not be any observations beyond the inner and outer fences. Therefore, they are probably outliers.

5.43 a. The proportion of measurements that one would expect to fall in the interval $\mu \pm \sigma$ is about .68.

b. The proportion of measurements that one would expect to fall in the interval $\mu \pm 2\sigma$ is about .95.

c. The proportion of measurements that one would expect to fall in the interval $\mu \pm 3\sigma$ is about 1.00.

5.45 If the data are normally distributed, then the normal probability plot should be an approximate straight line. Of the three plots, only plot c implies that the data are normally distributed. The data points in plot c form an approximately straight line. In both plots a and b, the plots of the data points do not form a straight line.

5.47 a. $\bar{x} \pm s \Rightarrow 48.16 \pm 6.015 \Rightarrow (42.145, 54.175)$ From the data, there are 38 ages in the interval or $38/50 = .76$. From the Empirical Rule, if the data are normal, there should be approximately .68 of the ages in this interval.

$\bar{x} \pm 2s \Rightarrow 48.16 \pm 2(6.015) \Rightarrow 48.16 \pm 12.030 \Rightarrow (36.130, 60.190)$ From the data, there are 47 ages in the interval or $47/50 = .94$. From the Empirical Rule, if the data are normal, there should be approximately .95 of the ages in this interval.

$\bar{x} \pm 3s \Rightarrow 48.16 \pm 3(6.015) \Rightarrow 48.16 \pm 18.045 \Rightarrow (30.115, 66.205)$ From the data, there are 49 ages in the interval or $49/50 = .98$. From the Empirical Rule, if the data are normal, there should be approximately all of the ages in this interval.

Using the Empirical Rule, there are more observations within one standard deviation than what you would expect if the data are normal. Thus, the data may not be normal.

In addition, if the data are normal, then IQR/s ≈ 1.3. From the printout, IQR = Q3 – Q1 = 51.25 – 45.00 = 6.25. IQR / s = 6.25 / 6.015 = 1.039. This is not particularly close 1.3, indicating the data may not be normal.

b. Using MINITAB, a relative frequency histogram of the data is:

```
Histogram of Age    N = 50

Midpoint          Count
    36              2    **
    40              3    ***
    44             11    **********
    48             17    *****************
    52             11    **********
    56              2    **
    60              2    **
    64              1    *
    68              1    *
```

From the histogram, the data are fairly mound-shaped. However, the data are somewhat skewed to the right. Thus, the data may not be normal.

5.49 a. Using MINITAB, the histogram for the data is:

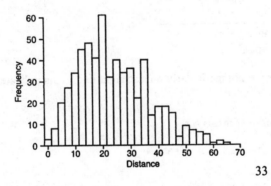

From the plot, the data appear to be somewhat mound-shaped, but it also appears that the data are somewhat skewed to the right.

b. $\bar{x} \pm s \Rightarrow 24.4 \pm 12.8 \Rightarrow (11.6, 37.2)$

$\bar{x} \pm 2s \Rightarrow 24.4 \pm 2(12.8) \Rightarrow 24.2 \pm 25.6 \Rightarrow (-1.4, 49.8)$

$\bar{x} \pm 3s \Rightarrow 24.4 \pm 3(12.8) \Rightarrow 24.2 \pm 38.4 \Rightarrow (-14.2, 62.6)$

Of the 590 measurements, 405 are in the interval 11.6 – 37.2. The proportion is 405/590 = .686. This is close to the proportion stated by the Empirical Rule.

Of the 590 measurements, 560 are in the interval –1.4 – 49.8. The proportion is 560/590 = .949. This is close to the proportion stated by the Empirical Rule.

Of the 590 measurements, 589 are in the interval –14.2 – 62.6. The proportion is 589/590 = .998. This is close to the proportion stated by the Empirical Rule.

This would imply that the data are approximately normal.

$IQR = Q_U - Q_L = 34 - 14 = 20$. $IQR/s = 20/12.8 = 1.5625$. If the data are normally distributed, this ratio should be close to 1.3. Since 1.5625 is fairly close to 1.3, this indicates that the data are approximately normal.

c. Using MINITAB, the normal probability plot is:

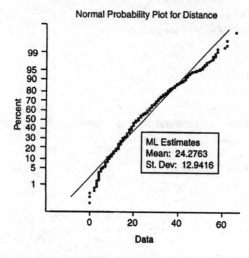

Normal Probability Plot for Distance

34

The data at the extremes are not particularly on a straight line. This indicates that the data are nor normally distributed.

5.51 Using MINITAB, the stem-and-leaf display is:

```
Stem-and-leaf of St. Louis    N = 58
Leaf Unit = 0.10

     2   1   00
     7   2   00000
    15   3   00000000
    20   4   00000
    26   5   000000
   (11)  6   00000000000
    21   7   0000
    17   8   00000000
     9   9   00
     7  10   00
     5  11   00
     3  12
     3  13   0
     2  14   0
     1  15   0
```

The data are somewhat mound-shaped, but also appear to be somewhat skewed to the right.

Using MINITAB, the descriptive statistics are:

Variable	N	Mean	Median	TrMean	StDev	SE Mean
St. Louis	58	5.966	6.000	5.769	3.129	0.4110

Variable	Minimum	Maximum	Q1	Q3
St. Louis	1.000	15.000	3.0000	8.000

$\bar{x} \pm s \Rightarrow 5.966 \pm 3.129 \Rightarrow (2.837, 9.095)$

$\bar{x} \pm 2s \Rightarrow 5.966 \pm 2(3.129) \Rightarrow 5.966 \pm 6.258 \Rightarrow (-0.292, 12.224)$

$\bar{x} \pm 3s \Rightarrow 5.966 \pm 3(3.129) \Rightarrow 5.966 \pm 9.387 \Rightarrow (-3.421, 15.353)$

Of the 58 measurements, 44 are in the interval 2.837 – 9.095. The proportion is 44/58 = .759. This is somewhat larger than the proportion stated by the Empirical Rule.

Of the 58 measurements, 55 are in the interval, –0.292 – 12.224. The proportion is 55/58 = .948. This is close to the proportion stated by the Empirical Rule.

Of the 58 measurements, 58 are in the interval, –3.421 – 15.353. The proportion is 58/58 = 1.000. This is close to the proportion stated by the Empirical Rule.

This would imply that the data are approximately normal.

IQR = $Q_U - Q_L = 8 - 3 = 5$. IQR/s = 5/3.129 = 1.598. If the data are normally distributed, this ratio should be close to 1.3. Since 1.598 is fairly close to 1.3, this indicates that the data are approximately normal.

Using MINITAB, the normal probability plot is:

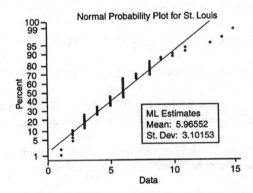

The data at the extremes are not particularly on a straight line. This indicates that the data are not normally distributed.

Several of the checks indicate that the data may not be normal.

b.   Using MINITAB, the stem-and-leaf display is:

```
Stem-and-leaf of Chicago     N = 54
Leaf Unit = 0.10

    2   1  00
    6   2  0000
   10   3  0000
   16   4  000000
   26   5  0000000000
   (8)  6  00000000
   20   7  000
   17   8  00000
   12   9  00000
    7  10  000
    4  11  00
    2  12
    2  13  0
    1  14
    1  15  0
```

The data is fairly mound shaped.

Using MINITAB, the descriptive statistics are:

Variable	N	Mean	Median	TrMean	StDev	SE Mean
Chicago	54	6.111	6.000	5.979	3.007	0.409

Variable	Minimum	Maximum	Q1	Q3
Chicago	1.000	15.000	4.000	8.000

$\bar{x} \pm s \Rightarrow 6.111 \pm 3.007 \Rightarrow (3.104, 9.118)$

$\bar{x} \pm 2s \Rightarrow 6.111 \pm 2(3.007) \Rightarrow 6.111 \pm 6.014 \Rightarrow (0.097, 12.125)$

$\bar{x} \pm 3s \Rightarrow 6.111 \pm 3(3.007) \Rightarrow 6.111 \pm 9.021 \Rightarrow (-2.91, 15.132)$

Of the 54 measurements, 37 are in the interval $3.104 - 9.118$. The proportion is 37/54 = .685. This is very close to the proportion stated by the Empirical Rule.

Of the 54 measurements, 52 are in the interval $0.097 - 12.125$. The proportion is 52/54 = .963. This is close to the proportion stated by the Empirical Rule.

Of the 54 measurements, 54 are in the interval $-2.91 - 15.132$. The proportion is 54/54 = 1.000. This is close to the proportion stated by the Empirical Rule.

This would imply that the data are approximately normal.

IQR $= Q_U - Q_L = 8 - 4 = 4$. IQR$/s = 4/3.007 = 1.33$. If the data are normally distributed, this ratio should be close to 1.3. Since 1.33 is very close to 1.3, this indicates that the data are approximately normal.

Using MINITAB, the normal probability is:

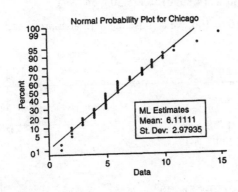

Normal Probability Plot for Chicago

ML Estimates
Mean: 6.11111
St. Dev: 2.97935

35

The data are very close to a straight line. This indicates that the data are normally distributed.
All of the checks are consistent with the data being normal.

5.53   The lowest score possible is 0. The sample mean is 7.62 and the standard deviation is 8.91. Since the lower limit of 0 is less than one standard deviation below the mean, this implies that the data are not normal, but skewed to the right.

5.55   a.   In order to approximate the binomial distribution with the normal distribution, the interval $\mu \pm 3\sigma \Rightarrow np \pm 3\sqrt{npq}$ should lie in the range 0 to n.

When $n = 25$ and $p = .4$,
$$np \pm 3\sqrt{npq} \Rightarrow 25(.4) \pm 3\sqrt{25(.4)(1-.4)}$$
$$\Rightarrow 10 \pm 3\sqrt{6} \Rightarrow 10 \pm 7.3485 \Rightarrow (2.6515, 17.3485)$$

Since the interval calculated does lie in the range 0 to 25, we can use the normal approximation.

b. $\mu = np = 25(.4) = 10$
$\sigma^2 = npq = 25(.4)(.6) = 6$

c. $P(x \geq 9) = 1 - P(x \leq 8) = 1 - .274 = .726$ (Table II, Appendix A)

d. $P(x \geq 9) \approx P\left(z \geq \dfrac{(9-.5)-10}{\sqrt{6}}\right)$
$= P(z \geq -.61)$
$= .5000 + .2291 = .7291$
(Using Table IV in Appendix A.)

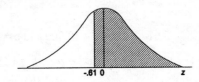

5.57 x is a binomial random variable with $n = 100$ and $p = .4$.

$$\mu \pm 3\sigma \Rightarrow np \pm 3\sqrt{npq} \Rightarrow 100(.4) \pm 3\sqrt{100(.4)(1-.4)}$$
$$\Rightarrow 40 \pm 3(4.8990) \Rightarrow (25.303, 54.697)$$

Since the interval lies in the range 0 to 100, we can use the normal approximation to approximate the probabilities.

a. $P(x \le 35) \approx P\left(z \le \dfrac{(35+.5)-40}{4.899}\right)$

$= P(z \le -.92)$

$= .5000 - .3212 = .1788$

(Using Table IV in Appendix A.)

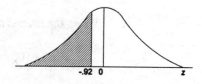

b. $P(40 \le x \le 50)$

$\approx P\left(\dfrac{(40-.5)-40}{4.899} \le z \le \dfrac{(50+.5)-40}{4.899}\right)$

$= P(-.10 \le z \le 2.14)$

$= P(-.10 \le z \le 0) + P(0 \le z \le 2.14)$

$= .0398 + .4838 = .5236$

(Using Table IV in Appendix A.)

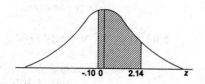

c. $P(x \ge 38) \approx P\left(z \ge \dfrac{(38-.5)-40}{4.899}\right)$

$= P(z \ge -.51)$

$= .5000 + .1950 = .6950$

(Using Table IV in Appendix A.)

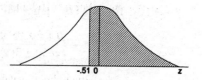

5.59   a. $\mu = E(x) = np = 200(.05) = 10$. This is the same value that was found in Exercise 4.43 c.

b. $\sigma = \sqrt{npq} = \sqrt{200(.05)(1-.05)} = \sqrt{9.5} = 3.082$

c. $z = \dfrac{x-\mu}{\sigma} = \dfrac{10.5-10}{3.082} = .16$

d. $P(x \ge 11) \approx P\left(z \ge \dfrac{10.5-10}{3.082}\right) = P(z \ge .16) = .5 - .0636 = .4364$

(Using Table IV, Appendix A)

5.61   a. $n = 10,000$ and $p = .60$

$E(x) = \mu = np = 10,000(.60) = 6000$

$\sigma^2 = npq = 10,000(.60)(.40) = 2400$

b. $\sigma = \sqrt{2400} = 48.99$

Using the normal approximation to the binomial,

$$P(x > 6100) \approx P\left(z > \frac{(6100 + .5) - 6000}{48.99}\right) = P(z > 2.05) = .5 - .4798 = .0202$$

(from Table IV, Appendix A)

c. Using the normal approximation to the binomial,

$$P(x > 6500) = P\left(z > \frac{(6500 + .5) - 6000}{48.99}\right) \; P(z > 10.21) \approx .5 - .5 = 0$$

Since this probability is essentially 0, we would not expect to see more than 6500 deaths in any one year.

5.63   a.   For $n = 100$ and $p = .01$:

$$\mu \pm 3\sigma \Rightarrow np \pm 3\sqrt{npq} \Rightarrow 100(.01) \pm 3\sqrt{100(.01)(.99)}$$
$$\Rightarrow 1 \pm 3(.995) \Rightarrow 1 \pm 2.985 \Rightarrow (-1.985, 3.985)$$

Since the interval does not lie in the range 0 to 100, we cannot use the normal approximation to approximate the probabilities.

b.   For $n = 100$ and $p = .5$:

$$\mu \pm 3\sigma \Rightarrow np \pm 3\sqrt{npq} \Rightarrow 100(.5) \pm 3\sqrt{100(.5)(.5)}$$
$$\Rightarrow 50 \pm 3(5) \Rightarrow 50 \pm 15 \Rightarrow (35, 65)$$

Since the interval lies in the range 0 to 100, we can use the normal approximation to approximate the probabilities.

c.   For $n = 100$ and $p = .9$:

$$\mu \pm 3\sigma \Rightarrow np \pm 3\sqrt{npq} \Rightarrow 100(.9) \pm 3\sqrt{100(.9)(.1)}$$
$$\Rightarrow 90 \pm 3(3) \Rightarrow 90 \pm 9 \Rightarrow (81, 99)$$

Since the interval lies in the range 0 to 100, we can use the normal approximation to approximate the probabilities.

5.65   a.   Let x = number of abused women in a sample of 150. The random variable x is a binomial random variable with $n = 150$ and $p = 1/3$. Thus, for the normal approximation,

$$\mu = np = 150(1/3) = 50 \text{ and } \sigma = \sqrt{npq} = \sqrt{150(1/3)(2/3)} = 5.7735$$
$$\mu \pm 3\sigma \Rightarrow 50 \pm 3(5.7735) \Rightarrow 50 \pm 17.3205 \Rightarrow (32.6795, 67.3205)$$

Since this interval lies in the range from 0 to $n = 150$, the normal approximation is appropriate.

$$P(x > 75) \approx P\left(z > \frac{(75 + .5) - 50}{5.7735}\right) = P(z > 4.42) \approx .5 - .5 = 0$$

(Using Table IV, Appendix A.)

b. $P(x < 50) \approx P\left(z < \dfrac{(50-.5)-50}{5.7735}\right) = P(z < -.09) \approx .5 - .0359 = .4641$

c. $P(x < 30) \approx P\left(z < \dfrac{(30-.5)-50}{5.7735}\right) = P(z < -3.55) \approx .5 - .5 = 0$

Since the probability of seeing fewer than 30 abused women in a sample of 150 is so small ($p \approx 0$), it would be very unlikely to see this event.

5.67   Let x = number of defective CDs in a sample of 1600. Then x is a binomial random variable with $n = 1600$ and $p = .006$. To see if the normal approximation is appropriate, we check:

$$np \pm 3\sqrt{npq} \Rightarrow 1600(.006) \pm 3\sqrt{1600(.006)(.994)}$$
$$\Rightarrow 9.6 \pm 3(3.089) \Rightarrow 9.6 \pm 9.267 \Rightarrow (.333, 18.867)$$

Since this interval falls in the interval 0 to $n = 1600$, the normal approximation is appropriate.

$$P(x \geq 12) \approx P\left(z \geq \dfrac{(12-.5)-9.6}{3.089}\right)$$
$$= P(z \geq .62)$$
$$= .5 - .2324 = .2676$$

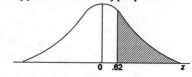

Since the probability of seeing 12 or more defective disks is not small ($p = .2676$), it would not be unusual if 99.4% of the disks are defect free. This does not cast doubt on the manufacturer's claim.

5.69   a. If 80% of the passengers pass through without their luggage being inspected, then 20% will be detained for luggage inspection. The expected number of passengers detained will be:

$$E(x) = np = 1,500(.2) = 300$$

b. For $n = 4,000$, $E(x) = np = 4,000(.2) = 800$

c. $P(x > 600) \approx P\left(z > \dfrac{(600+.5)-800}{\sqrt{4000(.2)(.8)}}\right) = P(z > -7.89) = .5 + .5 = 1.0$

5.71   For $\theta = 3$, $f(x) = \dfrac{1}{3}e^{-x/3}$. Using Table V, Appendix A:

x	$f(x)$
0	.333
1	.239
2	.171
3	.123
4	.088
5	.063
6	.045

For $\theta = 1, f(y) = e^{-y}$. Using Table V, Appendix A:

x	$f(x)$
0	1.000
1	.368
2	.135
3	.050
4	.018
5	.007
6	.002

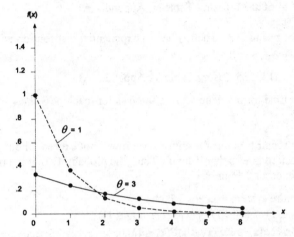

5.73 $P(x \geq a) = e^{-a/\theta} = e^{-a/1}$. Using Table V, Appendix A:

a. $P(x > 1) = e^{-1/1} = e^{-1} = .367879$

b. $P(x \leq 3) = 1 - P(x > 3) = 1 - e^{-3/1} = 1 - e^{-3} = 1 - .049787 = .950213$

c. $P(x > 1.5) = e^{-1.5/1} = e^{-1.5} = .223130$

d. $P(x \leq 5) = 1 - P(x > 5) = 1 - e^{-5/1} = 1 - e^{-5} = 1 - .006738 = .993262$

5.75 $\mu = \theta = 1, \sigma = \theta = 1$

$\mu \pm 2\sigma \Rightarrow 1 \pm 2(1) \Rightarrow 1 \pm 2 \Rightarrow (-1, 3)$

$P(-1 < x < 3) = P(0 < x < 3) = 1 - P(x \geq 3) = 1 - e^{-3/1}$
$$= 1 - e^{-3} = 1 - .049787 = .950213$$
(from Table V, Appendix A)

5.77   a. $P(x > 2) = e^{-2/2.5} = e^{-.8} = .449329$ (using Table V, Appendix A)

b. $P(x < 5) = 1 - P(x \geq 5) = 1 - e^{-5/2.5} = 1 - e^{-2} = 1 - .135335 = .864665$
(Using Table V, Appendix A)

5.79   Let x = breast height diameter of the western hemlock. Then x has an exponential distribution with a mean of 30 cm.

$$P(x > 25) + e^{-25/30} = e^{-.833333} = .434598$$

5.81   a.   Let x_1 = repair time for machine 1. Then x_1 has an exponential distribution with μ_1 = 1 hour.

$$P(x_1 > 1) = e^{-1/1} = e^{-1} = .367879 \text{ (using Table V, Appendix A)}$$

b.   Let x_2 = repair time for machine 2. Then x_2 has an exponential distribution with μ_2 = 2 hours.

$$P(x_2 > 1) = e^{-1/2} = e^{-.5} = .606531 \text{ (using Table V, Appendix A)}$$

c.   Let x_3 = repair time for machine 3. Then x_3 has an exponential distribution with μ_3 = .5 hours.

$$P(x_3 > 1) = e^{-1/.5} = e^{-2} = .135335 \text{ (using Table V, Appendix A)}$$

Since the mean repair time for machine 4 is the same as for machine 3, $P(x_4 > 1)$ = $P(x_3 > 1)$ = .135335.

d.   The only way that the repair time for the entire system will not exceed 1 hour is if all four machines are repaired in less than 1 hour. Thus, the probability that the repair time for the entire system exceeds 1 hour is:

$$P(\text{Repair time entire system exceeds 1 hour})$$
$$= 1 - P((x_1 \le 1) \cap (x_2 \le 1) \cap (x_3 \le 1) \cap (x_4 \le 1))$$
$$= 1 - P(x_1 \le 1)P(x_2 \le 1)P(x_3 \le 1)P(x_4 \le 1)$$
$$= 1 - (1 - .367879)(1 - .606531)(1 - .135335)(1 - .135335)$$
$$= 1 - (.632121)(.393469)(.864665)(.864665) = 1 - .185954 = .814046$$

5.83   a.   For $\mu = 17 = \theta$. To graph the distribution, we will pick several values of x and find the value of $f(x)$, where x = time between arrivals of the smaller craft at the pier.

$$f(x) = \frac{1}{\theta} e^{-x/\theta} = \frac{1}{17} e^{-x/17}$$

$$f(1) = \frac{1}{17} e^{-1/17} = .0555$$

$$f(3) = \frac{1}{17} e^{-3/17} = .0493$$

$$f(5) = \frac{1}{17} e^{-5/17} = .0438$$

$$f(7) = \frac{1}{17} e^{-7/17} = .0390$$

$$f(10) = \frac{1}{17} e^{-10/17} = .0327$$

$$f(15) = \frac{1}{17} e^{-15/17} = .0243$$

$$f(20) = \frac{1}{17}e^{-20/17} = .0181$$

$$f(25) = \frac{1}{17}e^{-25/17} = .0135$$

The graph is:

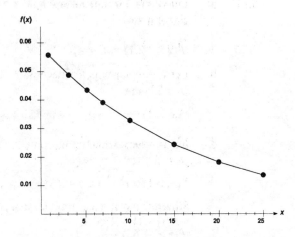

b. We want to find the probability that the time between arrivals is less than 15 minutes.

$$P(x < 15) = 1 - P(x \geq 15) = 1 - e^{-15/17} = 1 - .4138 = .5862$$

5.85   a.   For $\theta = 250$, $P(x > a) = e^{-a/250}$

For $a = 300$ and $b = 200$, show $P(x > a + b) \geq P(x > a)P(x > b)$

$$P(x > 300 + 200) = P(x > 500) = e^{-500/250} = e^{-2} = .1353$$

$$P(x > 300) P(x > 200) = e^{-300/250} e^{-200/250} = e^{-1.2} e^{-.8} = .3012(.4493) = .1353$$

Since $P(x > 300 + 200) = P(x > 300) P(x > 200)$, then
$P(x > 300 + 200) \geq P(x > 300) P(x > 200)$

Also, show $P(x > 300 + 200) \leq P(x > 300) P(x > 200)$. Since we already showed that
$P(x > 300 + 200) = P(x > 300) P(x > 200)$,
then $P(x > 300 + 200) \leq P(x > 300) P(x > 200)$.

b.   Let $a = 50$ and $b = 100$. Show $P(x > a + b) \geq P(x > a) P(x > b)$

$$P(x > 50 + 100) = P(x > 150) = e^{-150/250} = e^{-.6} = .5488$$

$$P(x > 50) P(x > 100) = e^{-50/250} e^{-100/250} = e^{-.2}e^{-.4} = .8187(.6703) = .5488$$

Since $P(x > 50 + 100) = P(x > 50) P(x > 100)$, then
$P(x > 50 + 100) \geq P(x > 50) P(x > 100)$

Also, show $P(x > 50 + 100) \leq P(x > 50) \, P(x > 100)$. Since we already showed that
$P(x > 50 + 100) = P(x > 50) \, P(x > 100)$,
then $P(x > 50 + 100) \leq P(x > 50) \, P(x > 100)$.

c. Show $P(x > a + b) \geq P(x > a) \, P(x > b)$
$P(x > a + b) = e^{-(a+b)/250} = e^{-a/250} \, e^{-b/250} = P(x > a) \, P(x > b)$

5.87 a. $P(z \leq -2.1) = A_1 + A_2$
$\qquad\qquad = .5 + .4821$
$\qquad\qquad = .9821$

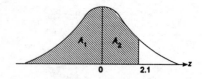

b. $P(z \geq 2.1) = A_2 = .5 - A_1$
$\qquad\qquad = .5 - .4821$
$\qquad\qquad = .0179$

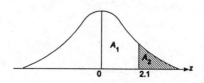

c. $P(z \geq -1.65) = A_1 + A_2$
$\qquad\qquad = .4505 + .5000$
$\qquad\qquad = .9505$

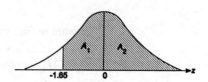

d. $P(-2.13 \leq z \leq -.41)$
$\qquad = P(-2.13 \leq z \leq 0) - P(-.41 \leq z \leq 0)$
$\qquad = .4834 - .1591$
$\qquad = .3243$

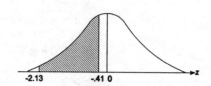

e. $P(-1.45 \leq z \leq 2.15) = A_1 + A_2$
$\qquad\qquad = .4265 + .4842$
$\qquad\qquad = .9107$

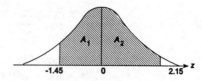

f. $P(z \leq -1.43) = A_1 = .5 - A_2$
$\qquad\qquad = .5000 - .4236$
$\qquad\qquad = .0764$

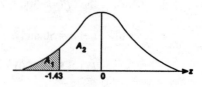

5.89 a. $P(x \leq 75) = P\left(z \leq \dfrac{75 - 70}{10} \right)$

$\qquad\qquad = P(z \leq .50)$
$\qquad\qquad = .5 + .1915 = .6915$

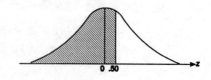

b. $P(x \geq 90) = P\left(z \geq \dfrac{90 - 70}{10}\right)$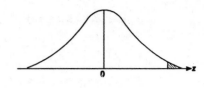

$= P(z \geq 2.00)$

$= .5000 - .4772 = .0228$

c. $P(60 \leq x \leq 75) = P\left(\dfrac{60 - 70}{10} \leq z \leq \dfrac{75 - 70}{10}\right)$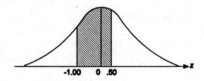

$= P(-1.00 \leq z \leq .50)$

$= .3413 + .1915 = .5328$

d. This is the probability of the complement of the event in part **a**. Therefore:

$P(x > 75) = 1 - P(x \leq 75) = 1 - .6915 = .3085$

e. $P(x = 75) = 0$ (True for any continuous random variable.)

f. $P(x \leq 95) = P\left(z \leq \dfrac{95 - 70}{10}\right)$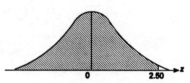

$= P(z \leq 2.50)$

$= .5000 + .4938 = .9938$

5.91 $\mu = np = 100(.5) = 50, \ \sigma = \sqrt{npq} = \sqrt{100(.5)(.5)} = 5$

a. $P(x \leq 48) = P\left(z \leq \dfrac{(48 + .5) - 50}{5}\right)$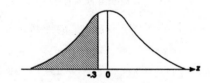

$= P(z \leq -.30)$

$= .5 - .1179 = .3821$

b. $P(50 \leq x \leq 65)$

$= P\left(\dfrac{(50 - .5) - 50}{5} \leq z \leq \dfrac{(65 + .5) - 50}{5}\right)$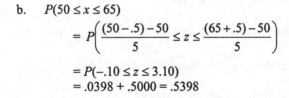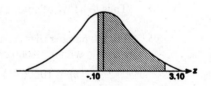

$= P(-.10 \leq z \leq 3.10)$

$= .0398 + .5000 = .5398$

c. $P(x \geq 70) = P\left(z \geq \dfrac{(70 - .5) - 50}{5}\right)$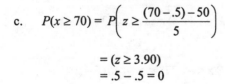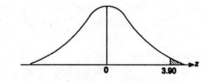

$= (z \geq 3.90)$

$= .5 - .5 = 0$

d. $P(55 \leq x \leq 58)$

$$= P\left(\frac{(55-.5)-50}{5} \leq z \leq \frac{(58+.5)-50}{5}\right)$$

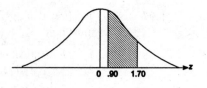

$= P(.90 \leq z \leq 1.70)$
$= P(0 \leq z \leq 1.70) - P(0 \leq z \leq .90)$
$= .4554 - .3159 = .1395$

e. $P(x = 62)$

$$= P\left(\frac{(62-.5)-50}{5} \leq z \leq \frac{(62+.5)-50}{5}\right)$$

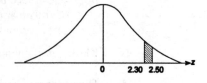

$= P(2.30 \leq z \leq 2.50)$
$= P(0 \leq z \leq 2.50) - (0 \leq z \leq 2.30)$
$= .4938 - .4893 = .0045$

f. $P(x \leq 49$ or $x \geq 72)$

$$= P\left(z \leq \frac{(49+.5)-50}{5}\right) + P\left(z \geq \frac{(72-.5)-50}{5}\right)$$

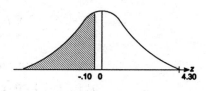

$= P(z \leq -.10) + P(z \geq 4.30)$
$= (.5 - .0398) + (.5 - .5) = .4602$

5.93   Let x = score for first time students on mathematics achievement test. Then x is a normal random variable with $\mu = 77$ and $\sigma = 7.3$.

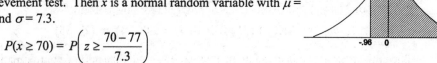

$$P(x \geq 70) = P\left(z \geq \frac{70-77}{7.3}\right)$$

$= P(z \geq -.96)$
$= .3315 + .5000 = .8315$

Thus, 83.15% of students will pass the test the first time.

5.95   Let x = number of American blacks with sickle-cell anemia. Then x is a binomial random variable with $n = 1000$ and $p = .16$.

$$\mu = np = 1000(.16) = 160 \text{ and } \sigma = \sqrt{npq} = \sqrt{1000(.16)(.84)} = 11.593$$

Using the normal approximation to the binomial and Table IV,

a. $P(x > 175) \approx P\left(z > \frac{(175+.5)-160}{11.593}\right) = P(z > 1.34) = .5 - .4099 = .0901$

b. $P(x < 140) \approx P\left(z < \frac{(140-.5)-160}{11.593}\right) = P(z < -1.77) = .5 - .4616 = .0384$

c. $P(130 \leq x \leq 180) \approx P\left(\frac{(130-.5)-160}{11.593} \leq z \leq \frac{(180+.5)-160}{11.593}\right)$

$$= P(-2.63 \leq z \leq 1.77) = .4957 + .4616 = .9573$$

5.97　a.　To show that the random variable has a binomial distribution, the five characteristics must hold:

    1.　n identical trials—There are 4000 identical jumps or trials.
    2.　Each trial results in one of two possible outcomes—On each trial, the main chute either fails (Success) or does not fail (Failure).
    3.　The probability of a success is constant from trial to trial—The probability that a main chute fails $= p = 1/1000 = .001$ and $q = .999$.
    4.　The trials are independent—The outcome of one chute does not affect the outcome of any other chute.
    5.　The random variable is $x =$ number of times the main chute fails in 4000 trials.

Thus, x is a binomial random variable.

b.　For the normal approximation, $\mu = np = 4000(.001) = 4$, $\sigma^2 = npq = 4000(.001)(.999)$ $= 3.996$, and $\sigma = \sqrt{3.996} = 1.999$.

$$P(x \geq 1) \approx P\left(z \geq \frac{(1-.5)-4}{1.999} \right) = P(x \geq -1.75) = .5 + .4599 = .9599$$

(Using Table IV, Appendix A.)

c.　For the Poisson approximation, the mean is $\mu = np = 4000(.001) = 4$.

$$P(x \geq 1) = 1 - P(x = 0) = 1 - .018 = .982 \text{ (Using Table III, Appendix A.)}$$

5.99　Let $x =$ number of white-collar employees in good shape who will develop stress related illnesses in a sample of 400. Then x is a binomial random variable with $n = 400$ and $p = .10$. To see if the normal approximation is appropriate for this problem:

$$np \pm 3\sqrt{npq} \Rightarrow 400(.1) \pm 3\sqrt{400(.1)(.9)} \Rightarrow 40 \pm 18 \Rightarrow (22, 58)$$

Since this interval is contained in the interval 0, $n = 400$, the normal approximation is appropriate.

$$P(x > 60) \approx P\left(z > \frac{(60+.5)-40}{6} \right)$$
$$= P(z > 3.42) \approx .5000 - .5000 = 0$$

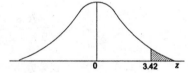

5.101　$c = 0$, $d = 4$

$$f(x) = \begin{cases} \dfrac{1}{d-c} = \dfrac{1}{4-0} = \dfrac{1}{4} & 0 \leq x \leq 4 \\ 0 & \text{otherwise} \end{cases}$$

$$\mu = \frac{c+d}{2} = \frac{0+4}{2} = 2, \ \sigma = \frac{d-c}{\sqrt{12}} = \frac{4-0}{\sqrt{12}} = 1.155$$

$$15 \text{ seconds} = \frac{15}{60} = .25 \text{ minutes}$$

It takes .25 minutes to go from floor 2 to floor 1. Thus, we must find the probability that the waiting time is less than $1.5 - .25 = 1.25$.

$$P(x < 1.25) = (1.25 - 0)\frac{1}{4} = 1.25\left(\frac{1}{4}\right) = .3125$$

5.103 Let x = velocity of a galaxy located within the galaxy cluster A2142. Then x is a normal random variable with mean $\mu = 27{,}117$ km/s and standard deviation $\sigma = 1{,}280$ km/s.

$$P(x \leq 24{,}350) = P\left(z \leq \frac{24{,}350 - 27{,}117}{1{,}280}\right) = P(z \leq -2.16) = .5 - .4846 = .0154$$

Since the probability of observing a galaxy from the galaxy cluster A2142 with a velocity of 24,350 km/s or slower is so small ($p = .0154$), it would be very unlikely that the galaxy observed came from the galaxy cluster A2142.

5.105 a. Using Table IV, Appendix A, with $\mu = 76.8$ and $\sigma = 9.2$.

$$P(x > 90) = P\left(z > \frac{90 - 76.8}{9.2}\right) = P(z > 1.43) = .5 - P(0 < z < 1.43)$$
$$= .5 - .4236 = .0764$$

Thus, the percentage of compounds formed from the plant extract that have a collagen amount greater than 90 grams is 7.64%.

  b. $P(x_L < x < x_U) = .8000$. Suppose we pick x_L and x_U so that the interval is symmetric around the mean.

If $P(x_L < x < x_U) = .8000$, then $P\left(\dfrac{x_L - 76.8}{9.2} < z < \dfrac{x_U - 76.8}{9.2}\right)$
$$= P(z_L < z < z_U) = .8000$$

Then $P(z_L < z < 0) = P(0 < z < z_U) = .4000$. Looking up .4000 in Table IV, the z-score is 1.28. Thus, $z_L = -1.28$ and $z_U = 1.28$. Now, we must convert z_L and z_U back to x scores.

$$z_L = \frac{x_L - 76.8}{9.2} \Rightarrow -1.28 = \frac{x_L - 76.8}{9.2} \Rightarrow x_L = 9.2(-1.28) + 76.8 = 65.024$$

$$z_U = \frac{x_U - 76.8}{9.2} \Rightarrow 1.28 = \frac{x_U - 76.8}{9.2} \Rightarrow x_L = 9.2(1.28) + 76.8 = 88.576$$

Thus, a symmetric interval that contains 80% of the compounds is 65.024 to 88.576.

5.107 From Table IV, Appendix A, with $\mu = 6.3$ and $\sigma = .6$

  a. $P(x < 5) = P\left(z < \dfrac{5 - 6.3}{.6}\right) = P(z < -2.17) = .5 - .4850 = .0150$

Thus, the percentage of days when the oxygen content is undesirable is 1.5%.

  b. We would expect the oxygen contents to fall within 2 standard deviations of the mean, or

$$\mu \pm 2\sigma \Rightarrow 6.3 \pm 2(.6) \Rightarrow 6.3 \pm 1.2 \Rightarrow (5.1, 7.5)$$

5.109 a. (i) $P(0 < z < 1.2) \approx z(4.4 - z)/10 = 1.2(4.4 - 1.2)/10 = .384$
          (ii) $P(0 < z < 2.5) \approx .49$
          (iii) $P(z > .8) = .5 - P(0 < z < .8) \approx .5 - .8(4.4 - .8)/10 = .5 - .288 = .212$
          (iv) $P(z < 1.0) = .5 + P(0 < z < 1.0) = .5 + 1(4.4 - 1)/10 = .5 + .34 = .84$

       b. Using Table IV, Appendix A:
          (i) $P(0 < z < 1.2) = .3849$
          (ii) $P(0 < z < 2.5) = .4938$
          (iii) $P(z > .8) = .5 - P(0 < z < .8) = .5 - .2881 = .2119$
          (iv) $P(z < 1.0) = .5 + P(0 < z < 1.0) = .5 + .3413 = .8413$

       c. For each part, the absolute error is:
          (i) Error $= | .384 - .3869 | = .0029$
          (ii) Error $= | .49 - .4938 | = .0038$
          (iii) Error $= | .212 - .2119 | = .0001$
          (iv) Error $= | .84 - .8413 | = .0013$

          In all cases, the absolute error is less than .0052.

5.111 Since we know for a mound-shaped distribution that approximately 95% of the data lies within two standard deviations of the mean, by symmetry approximately 2.5% would fall below 2 standard deviations below the mean and approximately 2.5% would fall more than 2 standard deviations above the mean. Therefore, the amount of time allotted should be two standard deviations above the mean. This is:

$$\mu + 2\sigma = 40 + 2(6) = 52 \text{ minutes}$$

5.113 Using Table IV, Appendix A, with $\mu = .27$ and $\sigma = .04$,

$$P(x \le .14) = P\left(z \le \frac{.14 - .27}{.04} \right) = P(z < -3.25) \approx .5 - .5 = .0000$$

       Since the probability of observing a reading of .14 or smaller is so small ($p \approx .0000$) if the pressure is .1 MPa, it would be extremely unlikely that this reading was obtained at a pressure of .1 MPa.

5.115 a. Define x = the number of serious accidents per month. Then x has a *Poisson* distribution with $\lambda = 2$. If we define y = the time between adjacent serious accidents, then y has an exponential distribution with $\mu = 1/\lambda = 1/2$. If an accident occurs today, the probability that the next serious accident will **not** occur during the next month is:

$$P(y > 1) = e^{-1(2)} = e^{-2} = .135335$$

          Alternatively, we could solve the problem in terms of the random variable x by noting that the probability that the next serious accident will **not** occur during the next month is the same as the probability that the number of serious accident next month is zero, i.e.,

$$P(y > 1) = P(x = 0) = \frac{e^{-2} 2^0}{0!} = e^{-2} = .135335$$

       b. $P(x > 1) = 1 - P(x \le 1) = 1 - .406 = .594$
                          (Using Table III in Appendix A with $\lambda = 2$)

Sampling Distributions

6.1   a–b.   The different samples of $n = 2$ with replacement and their means are:

Possible Samples	$\bar{x}$	Possible Samples	$\bar{x}$
0, 0	0	4, 0	2
0, 2	1	4, 2	3
0, 4	2	4, 4	4
0, 6	3	4, 6	5
2, 0	1	6, 0	3
2, 2	2	6, 2	4
2, 4	3	6, 4	5
2, 6	4	6, 6	6

   c.   Since each sample is equally likely, the probability of any 1 being selected is

$$\frac{1}{4}\left(\frac{1}{4}\right) = \frac{1}{16}$$

   d. $P(\bar{x} = 0) = \dfrac{1}{16}$

$P(\bar{x} = 1) = \dfrac{1}{16} + \dfrac{1}{16} = \dfrac{2}{16}$

$P(\bar{x} = 2) = \dfrac{1}{16} + \dfrac{1}{16} + \dfrac{1}{16} = \dfrac{3}{16}$

$P(\bar{x} = 3) = \dfrac{1}{16} + \dfrac{1}{16} + \dfrac{1}{16} + \dfrac{1}{16} = \dfrac{4}{16}$

$P(\bar{x} = 4) = \dfrac{1}{16} + \dfrac{1}{16} + \dfrac{1}{16} = \dfrac{3}{16}$

$P(\bar{x} = 5) = \dfrac{1}{16} + \dfrac{1}{16} = \dfrac{2}{16}$

$P(\bar{x} = 6) = \dfrac{1}{16}$

$\bar{x}$	$p(\bar{x})$
0	1/16
1	2/16
2	3/16
3	4/16
4	3/16
5	2/16
6	1/16

   e.

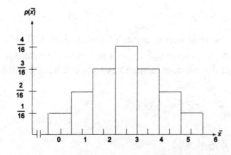

6.3   If the observations are independent of each other, then

$$P(1, 1) = p(1)p(1) = .2(.2) = .04$$
$$P(1, 2) = p(1)p(2) = .2(.3) = .06$$
$$P(1, 3) = p(1)p(3) = .2(.2) = .04$$

etc.

a.

Possible Samples	$\bar{x}$	$p(\bar{x})$	Possible Samples	$\bar{x}$	$p(\bar{x})$
1, 1	1	.04	3, 4	3.5	.04
1, 2	1.5	.06	3, 5	4	.02
1, 3	2	.04	4, 1	2.5	.04
1, 4	2.5	.04	4, 2	3	.06
1, 5	3	.02	4, 3	3.5	.04
2, 1	1.5	.06	4, 4	4	.04
2, 2	2	.09	4, 5	4.5	.02
2, 3	2.5	.06	5, 1	3	.02
2, 4	3	.06	5, 2	3.5	.03
2, 5	3.5	.03	5, 3	4	.02
3, 1	2	.04	5, 4	4.5	.02
3, 2	2.5	.06	5, 5	5	.01
3, 3	3	.04			

Summing the probabilities, the probability distribution of $\bar{x}$ is:

$\bar{x}$	$p(\bar{x})$
1	.04
1.5	.12
2	.17
2.5	.20
3	.20
3.5	.14
4	.08
4.5	.04
5	.01

b.

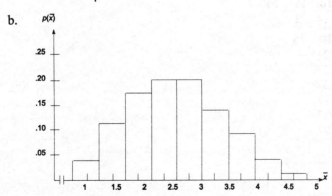

c. $P(\bar{x} \geq 4.5) = .04 + .01 = .05$

d.   No. The probability of observing $\bar{x} = 4.5$ or larger is small (.05).

6.5   a.   For a sample of size $n = 2$, the sample mean and sample median are exactly the same. Thus, the sampling distribution of the sample median is the same as that for the sample mean (see Exercise 6.3a).

      b.   The probability histogram for the sample median is identical to that for the sample mean (see Exercise 6.3b).

6.9   a. $\mu = \sum x p(x) = 2\left(\dfrac{1}{3}\right) + 4\left(\dfrac{1}{3}\right) + 9\left(\dfrac{1}{3}\right) = \dfrac{15}{3} = 5$

      b.   The possible samples of size $n = 3$, the sample means, and the probabilities are:

Possible Samples	$\bar{x}$	$p(\bar{x})$	m	Possible Samples	$\bar{x}$	$p(\bar{x})$	m
2, 2, 2	2	1/27	2	4, 4, 4	4	1/27	4
2, 2, 4	8/3	1/27	2	4, 4, 9	17/3	1/27	4
2, 2, 9	13/3	1/27	2	4, 9, 2	5	1/27	4
2, 4, 2	8/3	1/27	2	4, 9, 4	17/3	1/27	4
2, 4, 4	10/3	1/27	4	4, 9, 9	22/3	1/27	9
2, 4, 9	5	1/27	4	9, 2, 2	13/3	1/27	2
2, 9, 2	13/3	1/27	2	9, 2, 4	5	1/27	4
2, 9, 4	5	1/27	4	9, 2, 9	20/3	1/27	9
2, 9, 9	20/3	1/27	9	9, 4, 2	5	1/27	4
4, 2, 2	8/3	1/27	2	9, 4, 4	17/3	1/27	4
4, 2, 4	10/3	1/27	4	9, 4, 9	22/3	1/27	9
4, 2, 9	5	1/27	4	9, 9, 2	20/3	1/27	9
4, 4, 2	10/3	1/27	4	9, 9, 4	22/3	1/27	9
				9, 9, 9	9	1/27	9

The sampling distribution of is:

$\bar{x}$	$p(\bar{x})$
2	1/27
8/3	3/27
10/3	3/27
4	1/27
13/3	3/27
5	6/27
17/3	3/27
20/3	3/27
22/3	3/27
9	1/27
	27/27

$$E(\bar{x}) = \sum \bar{x}p(\bar{x}) = 2\left(\frac{1}{27}\right) + \frac{8}{3}\left(\frac{3}{27}\right) + \frac{10}{3}\left(\frac{3}{27}\right) + 4\left(\frac{1}{27}\right) + \frac{13}{3}\left(\frac{3}{27}\right)$$

$$+ 5\left(\frac{6}{27}\right) + \frac{17}{3}\left(\frac{3}{27}\right) + \frac{20}{3}\left(\frac{3}{27}\right) + \frac{22}{3}\left(\frac{3}{27}\right) + 9\left(\frac{1}{27}\right)$$

$$= \frac{2}{27} + \frac{8}{27} + \frac{10}{27} + \frac{4}{27} + \frac{13}{27} + \frac{30}{27} + \frac{17}{27} + \frac{20}{27} + \frac{22}{27} + \frac{9}{27}$$

$$= \frac{135}{27} = 5$$

Since $\mu = 5$ in part **a**, and $E(\bar{x}) = \mu = 5$, $\bar{x}$ is an unbiased estimator of μ.

c. The median was calculated for each sample and is shown in the table in part **b**. The sampling distribution of m is:

m	$p(m)$
2	7/27
4	13/27
9	7/27
	27/27

$$E(m) = \sum mp(m) = 2\left(\frac{7}{27}\right) + 4\left(\frac{13}{27}\right) + 9\left(\frac{7}{27}\right) = \frac{14}{27} + \frac{52}{27} + \frac{63}{27} = \frac{129}{27} = 4.778$$

The $E(m) = 4.778 \neq \mu = 5$. Thus, m is a biased estimator of μ.

d. Use the sample mean, $\bar{x}$. It is an unbiased estimator.

6.13 a. Refer to the solution to Exercise 6.3. The values of s^2 and the corresponding probabilities are listed below:

$$s^2 = \frac{\sum (x^2) - \frac{\left(\sum x\right)^2}{n}}{n-1}$$

For sample 1, 1, $s^2 = \dfrac{2 - \dfrac{2^2}{2}}{1} = 0$

For sample 1, 2, $s^2 = \dfrac{5 - \dfrac{3^2}{2}}{1} = .5$

The rest of the values are calculated and shown:

s^2	$p(s^2)$	s^2	$p(s^2)$
0.0	.04	0.5	.04
0.5	.06	2.0	.02
2.0	.04	4.5	.04
4.5	.04	2.0	.06
8.0	.02	0.5	.04
0.5	.06	0.0	.04
0.0	.09	0.5	.02
0.5	.06	8.0	.02
2.0	.06	4.5	.03
4.5	.03	2.0	.02
2.0	.04	0.5	.02
0.5	.06	0.0	.01
0.0	.04		

The sampling distribution of s^2 is:

s^2	$p(s^2)$
0.0	.22
0.5	.36
2.0	.24
4.5	.14
8.0	.04

b. $\sigma^2 = \sum (x - \mu)^2 p(x) = (1 - 2.7)^2(.2) + (2 - 2.7)^2(.3) + (3 - 2.7)^2(.2)$
$$+ (4 - 2.7)^2(.2) + (5 - 2.7)^2(.1)$$
$$= 1.61$$

c. $E(s^2) = \sum s^2 p(s^2) = 0(.22) + .5(.36) + 2(.24) + 4.5(.14) + 8(.04) = 1.61$

d. The sampling distribution of s is listed below, where $s = \sqrt{s^2}$:

s	$p(s)$
0.000	.22
0.707	.36
1.414	.24
2.121	.14
2.828	.04

e. $E(s) = \sum s p(s) = 0(.22) + .707(.36) + 1.414(.24) + 2.121(.14) + 2.828(.04)$
$$= 1.00394$$

Since $E(s) = 1.00394$ is not equal to $\sigma = \sqrt{\sigma^2} = \sqrt{1.61} = 1.269$, s is a biased estimator of σ.

6.15 a. $\mu_{\bar{x}} = \mu = 100, \sigma_{\bar{x}} = \dfrac{\sigma}{\sqrt{n}} = \dfrac{\sqrt{100}}{\sqrt{4}} = 5$

  b. $\mu_{\bar{x}} = \mu = 100, \sigma_{\bar{x}} = \dfrac{\sigma}{\sqrt{n}} = \dfrac{\sqrt{100}}{\sqrt{25}} = 2$

  c. $\mu_{\bar{x}} = \mu = 100, \sigma_{\bar{x}} = \dfrac{\sigma}{\sqrt{n}} = \dfrac{\sqrt{100}}{\sqrt{100}} = 1$

  d. $\mu_{\bar{x}} = \mu = 100, \sigma_{\bar{x}} = \dfrac{\sigma}{\sqrt{n}} = \dfrac{\sqrt{100}}{\sqrt{50}} = 1.414$

  e. $\mu_{\bar{x}} = \mu = 100, \sigma_{\bar{x}} = \dfrac{\sigma}{\sqrt{n}} = \dfrac{\sqrt{100}}{\sqrt{500}} = .447$

  f. $\mu_{\bar{x}} = \mu = 100, \sigma_{\bar{x}} = \dfrac{\sigma}{\sqrt{n}} = \dfrac{\sqrt{100}}{\sqrt{1000}} = .316$

6.17 a. $\mu = \sum xp(x) = 1(.1) + 2(.4) + 3(.4) + 8(.1) = 2.9$

   $\sigma^2 = \sum (x - \mu)^2 p(x) = (1 - 2.9)^2(.1) + (2 - 2.9)^2(.4) + (3 - 2.9)^2(.4) + (8 - 2.9)^2(.1)$
         $= .361 + .324 + .004 + 2.601 = 3.29$

   $\sigma = \sqrt{3.29} = 1.814$

  b. The possible samples, values of $\bar{x}$, and associated probabilities are listed:

Possible Samples	$\bar{x}$	$p(\bar{x})$	Possible Samples	$\bar{x}$	$p(\bar{x})$
1, 1	1	.01	3, 1	2	.04
1, 2	1.5	.04	3, 2	2.5	.16
1, 3	2	.04	3, 3	3	.16
1, 8	4.5	.01	3, 8	5.5	.04
2, 1	1.5	.04	8, 1	4.5	.01
2, 2	2	.16	8, 2	5	.04
2, 3	2.5	.16	8, 3	5.5	.04
2, 8	5	.04	8, 8	8	.01

   $P(1, 1) = p(1)p(1) = .1(.1) = .01$
   $P(1, 2) = p(1)p(2) = .1(.4) = .04$
   $P(1, 3) = p(1)p(3) = .1(.4) = .04$
    etc.

The sampling distribution of is:

$\bar{x}$	$p(\bar{x})$
1	.01
1.5	.08
2	.24
2.5	.32
3	.16
4.5	.02
5	.08
5.5	.08
8	.01
	1.00

c. $\mu_{\bar{x}} = E(\bar{x}) = \sum \bar{x}p(\bar{x}) = 1(.01) + 1.5(.08) + 2(.24) + 2.5(.32) + 3(.16) + 4.5(.02)$
$$+ 5(.08) + 5.5(.08) + 8(.01)$$
$$= 2.9 = \mu$$

$\sigma_{\bar{x}}^2 = \sum (\bar{x} - \mu_{\bar{x}})^2 p(\bar{x}) = (1 - 2.9)^2(.01) + (1.5 - 2.9)^2(.08) + (2 - 2.9)^2(.24)$
$$+ (2.5 - 2.9)^2(.32) + (3 - 2.9)^2(.16) + (4.5 - 2.9)^2(.02)$$
$$+ (5 - 2.9)^2(.08) + (5.5 - 2.9)^2(.08) + (8 - 2.9)^2(.01)$$
$$= .0361 + .1568 + .1944 + .0512 + .0016 + .0512 + .3528$$
$$+ .5408 + .2601$$
$$= 1.645$$

$\sigma_{\bar{x}} = \sqrt{1.645} = 1.283$

$\sigma_{\bar{x}} = \sigma / \sqrt{n} = 1.814 / \sqrt{2} = 1.283$

6.19   a. $\mu_{\bar{x}} = \mu = 20$, $\sigma_{\bar{x}} = \sigma / \sqrt{n} = 16 / \sqrt{64} = 2$

b.   By the Central Limit Theorem, the distribution of $\bar{x}$ is approximately normal. In order for the Central Limit Theorem to apply, n must be sufficiently large. For this problem, $n = 64$ is sufficiently large.

c. $z = \dfrac{\bar{x} - \mu_{\bar{x}}}{\sigma_{\bar{x}}} = \dfrac{15.5 - 20}{2} = -2.25$

d. $z = \dfrac{\bar{x} - \mu_{\bar{x}}}{\sigma_{\bar{x}}} = \dfrac{23 - 20}{2} = 1.50$

6.21   By the Central Limit Theorem, the sampling distribution of is approximately normal with $\mu_{\bar{x}} = \mu = 30$ and $\sigma_{\bar{x}} = \sigma / \sqrt{n} = 16 / \sqrt{100} = 1.6$. Using Table IV, Appendix A:

a. $P(\bar{x} \geq 28) = P\left(z \geq \dfrac{28 - 30}{1.6} \right) = P(z \geq -1.25) = .5 + .3944 = .8944$

b. $P(22.1 \le \bar{x} \le 26.8) = P\left(\dfrac{22.1-30}{1.6} \le z \le \dfrac{26.8-30}{1.6}\right) = P(-4.94 \le z \le -2)$

$$= .5 - .4772 = .0228$$

c. $P(\bar{x} \le 28.2) = P\left(z \le \dfrac{28.2-30}{1.6}\right) = P(z \le -1.13) = .5 - .3708 = .1292$

d. $P(\bar{x} \ge 27.0) = P\left(z \ge \dfrac{27.0-30}{1.6}\right) = P(z \ge -1.88) = .5 + .4699 = .9699$

6.25   a. $\mu_{\bar{x}} = \mu = 89.34;\ \sigma_{\bar{x}} = \dfrac{\sigma}{\sqrt{n}} = \dfrac{7.74}{\sqrt{35}} = 1.3083$

      b.

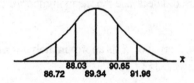

                         86.72 88.03 89.34 90.65 91.96

c. $P(\bar{x} > 88) = P\left(z > \dfrac{88-89.34}{1.3083}\right) = P(z > -1.02) = .5 + .3461 = .8461$

                                              (using Table IV, Appendix A)

d. $P(\bar{x} < 87) = P\left(z < \dfrac{87-89.34}{1.3083}\right) = P(z < -1.79) = .5 - .4633 = .0367$

                                              (using Table IV, Appendix A)

6.27   Let $\bar{x}$ = sample mean score change. By the Central Limit Theorem, the sampling distribution of $\bar{x}$ is approximately normal with

         with $\mu_{\bar{x}} = \mu = 19$ and $\sigma_{\bar{x}} = \dfrac{\sigma}{\sqrt{n}} = \dfrac{65}{\sqrt{100}} = 6.5.$

      b. $P(\bar{x} < 10) = P\left(z < \dfrac{10-19}{6.5}\right) = P(z < -1.38) = .5 - .4162 = .0838$

                                            (using Table IV, Appendix A)

6.29   a. $\mu_{\bar{x}} = \mu = 5.1;\ \sigma_{\bar{x}} = \dfrac{\sigma}{\sqrt{n}} = \dfrac{6.1}{\sqrt{150}} = .4981$

      b.   Because the sample size is large, $n = 150$, the Central Limit Theorem says that the sampling distribution of $\bar{x}$ is approximately normal.

c. $P(\bar{x} > 5.5) = P\left(z > \dfrac{5.5-5.1}{.4981}\right) = P(z > .80) = .5 - .2881 = .2119$

                                            (using Table IV, Appendix A)

d. $P(4 < \bar{x} < 5) = P\left(\dfrac{4-5.1}{.4981} < z < \dfrac{5-5.1}{.4981}\right) = P(-2.21 < z < -.20)$

$$= .4864 - .0793 = .4071$$

(using Table IV, Appendix A)

6.31 By the Central Limit Theorem, the sampling distribution of is approximately normal.

$$\mu_{\bar{x}} = \mu = 7; \ \sigma_{\bar{x}} = \frac{\sigma}{\sqrt{n}} = \frac{2}{\sqrt{100}} = .2$$

a. $P(\bar{x} \le 6.4) = P\left(z \le \dfrac{6.4-7}{.2}\right) = P(z \le -3.00) = .5 - .4987 = .0013$

b. Since the probability of observing the sample mean 6.4 or less is only .0013, it would be reasonable to conclude that the program did decrease the mean number of sick days taken by the company's employees.

6.33 a. By the Central Limit Theorem, the distribution of $\bar{x}$ is approximately normal, with $\mu_{\bar{x}} = \mu = 157$ and $\sigma_{\bar{x}} = \sigma/\sqrt{n} = 3/\sqrt{40} = .474$.

The sample mean is 1.3 psi below 157 or $\bar{x} = 157 - 1.3 = 155.7$

$$P(\bar{x} \le 155.7) = P\left(z \le \frac{155.7-157}{.474}\right) = P(z \le -2.74) = .5 - .4969 = .0031$$

(using Table IV, Appendix A)

If the claim is true, it is very unlikely (probability = .0031) to observe a sample mean 1.3 psi below 157 psi. Thus, the actual population mean is probably not 157 but something lower.

b. $P(\bar{x} \le 155.7) = P\left(z \le \dfrac{155.7-156}{.474}\right) = P(z \le -.63) = .5 - .2357 = .2643$

(using Table IV, Appendix A)

The observed sample is more likely if $\mu = 156$ rather than 157.

$$P(\bar{x} \le 155.7) = P\left(z \le \frac{155.7-158}{.474}\right) = P(z \le -4.85) = .5 - .5 = 0$$

The observed sample is less likely if $\mu = 158$ rather than 157.

c. $\sigma_{\bar{x}} = \sigma/\sqrt{n} = 2/\sqrt{40} = .316 \qquad \mu_{\bar{x}} = 157$

$$P(\bar{x} \le 155.7) = P\left(z \le \frac{155.7-157}{.316}\right) = P(z \le -4.11) = .5 - .5 = 0$$

(using Table IV, Appendix A)

The observed sample is less likely if $\sigma = 2$ rather than 3.

$$\sigma_{\bar{x}} = \sigma / \sqrt{n} = 6 / \sqrt{40} = .949 \qquad \mu_{\bar{x}} = 157$$

$$P(\bar{x} \le 155.7) = P\left(z \le \frac{155.7 - 157}{.949}\right) = P(z \le -1.37) = .5 - .4147 = .0853$$

(using Table IV, Appendix A)

The observed sample is more likely if $\sigma = 6$ rather than 3.

6.35   a.   "The sampling distribution of the sample statistic A" is the probability distribution of the variable A.

        b.   "A" is an unbiased estimator of α because the mean of the sampling distribution of A is α.

        c.   If both A and B are unbiased estimators of α, then the statistic whose standard deviation is smaller is a better estimator of α.

        d.   No. The Central Limit Theorem applies only to the sample mean. If A is the sample mean, $\bar{x}$, and n is sufficiently large, then the Central Limit Theorem will apply. However, both A and B cannot be sample means. Thus, we cannot apply the Central Limit Theorem to both A and B.

6.37   By the Central Limit Theorem, the sampling distribution of is approximately normal.

$$\mu_{\bar{x}} = \mu = 19.6, \ \sigma_{\bar{x}} = \frac{3.2}{\sqrt{68}} = .388$$

a. $P(\bar{x} \le 19.6) = P\left(z \le \frac{19.6 - 19.6}{.388}\right) = P(z \le 0) = .5$ (using Table IV, Appendix A)

b. $P(\bar{x} \le 19) = P\left(z \le \frac{19 - 19.6}{.388}\right) = P(z \le -1.55) = .5 - .4394 = .0606$

(using Table IV, Appendix A)

c. $P(\bar{x} \ge 20.1) = P\left(z \ge \frac{20.1 - 19.6}{.388}\right) = P(z \ge 1.29) = .5 - .4015 = .0985$

(using Table IV, Appendix A)

d. $P(19.2 < \bar{x} < 20.6) = P\left(\frac{19.2 - 19.6}{.388} < z < \frac{20.6 - 19.6}{.388}\right)$

$$= P(-1.03 < z < 2.58) = .3485 + .4951 = .8436$$

(using Table IV, Appendix A)

6.41   Given: $\mu = 100$ and $\sigma = 10$

n	1	5	10	20	30	40	50
$\dfrac{\sigma}{\sqrt{n}}$	10	4.472	3.162	2.236	1.826	1.581	1.414

The graph of $\sigma/\sqrt{n}$ against n is given here:

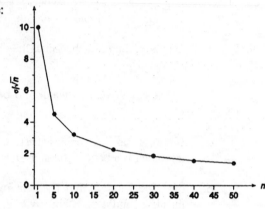

6.43   a. $\mu_{\bar{x}} = \mu = 1.3$, $\sigma_{\bar{x}} = \sigma/\sqrt{n} = 1.7/\sqrt{50} = .240$

b.   Yes. By the Central Limit Theorem, the distribution of is approximately normal for n sufficiently large. For this problem, $n = 50$ is sufficiently large.

c. $P(\bar{x} < 1) = P\left(z < \dfrac{1-1.3}{.24}\right) = P(z < -1.25) = .5 - .3944 = .1056$

(using Table IV, Appendix A)

d. $P(\bar{x} > 1.9) = P\left(z > \dfrac{1.9-1.3}{.24}\right) = P(z > 2.5) = .5 - .4938 = .0062$

(using Table IV, Appendix A)

6.45   By the Central Limit Theorem, the sampling distribution of is approximately normal.

$\mu_{\bar{x}} = \mu = 75$; $\sigma_{\bar{x}} = \dfrac{\sigma}{\sqrt{n}} = \dfrac{10}{\sqrt{36}} = 1.6667$

a. $P(\bar{x} > 79) = P\left(z > \dfrac{79-75}{1.6667}\right) = P(z > 2.4) = .5 - .4918 = .008$

(using Table IV, Appendix A)

b.   We must assume the sample size of $n = 36$ is sufficiently large so that the Central Limit Theorem applies.

6.47 a. Discrete. There are only 3 possible outcomes for the response: 1, 1.5, or 2.

b. By the Central Limit Theorem, the sampling distribution of is approximately normal ($n = 651$) with $\mu_{\bar{x}} = \mu = 1.50$ and $\sigma_{\bar{x}} = \sigma/\sqrt{n} = .45/\sqrt{651} = .01764$. There are no assumptions necessary.

c. $P(\bar{x} \leq 1.45) = P\left(z \leq \dfrac{1.45 - 1.5}{.01764}\right) = P(z \leq -2.83) = .5 - .4977 = .0023$

(using Table IV, Appendix A)

d. $P(\bar{x} \leq 1.36) = P\left(z \leq \dfrac{1.36 - 1.5}{.01764}\right) = P(z \leq -7.94) \approx .5 - .5 = 0$

(using Table IV, Appendix A)

Since this probability is so small (≈ 0), it is very unlikely that the population mean is 1.50, but probably is less than 1.50.

6.49 a. By the Central Limit Theorem, the sampling distribution of is approximately normal, regardless of the shape of the distribution of the verbal IQ scores. The mean is $\mu_{\bar{x}} = \mu = 107$, and the standard deviation is $\sigma_{\bar{x}} = \sigma/\sqrt{n} = 15/\sqrt{84} = 1.637$.

b. $P(\bar{x} \geq 110) = P\left(z \geq \dfrac{110 - 107}{1.637}\right) = P(z \geq 1.83) = .5 - .4664 = .0336$

(using Table IV, Appendix A)

c. No. If the mean and standard deviation for the nondelinquent juveniles were the same as those for all juveniles, it would be very unlikely (probability = .0336) to observe a sample mean of 110 or higher.

6.51 Even though the number of flaws per piece of siding has a Poisson distribution, the Central Limit Theorem implies that the distribution of the sample mean will be approximately normal with $\mu_{\bar{x}} = \mu = 2.5$ and $\sigma_{\bar{x}} = \dfrac{\sigma}{\sqrt{n}} = \dfrac{\sqrt{2.5}}{\sqrt{35}} = .2673$. Therefore,

$$P(\bar{x} > 2.1) + P\left(z > \dfrac{2.1 - 2.5}{\sqrt{2.5}/\sqrt{35}}\right) = P(z > -1.50) = .5 + .4332 = .9332$$

(using Table IV, Appendix A)

Inferences Based on a Single Sample: Estimation with Confidence Intervals

7.1   a.   For $\alpha = .10$, $\alpha/2 = .10/2 = .05$. $z_{\alpha/2} = z_{.05}$ is the z-score with .05 of the area to the right of it. The area between 0 and $z_{.05}$ is $.5 - .05 = .4500$. Using Table IV, Appendix A, $z_{.05} = 1.645$.

    b.   For $\alpha = .01$, $\alpha/2 = .01/2 = .005$. $z_{\alpha/2} = z_{.005}$ is the z-score with .005 of the area to the right of it. The area between 0 and $z_{.005}$ is $.5 - .005 = .4950$. Using Table IV, Appendix A, $z_{.005} = 2.58$.

    c.   For $\alpha = .05$, $\alpha/2 = .05/2 = .025$. $z_{\alpha/2} = z_{.025}$ is the z-score with .025 of the area to the right of it. The area between 0 and $z_{.025}$ is $.5 - .025 = .4750$. Using Table IV, Appendix A, $z_{.025} = 1.96$.

    d.   For $\alpha = .20$, $\alpha/2 = .20/2 = .10$. $z_{\alpha/2} = z_{.10}$ is the z-score with .10 of the area to the right of it. The area between 0 and $z_{.10}$ is $.5 - .10 = .4000$. Using Table IV, Appendix A, $z_{.10} = 1.28$.

7.3   a. $s^2 = \dfrac{\sum(x_i - \bar{x})^2}{n-1} = \dfrac{4238}{64-1} = 67.2698$, $s = \sqrt{67.2698} = 8.2018$

$$\bar{x} = \frac{\sum x}{n} = \frac{700}{64} = 10.9375$$

For confidence coefficient .95, $\alpha = .05$ and $\alpha/2 = .05/2 = .025$. From Table IV, Appendix A, $z_{.025} = 1.96$. The confidence interval is:

$$\bar{x} \pm z_{\alpha/2}\frac{s}{\sqrt{n}} \Rightarrow 10.9375 \pm 1.96\frac{8.2018}{\sqrt{64}} \Rightarrow 10.9375 \pm 2.0094 \Rightarrow (8.9281, 12.9469)$$

    b.   We are 95% confident the true mean is between 8.9281 and 12.9469.

7.5   a.   For confidence coefficient .95, $\alpha = .05$ and $\alpha/2 = .05/2 = .025$. From Table IV, Appendix A, $z_{.025} = 1.96$. The confidence interval is:

$$\bar{x} \pm z_{\alpha/2}\frac{s}{\sqrt{n}} \Rightarrow 83.2 \pm 1.96\frac{6.4}{\sqrt{100}} \Rightarrow 83.2 \pm 1.25 \Rightarrow (81.95, 84.45)$$

    b.   The confidence coefficient of .95 means that in repeated sampling, 95% of all confidence intervals constructed will include μ.

c. For confidence coefficient .99, $\alpha = .01$ and $\alpha/2 = .01/2 = .005$. From Table IV, Appendix A, $z_{.005} = 2.58$. The confidence interval is:

$$\bar{x} \pm z_{\alpha/2}\frac{s}{\sqrt{n}} \Rightarrow 83.2 \pm 2.58\frac{6.4}{\sqrt{100}} \Rightarrow 83.2 \pm 1.65 \Rightarrow (81.55, 84.85)$$

d. As the confidence coefficient increases, the width of the confidence interval also increases.

e. Yes. Since the sample size is 100, the Central Limit Theorem applies. This ensures the distribution of $\bar{x}$ is normal, regardless of the original distribution.

7.7 An interval estimator estimates μ with a range of values, while a point estimator estimates μ with a single point.

7.9 Yes. As long as the sample size is sufficiently large, the Central Limit Theorem says the distribution of $\bar{x}$ is approximately normal regardless of the original distribution.

7.11 a. The 95% confidence interval is (19.7, 20.3). We are 95% confident that the mean number of cigarettes smoked per day by all smokers is between 19.7 and 20.3.

b. Since the sample size is so large ($n = 11,000$), no assumptions are necessary. The Central Limit Theorem indicates that the sampling distribution of $\bar{x}$ is approximately normal.

c. Since the entire 95% confidence interval is above the value of 15, the claim made by the tobacco industry researcher is probably not true.

7.13 a. For confidence coefficient .95, $\alpha = .05$ and $\alpha/2 = .05/2 = .025$. From Table IV, Appendix   A, $z_{.025} = 1.96$. The confidence interval is:

$$\bar{x} \pm z_{\alpha/2}\frac{s}{\sqrt{n}} \Rightarrow 19 \pm 1.96\frac{65}{\sqrt{265}} \Rightarrow 19 \pm 7.826 \Rightarrow (11.174, \ 26.826)$$

b. For confidence coefficient .95, $\alpha = .05$ and $\alpha/2 = .05/2 = .025$. From Table IV, Appendix   A, $z_{.025} = 1.96$. The confidence interval is:

$$\bar{x} \pm z_{\alpha/2}\frac{s}{\sqrt{n}} \Rightarrow 7 \pm 1.96\frac{49}{\sqrt{265}} \Rightarrow 7 \pm 5.90 \Rightarrow (1.10, \ 12.90)$$

c. The SAT-Mathematics test would be more likely to have a mean change of 15 because 15 is in the 95% confidence interval for the mean change in SAT-Mathematics. Since 15 is in the confidence interval, it is a likely value. The value 15 is not in the 95% confidence interval for the SAT-Verbal. Thus, it is not a likely value.

7.15   a.   For confidence coefficient .99, $\alpha = .01$ and $\alpha/2 = .01/2 = .005$. From Table IV,
              Appendix   A, $z_{.005} = 2.58$. The confidence interval is:

$$\bar{x} \pm z_{\alpha/2} \frac{s}{\sqrt{n}} \Rightarrow 1.13 \pm 2.58 \frac{2.21}{\sqrt{72}} \Rightarrow 1.13 \pm .672 \Rightarrow (.458, 1.802)$$

We are 99% confident that the true mean number of pecks made by chickens pecking at blue string is between .458 and 1.802.

        b.   Yes, there is evidence that chickens are more apt to peck at white string. The mean
              number of pecks for white string is 7.5. Since 7.5 is not in the 99% confidence interval
              for the mean number of pecks at blue string, it is not a likely value for the true mean for
              blue string.

7.17   From the printout, the 95% confidence interval is (.3526, .4921). We are 95% confident that
        the true mean correlation coefficient between appraisal participation and a subordinate's
        satisfaction with the appraisal is between .3526 and .4921.

7.19   a.   For confidence coefficient .95, $\alpha = .05$ and $\alpha/2 = .025$. From Table IV, Appendix A,
              $z_{.025} = 1.96$. The confidence interval is:

$$\bar{x} \pm z_{\alpha/2} \frac{s}{\sqrt{n}}$$

Younger: $4.17 \pm 1.96 \frac{.75}{\sqrt{241}} \Rightarrow 4.17 \pm .095 \Rightarrow (4.075, 4.265)$

We are 95% confident that the mean job satisfaction score for all adults in the younger age group is between 4.075 and 4.265.

Middle-Age: $4.04 \pm 1.96 \frac{.81}{\sqrt{768}} \Rightarrow 4.04 \pm .057 \Rightarrow (3.983, 4.097)$

We are 95% confident that the mean job satisfaction score for all adults in the middle-age age group is between 3.983 and 4.097.

Older: $4.31 \pm 1.96 \frac{.82}{\sqrt{677}} \Rightarrow 4.31 \pm .062 \Rightarrow (4.248, 4.372)$

We are 95% confident that the mean job satisfaction score for all adults in the older age group is between 4.248 and 4.372.

        b.   Let y = number of 95% confidence intervals that do not contain the population mean in
              3 trials. Then y is a binomial random variable with $n = 3$ and $p = .05$.

$$P(y \geq 1) = 1 - P(y = 0) = 1 - \binom{3}{0}.05^0.95^3 = 1 - .857375 = .142625$$

Thus, it is more likely that at least one of three intervals will not contain the population mean than it is for a single confidence interval to miss the population mean.

The probability that a single confidence interval will not contain the population mean is .05.

7.21   a.   If x is normally distributed, the sampling distribution of is normal, regardless of the sample size.

b.   If nothing is known about the distribution of x, the sampling distribution of is approximately normal if n is sufficiently large. If n is not large, the distribution of is unknown if the distribution of x is not known.

7.23   a. $P(t \geq t_0) = .025$ where df $= 10$

$$t_0 = 2.228$$

b. $P(t \geq t_0) = .01$ where df $= 17$

$$t_0 = 2.567$$

c. $P(t \leq t_0) = .005$ where df $= 6$

Because of symmetry, the statement can be rewritten

$$P(t \geq -t_0) = .005 \text{ where df} = 6$$
$$t_0 = -3.707$$

d. $P(t \leq t_0) = .05$ where df $= 13$
$$t_0 = -1.771$$

7.25   First, we must compute and s.

$$\bar{x} = \frac{\sum x}{n} = \frac{30}{6} = 5$$

$$s^2 = \frac{\sum x^2 - \frac{\left(\sum x\right)^2}{n}}{n-1} = \frac{176 - \frac{(30)^2}{6}}{6-1} = \frac{26}{5} = 5.2$$

$$s = \sqrt{5.2} = 2.2804$$

a.   For confidence coefficient .90, $\alpha = 1 - .90 = .10$ and $\alpha/2 = .10/2 = .05$. From Table VI, Appendix A, with df $= n - 1 = 6 - 1 = 5$, $t_{.05} = 2.015$. The 90% confidence interval is:

$$\bar{x} \pm t_{.05} \frac{s}{\sqrt{n}} \Rightarrow 5 \pm 2.015 \frac{2.2804}{\sqrt{6}} \Rightarrow 5 \pm 1.876 \Rightarrow (3.124, 6.876)$$

b.   For confidence coefficient .95, $\alpha = 1 - .95 = .05$ and $\alpha/2 = .05/2 = .025$. From Table VI, Appendix A, with df $= n - 1 = 6 - 1 = 5$, $t_{.025} = 2.571$. The 95% confidence interval is:

$$\bar{x} \pm t_{.025} \frac{s}{\sqrt{n}} \Rightarrow 5 \pm 2.571 \frac{2.2804}{\sqrt{6}} \Rightarrow 5 \pm 2.394 \Rightarrow (2.606, 7.394)$$

c. For confidence coefficient .99, $\alpha = 1 - .99 = .01$ and $\alpha/2 = .01/2 = .005$. From Table VI, Appendix A, with df $= n - 1 = 6 - 1 = 5$, $t_{.005} = 4.032$. The 99% confidence interval is:

$$\bar{x} \pm t_{.005} \frac{s}{\sqrt{n}} \Rightarrow 5 \pm 4.032 \frac{2.2804}{\sqrt{6}} \Rightarrow 5 \pm 3.754 \Rightarrow (1.246, 8.754)$$

d. a) For confidence coefficient .90, $\alpha = 1 - .90 = .10$ and $\alpha/2 = .10/2 = .05$. From Table VI, Appendix A, with df $= n - 1 = 25 - 1 = 24$, $t_{.05} = 1.711$. The 90% confidence interval is:

$$\bar{x} \pm t_{.05} \frac{s}{\sqrt{n}} \Rightarrow 5 \pm 1.711 \frac{2.2804}{\sqrt{25}} \Rightarrow 5 \pm .780 \Rightarrow (4.220, 5.780)$$

b) For confidence coefficient .95, $\alpha = 1 - .95 = .05$ and $\alpha/2 = .05/2 = .025$. From Table VI, Appendix A, with df $= n - 1 = 25 - 1 = 24$, $t_{.025} = 2.064$. The 95% confidence interval is:

$$\bar{x} \pm t_{.025} \frac{s}{\sqrt{n}} \Rightarrow 5 \pm 2.064 \frac{2.2804}{\sqrt{25}} \Rightarrow 5 \pm .941 \Rightarrow (4.059, 5.941)$$

c) For confidence coefficient .99, $\alpha = 1 - .99 = .01$ and $\alpha/2 = .01/2 = .005$. From Table VI, Appendix A, with df $= n - 1 = 25 - 1 = 24$, $t_{.005} = 2.797$. The 99% confidence interval is:

$$\bar{x} \pm t_{.005} \frac{s}{\sqrt{n}} \Rightarrow 5 \pm 2.797 \frac{2.2804}{\sqrt{25}} \Rightarrow 5 \pm 1.276 \Rightarrow (3.724, 6.276)$$

Increasing the sample size decreases the width of the confidence interval.

7.27 a. First, we must compute some preliminary statistics:

$$\bar{x} = \frac{\sum x}{n} = \frac{28.856}{10} = 2.8856$$

$$s^2 = \frac{\sum x^2 - \frac{\left(\sum x\right)^2}{n}}{n-1} = \frac{221.90161 - \frac{(28.856)^2}{10}}{10-1} = 15.4039$$

$$s = \sqrt{s^2} = \sqrt{15.4039} = 3.925$$

For confidence coefficient .99, $\alpha = .01$ and $\alpha/2 = .01/2 = .005$. From Table VI, Appendix A, with df $= n - 1 = 10 - 1 = 9$, $t_{.005} = 3.250$. The confidence interval is:

$$\bar{x} \pm t_{.005} \frac{s}{\sqrt{n}} \Rightarrow 2.8856 \pm 3.250 \frac{3.925}{\sqrt{10}} \Rightarrow 2.8856 \pm 4.034 \Rightarrow (-1.148, 6.919)$$

b. First, we must compute some preliminary statistics:

$$\bar{x} = \frac{\sum x}{n} = \frac{4.083}{10} = .4083$$

$$s^2 = \frac{\sum x^2 - \frac{\left(\sum x\right)^2}{n}}{n-1} = \frac{2.227425 - \frac{(4.083)^2}{10}}{10-1} = .06226$$

$$s = \sqrt{s^2} = \sqrt{.06226} = .2495$$

For confidence coefficient .99, $\alpha = .01$ and $\alpha/2 = .01/2 = .005$. From Table VI, Appendix A, with df $= n - 1 = 10 - 1 = 9$, $t_{.005} = 3.250$. The confidence interval is:

$$\bar{x} \pm t_{.005} \Rightarrow .4083 \pm 3.250 \frac{.2495}{\sqrt{10}} \Rightarrow .4083 \pm .2564 \Rightarrow (.1519, .6647)$$

c. We are 99% confident that the mean lead level in water specimens from Crystal Lake Manors is between −1.148 and 6.919.

We are 99% confident that the mean copper level in water specimens from Crystal Lake Manors is between .1519 and .6647.

d. The phrase "99% confident" means that if repeated samples of size n were selected and 99% confidence intervals constructed for the mean, 99% of all intervals constructed would contain the mean.

7.29 We must assume that the weights of dry seed in the crop of the pigeons are normally distributed. From the printout, the 99% confidence interval is:

(0.61, 2.13)

We are 99% confident that the mean weight of dry seeds in the crop of the spinifex pigeons inhabiting the Western Australian desert is between 0.61 and 2.13 grams.

7.31 a. The population is the set of all DOT permanent count stations in the state of Florida.

b. Yes. There are several types of routes included in the sample. There are 3 recreational areas, 7 rural areas, 5 small cities, and 5 urban areas.

c. The 95% confidence interval for the mean traffic count at the 30th highest hour is (1,633, 2,779). We are 95% confident that the mean traffic count at the 30th highest hour is between 1,633 and 2,779.

d. We must assume that the distribution of the traffic counts at the 30th highest hour is normal. From the stem-and-leaf display, the data look fairly mound-shaped. Thus, the assumption of normality is probably met.

e.   The 95% confidence interval for the mean traffic count at the 100th highest hour is (1,533, 2,659). We are 95% confident that the mean traffic count at the 100th highest hour is between 1,533 and 2,659.

We must assume that the distribution of the traffic counts at the 100th highest hour is normal. From the stem-and-leaf display, the data look fairly mound-shaped. Thus, the assumption of normality is probably met.

7.33   a.   Some preliminary calculations are:

$$\bar{x} = \frac{\sum x}{n} = \frac{196}{11} = 17.82$$

$$s^2 = \frac{\sum x^2 - \frac{\left(\sum x\right)^2}{n}}{n-1} = \frac{3,734 - \frac{196^2}{11}}{11-1} = 24.1636$$

$$s = \sqrt{26.1636} = 4.92$$

For confidence coefficient .95, $\alpha = .05$ and $\alpha/2 = .05/2 = .025$. From Table VI, with df $= n - 1 = 11 - 1 = 10$, $t_{.025} = 2.228$. The 95% confidence interval is:

$$\bar{x} \pm t_{\alpha/2}\frac{s}{\sqrt{n}} \Rightarrow 17.82 \pm 2.228\frac{4.92}{\sqrt{11}} \Rightarrow 17.82 \pm 3.31 \Rightarrow (14.51,\ 21.13)$$

We are 95% confident that the mean FNE score of the population of bulimic female students is between 14.51 and 21.13.

b.   Some preliminary calculations are:

$$\bar{x} = \frac{\sum x}{n} = \frac{198}{14} = 14.14$$

$$s^2 = \frac{\sum x^2 - \frac{\left(\sum x\right)^2}{n}}{n-1} = \frac{3,164 - \frac{198^2}{14}}{14-1} = 27.9780$$

$$s = \sqrt{27.9780} = 5.29$$

For confidence coefficient .95, $\alpha = .05$ and $\alpha/2 = .05/2 = .025$. From Table VI, with df $= n - 1 = 14 - 1 = 13$, $t_{.025} = 2.160$. The 95% confidence interval is:

$$\bar{x} \pm t_{\alpha/2}\frac{s}{\sqrt{n}} \Rightarrow 14.14 \pm 2.160\frac{5.29}{\sqrt{14}} \Rightarrow 14.14 \pm 3.05 \Rightarrow (11.09,\ 17.19)$$

We are 95% confident that the mean FNE score of the population of normal female students is between 11.09 and 17.19.

c. We must assume that the populations of FNE scores for both the bulimic and normal female students are normally distributed.

Stem-and-leaf displays for the two groups are below:

```
Stem-and-leaf of Bulimia    N  = 11
Leaf Unit = 1.0

  1       1 0
  3       1 33
  4       1 4
  5       1 6
 (1)      1 9
  5       2 011
  2       2
  2       2 45
```

```
Stem-and-leaf of Normal     N  = 14
Leaf Unit = 1.0

  2       0 67
  3       0 8
  5       1 01
  7       1 33
  7       1 5
  6       1 6
  5       1 899
  2       2 0
  1       2 3
```

From both of these plots, the assumption of normality is questionable for both groups. Neither of the plots look mound-shaped.

7.35   By the Central Limit Theorem, the sampling distribution of $\hat{p}$ is approximately normal with mean $\mu_{\hat{p}} = p$ and standard deviation $\sigma_{\hat{p}} = \sqrt{\dfrac{pq}{n}}$.

7.37   a.   The sample size is large enough if $\hat{p} \pm 3\sigma_{\hat{p}}$ does not include 0 or 1.

$$\hat{p} \pm 3\sigma_{\hat{p}} \approx .64 \pm 3\sqrt{\frac{.64(.36)}{196}} \Rightarrow .64 \pm .013 \Rightarrow (.537, .743)$$

Since this interval does not contain 0 or 1, the sample size is sufficiently large.

b. For confidence coefficient .95, $\alpha = .05$ and $\alpha/2 = .025$. From Table IV, Appendix A, $z_{.025} = 1.96$. The confidence interval is:

$$\hat{p} \pm z_{.025}\sqrt{\frac{\hat{p}\hat{q}}{n}} \Rightarrow .64 \pm 1.96\sqrt{\frac{.64(.36)}{196}} \Rightarrow .64 \pm .067 \Rightarrow (.573, .707)$$

c. We are 95% confident the true value of p is between .372 and .468.

d. "95% confidence" means that if repeated samples of size 196 were selected from the population and 95% confidence intervals formed, 95% of all confidence intervals will contain the true value of p.

7.39   The sample size is sufficiently large if $\hat{p} \pm 3\sigma_{\hat{p}}$ does not include 0 or 1.

a. $\hat{p} \pm 3\sigma_{\hat{p}} \approx .05 \pm 3\sqrt{\frac{.05(.95)}{500}} \Rightarrow .05 \pm .029 \Rightarrow (.021, .079)$

Since this interval does not contain 0 or 1, the sample size is sufficiently large.

b. $\hat{p} \pm 3\sigma_{\hat{p}} \approx .05 \pm 3\sqrt{\frac{.05(.95)}{100}} \Rightarrow .05 \pm .065 \Rightarrow (-.015, .115)$

Since the interval contains 0, the sample size is not sufficiently large.

c. $\hat{p} \pm 3\sigma_{\hat{p}} \approx .5 \pm 3\sqrt{\frac{.5(.5)}{10}} \Rightarrow .5 \pm .474 \Rightarrow (.026, .974)$

Since this interval does not contain 0 or 1, the sample size is sufficiently large.

d. $\hat{p} \pm 3\sigma_{\hat{p}} \approx .3 \pm 3\sqrt{\frac{.3(.7)}{10}} \Rightarrow .3 \pm .435 \Rightarrow (-.135, .735)$

Since the interval contains 0, the sample size is not sufficiently large.

7.41   First we must compute the sample proportion:

$$\hat{p} = \frac{x}{n} = \frac{278}{500} = .556$$

We must check to see if the sample size is sufficiently large:

$$\hat{p} \pm 3\sigma_{\hat{p}} \approx \hat{p} \pm 3\sqrt{\frac{\hat{p}\hat{q}}{n}} \Rightarrow .556 \pm 3\sqrt{\frac{.556(.444)}{500}} \Rightarrow .556 \pm .067 \Rightarrow (.489, .623)$$

Since the interval is wholly contained in the interval (0, 1), we may conclude that the normal approximation is reasonable.

For confidence coefficient .95, $\alpha = .05$ and $\alpha/2 = .025$. From Table IV, Appendix A, $z_{.025} = 1.96$. The confidence interval is:

$$\hat{p} \pm z_{.025}\sqrt{\frac{\hat{p}\hat{q}}{n}} \Rightarrow .556 \pm 1.96\sqrt{\frac{.556(.444)}{500}} \Rightarrow .556 \pm .044 \Rightarrow (.512, .600)$$

7.43   First, we compute $\hat{p}$: $= \hat{p} = \dfrac{x}{n} = \dfrac{183}{837} = .219$

To see if the sample size is sufficiently large:

$$\hat{p} \pm 3\sigma_{\hat{p}} \approx \hat{p} \pm 3\sqrt{\frac{\hat{p}\hat{q}}{n}} \Rightarrow .219 \pm 3\sqrt{\frac{.219(.781)}{837}} \Rightarrow .219 \pm .043 \Rightarrow (.176, .262)$$

Since the interval is wholly contained in the interval $(0,1)$, we may conclude that the normal approximation is reasonable.

For confidence coefficient .90, $\alpha = .10$ and $\alpha/2 = .10/2 = .05$. From Table IV, Appendix A, $z_{.05} = 1.645$. The confidence interval is:

$$\hat{p} \pm z_{.05}\sqrt{\frac{pq}{n}} \approx \hat{p} \pm 1.645\sqrt{\frac{\hat{p}\hat{q}}{n}} \Rightarrow .219 \pm 1.645\sqrt{\frac{.219(.781)}{837}} \Rightarrow .219 \pm .024 \Rightarrow (.195, .243)$$

7.45   a.   First, we compute $\hat{p}$: $\hat{p} = \dfrac{x}{n} = \dfrac{14}{38} = .368$

To see if the sample size is sufficiently large:

$$\hat{p} \pm 3\sigma_{\hat{p}} \approx \hat{p} \pm 3\sqrt{\frac{\hat{p}\hat{q}}{n}} \Rightarrow .368 \pm 3\sqrt{\frac{.368(.632)}{38}} \Rightarrow .368 \pm .235 \Rightarrow (.133, .603)$$

Since the interval is wholly contained in the interval $(0,1)$, we may conclude that the normal approximation is reasonable.

For confidence coefficient .90, $\alpha = .10$ and $\alpha/2 = .10/2 = .05$. From Table IV, Appendix A, $z_{.05} = 1.645$. The confidence interval is:

$$\hat{p} \pm z_{.05}\sqrt{\frac{pq}{n}} \approx \hat{p} \pm 1.645\sqrt{\frac{\hat{p}\hat{q}}{n}} \Rightarrow .368 \pm 1.645\sqrt{\frac{.368(.632)}{38}} \Rightarrow .368 \pm .129 \Rightarrow (.239, .497)$$

We are 90% confident that the true dropout rate for exercises who vary their routine in workouts is between .239 and .497.

b.   First, we compute $\hat{p}$: $\hat{p} = \dfrac{x}{n} = \dfrac{23}{38} = .605$

To see if the sample size is sufficiently large:

$$\hat{p} \pm 3\sigma_{\hat{p}} \approx \hat{p} \pm 3\sqrt{\dfrac{\hat{p}\hat{q}}{n}} \Rightarrow .605 \pm 3\sqrt{\dfrac{.605(.395)}{38}} \Rightarrow .605 \pm .238 \Rightarrow (.367,\ .843)$$

Since the interval is wholly contained in the interval (0,1), we may conclude that the normal approximation is reasonable.

For confidence coefficient .90, $\alpha = .10$ and $\alpha/2 = .10/2 = .05$. From Table IV, Appendix A, $z_{.05} = 1.645$. The confidence interval is:

$$\hat{p} \pm z_{.05}\sqrt{\dfrac{pq}{n}} \approx \hat{p} \pm 1.645\sqrt{\dfrac{\hat{p}\hat{q}}{n}} \Rightarrow .605 \pm 1.645\sqrt{\dfrac{.605(.395)}{38}} \Rightarrow .605 \pm .130 \Rightarrow (.475,\ .735)$$

We are 90% confident that the true dropout rate for exercises who have no set schedule for their workouts is between .475 and .735.

7.47   a.   The point estimate of p is $\hat{p} = x/n = 35/55 = .636$.

b.   We must check to see if the sample size is sufficiently large:

$$\hat{p} \pm 3\sigma_{\hat{p}} \approx \hat{p} \pm 3\sqrt{\dfrac{\hat{p}\hat{q}}{n}} \Rightarrow .636 \pm 3\sqrt{\dfrac{.636(.364)}{55}} \Rightarrow .636 \pm .195 \Rightarrow (.441,\ .831)$$

Since the interval is wholly contained in the interval (0, 1) we may assume that the normal approximation is reasonable.

For confidence coefficient, .99, $\alpha = .01$ and $\alpha/2 = .01/2 = .005$. From Table IV, Appendix A, $z_{.005} = 2.58$. The confidence interval is:

$$\hat{p} \pm z_{.005}\sqrt{\dfrac{\hat{p}\hat{q}}{n}} \Rightarrow .636 \pm 2.58\sqrt{\dfrac{.636(.364)}{55}} \Rightarrow .636 \pm .167 \Rightarrow (.469,\ .803)$$

c.   We are 99% confident that the true proportion of fatal accidents involving children is between .469 and .803.

d.   The sample proportion of children killed by air bags who were not wearing seat belts or were improperly restrained is 24/35 = .686. This is rather large proportion. Whether a child is killed by an airbag could be related to whether or not he/she was properly restrained or not. Thus, the number of children killed by air bags could possibly be reduced if the child were properly restrained.

7.49 a. First we must compute $\hat{p}$: $\hat{p} = \dfrac{x}{n} = \dfrac{97}{185} = .524$

To see if the sample size is sufficiently large:

$$\hat{p} \pm 3\sigma_{\hat{p}} \approx \hat{p} \pm 3\sqrt{\dfrac{\hat{p}\hat{q}}{n}} \Rightarrow .524 \pm 3\sqrt{\dfrac{.524(.476)}{185}} \Rightarrow .524 \pm .110 \Rightarrow (.414, .634)$$

Since this interval is wholly contained in the interval $(0, 1)$, we may conclude that the normal approximation is reasonable.

For confidence coefficient .99, $\alpha = .01$ and $\alpha/2 = .01/2 = .005$. From Table IV, Appendix A, $z_{.005} = 2.58$. The confidence interval is:

$$\hat{p} \pm z_{.005}\sqrt{\dfrac{pq}{n}} \approx \hat{p} \pm 2.58\sqrt{\dfrac{\hat{p}\hat{q}}{n}} \Rightarrow .524 \pm 2.58\sqrt{\dfrac{.524(.476)}{185}} \Rightarrow .524 \pm .095$$
$$\Rightarrow (.429, .619)$$

b. We are 99% confident that the true proportion of bottlenose dolphin signature whistles that are Type "a" whistles is between .429 and .619.

7.51 To compute the necessary sample size, use

$$n = \dfrac{(z_{\alpha/2})^2 \sigma^2}{B^2} \quad \text{where } \alpha = 1 - .95 = .05 \text{ and } \alpha/2 = .05/2 = .025$$

From Table IV, Appendix A, $z_{.025} = 1.96$. Thus,

$$n = \dfrac{(1.96)^2 (5.4)}{.2^2} = 518.616 \approx 519$$

You would need to take 519 samples.

7.53 a. Range $= 39 - 31 = 8$. $\sigma \approx \dfrac{\text{Range}}{4} = \dfrac{8}{4} = 2$

For confidence coefficient .90, $\alpha = .10$ and $\alpha/2 = .05$. From Table IV, Appendix A, $z_{.05} = 1.645$.

The sample size is $n = \dfrac{z_{\alpha/2}^2 \sigma^2}{B^2} = \dfrac{1.645^2 (2^2)}{.15^2} = 481.07 \approx 482$

b. $\sigma \approx \dfrac{\text{Range}}{6} = \dfrac{8}{6} = 1.333$

The sample size is $n = \dfrac{z_{\alpha/2}^2 \sigma^2}{B^2} = \dfrac{1.645^2 (1.333^2)}{.15^2} = 213.7 \approx 214$

7.55   For confidence coefficient .90, $\alpha = .10$ and $\alpha/2 = .05$. From Table IV, Appendix A, $z_{.05} = 1.645$.

We know $\hat{p}$ is in the middle of the interval, so $\hat{p} = \dfrac{.54 + .26}{2} = .4$

The confidence interval is $\hat{p} \pm z_{.05}\sqrt{\dfrac{\hat{p}\hat{q}}{n}} \Rightarrow .4 \pm 1.645\sqrt{\dfrac{.4(.6)}{n}}$

We know $.4 - 1.645\sqrt{\dfrac{.4(.6)}{n}} = .26$

$\Rightarrow .4 - \dfrac{.8059}{\sqrt{n}} = .26$

$\Rightarrow .4 - .26 = \dfrac{.8059}{\sqrt{n}} \Rightarrow \sqrt{n} = \dfrac{.8059}{.14} = 5.756$

$\Rightarrow n = 5.756^2 = 33.1 \approx 34$

7.57   a.   The width of a confidence interval is $2B = 2z_{\alpha/2}\dfrac{\sigma}{\sqrt{n}}$

For confidence coefficient .95, $\alpha = 1 - .95 = .05$ and $\alpha/2 = .05/2 = .025$. From Table IV, Appendix A, $z_{.025} = 1.96$.

For $n = 16$,

$$W = 2z_{\alpha/2}\frac{\sigma}{\sqrt{n}} = 2(1.96)\frac{1}{\sqrt{16}} = 0.98$$

For $n = 25$,

$$W = 2z_{\alpha/2}\frac{\sigma}{\sqrt{n}} = 2(1.96)\frac{1}{\sqrt{25}} = 0.784$$

For $n = 49$,

$$W = 2z_{\alpha/2}\frac{\sigma}{\sqrt{n}} = 2(1.96)\frac{1}{\sqrt{49}} = 0.56$$

For $n = 100$,

$$W = 2z_{\alpha/2}\frac{\sigma}{\sqrt{n}} = 2(1.96)\frac{1}{\sqrt{100}} = 0.392$$

For $n = 400$,

$$W = 2z_{\alpha/2}\frac{\sigma}{\sqrt{n}} = 2(1.96)\frac{1}{\sqrt{400}} = 0.196$$

b.

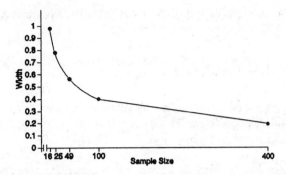

7.59　For confidence coefficient .95, $\alpha = .05$ and $\alpha/2 = .05/2 = .025$. From Table IV, Appendix A, $z_{.025} = 1.96$. For this study,

$$\hat{p} = \frac{190}{532} = .36. \text{ Using } \hat{p} \text{ to estimate } p, \text{ we get}$$

$$n = \frac{(z_{\alpha/2})^2 pq}{B^2} \approx \frac{1.96^2(.36)(.64)}{.03^2} = 983.4 \approx 984$$

No. The sample size needed to be 984 instead of 532.

7.61　For confidence coefficient .99, $\alpha = .01$ and $\alpha/2 = .01/2 = .005$. From Table IV, Appendix A, $z_{.005} = 2.58$. From the previous estimate, we will use $\hat{p} = .333$ to estimate p.

$$n = \frac{z_{\alpha/2}^2 pq}{B^2} = \frac{2.58^2(.333)(.667)}{.01^2} = 14,784.5966 \approx 14,785$$

7.63　From Exercise 7.28, $s = .2316$. For confidence coefficient .95, $\alpha = .05$ and $\alpha/2 = .05/2 = .025$. From Table IV, Appendix A, $z_{.025} = 1.96$.

$$n = \frac{z_{\alpha/2}^2 \alpha^2}{B^2} = \frac{1.96^2(.2316)^2}{.04^2} = 128.786 \approx 129$$

7.65　For confidence coefficient .90, $\alpha = .10$ and $\alpha/2 = .05$. From Table IV, Appendix A, $z_{.05} = 1.645$.

The sample size is $n = \dfrac{(z_{\alpha/2})^2 \sigma^2}{B^2} = \dfrac{(1.645)^2(10^2)}{1^2} = 270.6 \approx 271$

7.67　a.　For confidence coefficient .90, $\alpha = .10$ and $\alpha/2 = .05$. From Table IV, Appendix A, $z_{.05} = 1.645$.

The sample size is $n = \dfrac{z_{\alpha/2}^2 \sigma^2}{B^2} = \dfrac{1.645^2(2^2)}{.1^2} = 1082.4 \approx 1083$

　　　b.　In part **a**, we found $n = 1083$. If we used an n of only 100, the width of the confidence interval for μ would be wider since we would be dividing by a smaller number.

c. We know $B = \dfrac{z_{\alpha/2}\sigma}{\sqrt{n}} \Rightarrow z_{\alpha/2} = \dfrac{B\sqrt{n}}{\sigma} = \dfrac{.1\sqrt{100}}{2} = .5$

$P(-.5 \le z \le .5) = .1915 + .1915 = .3830$. Thus, the level of confidence is approximately 38.3%.

7.69 a. For a small sample from a normal distribution with unknown standard deviation, we use the t statistic. For confidence coefficient .95, $\alpha = 1 - .95 = .05$ and $\alpha/2 = .05/2 = .025$. From Table VI, Appendix A, with df $= n - 1 = 21 - 1 = 20$, $t_{.025} = 2.086$.

b. For a large sample from a distribution with an unknown standard deviation, we can estimate the population standard deviation with s and use the z statistic. For confidence coefficient .95, $\alpha = 1 - .95 = .05$ and $\alpha/2 = .05/2 = .025$. From Table IV, Appendix A, $z_{.025} = 1.96$.

c. For a small sample from a normal distribution with known standard deviation, we use the z statistic. For confidence coefficient .95, $\alpha = 1 - .95 = .05$ and $\alpha/2 = .05/2 = .025$. From Table IV, Appendix A, $z_{.025} = 1.96$.

d. For a large sample from a distribution about which nothing is known, we can estimate the population standard deviation with s and use the z statistic. For confidence coefficient .95, $\alpha = 1 - .95 = .05$ and $\alpha/2 = .05/2 = .025$. From Table IV, Appendix A, $z_{.025} = 1.96$.

e. For a small sample from a distribution about which nothing is known, we can use neither z nor t.

7.71 The parameters of interest for the problems are:

(1) The question requires a categorical response. One parameter of interest might be the proportion, p, of all Americans over 18 years of age who think their health is generally very good or excellent.
(2) A parameter of interest might be the mean number of days, μ, in the previous 30 days that all Americans over 18 years of age felt that their physical health was not good because of injury or illness.
(3) A parameter of interest might be the mean number of days, μ, in the previous 30 days that all Americans over 18 years of age felt that their mental health was not good because of stress, depression, or problems with emotions.
(4) A parameter of interest might be the mean number of days, μ, in the previous 30 days that all Americans over 18 years of age felt that their physical or mental health prevented them from performing their usual activities.

7.73 a. First we must compute $\hat{p}$: $\hat{p} = \dfrac{x}{n} = \dfrac{11{,}239}{43{,}732} = .257$

To see if the sample size is sufficiently large:

$$\hat{p} \pm 3\sigma_{\hat{p}} \approx \hat{p} \pm 3\sqrt{\dfrac{\hat{p}\hat{q}}{n}} \Rightarrow .257 \pm 3\sqrt{\dfrac{.257(.743)}{43{,}732}} \Rightarrow .257 \pm .006 \Rightarrow (.251, .263)$$

Since this interval is wholly contained in the interval (0, 1), we may conclude that the normal approximation is reasonable.

For confidence coefficient .90, $\alpha = .10$ and $\alpha/2 = .10/2 = .05$. From Table IV, Appendix A, $z_{.05} = 1.645$. The confidence interval is:

$$\hat{p} \pm z_{.05}\sqrt{\frac{pq}{n}} \approx \hat{p} \pm 1.645\sqrt{\frac{\hat{p}\hat{q}}{n}} \Rightarrow .257 \pm 1.645\sqrt{\frac{.257(.743)}{43,732}} \Rightarrow .257 \pm .003$$

$$\Rightarrow (.254, .260)$$

We are 90% confident that the true percentage of U.S. adults who currently smoke is between 25.4% and 26.0%.

b. First we must compute $\hat{p}$: $\hat{p} = \dfrac{x}{n} = \dfrac{10,539}{43,732} = .241$

To see if the sample size is sufficiently large:

$$\hat{p} \pm 3\sigma_{\hat{p}} \approx \hat{p} \pm 3\sqrt{\frac{\hat{p}\hat{q}}{n}} \Rightarrow .241 \pm 3\sqrt{\frac{.241(.759)}{43,732}} \Rightarrow .241 \pm .006 \Rightarrow (.235, .247)$$

Since this interval is wholly contained in the interval (0, 1), we may conclude that the normal approximation is reasonable.

For confidence coefficient .90, $\alpha = .10$ and $\alpha/2 = .10/2 = .05$. From Table IV, Appendix A, $z_{.05} = 1.645$. The confidence interval is:

$$\hat{p} \pm z_{.05}\sqrt{\frac{pq}{n}} \approx \hat{p} \pm 1.645\sqrt{\frac{\hat{p}\hat{q}}{n}} \Rightarrow .241 \pm 1.645\sqrt{\frac{.241(.759)}{43,732}} \Rightarrow .241 \pm .003$$

$$\Rightarrow (.238, .244)$$

We are 90% confident that the true percentage of U.S. adults who are former smokers is between 23.8% and 24.4%.

7.75 For confidence coefficient .90, $\alpha = .10$ and $\alpha/2 = .05$. From Table IV, Appendix A, $z_{.05} = 1.645$. The confidence interval is:

$$\bar{x} \pm z_{\alpha/2}\frac{s}{\sqrt{n}} \Rightarrow 2 \pm 1.645\frac{1.7}{\sqrt{35}} \Rightarrow 2 \pm .47 \Rightarrow (1.53, 2.47)$$

We are 90% confident the mean number of cavities per child under age of 12 who lives in the sampled environment is between 1.53 and 2.47.

7.77 a. First, we compute $\hat{p}$: $\hat{p} = \dfrac{x}{n} = \dfrac{665}{1007} = .660$

To see if the sample size is sufficiently large:

$$\hat{p} \pm 3\sigma_{\hat{p}} \approx \hat{p} \pm 3\sqrt{\frac{\hat{p}\hat{q}}{n}} \Rightarrow .660 \pm 3\sqrt{\frac{.660(.340)}{1007}} \Rightarrow .660 \pm .045 \Rightarrow (.615, .705)$$

Since the interval is wholly contained in the interval (0,1), we may conclude that the normal approximation is reasonable.

For confidence coefficient .95, $\alpha = .05$ and $\alpha/2 = .05/2 = .025$. From Table IV, Appendix A, $z_{.025} = 1.96$. The confidence interval is:

$$\hat{p} \pm z_{.025}\sqrt{\frac{pq}{n}} \approx \hat{p} \pm 1.96\sqrt{\frac{\hat{p}\hat{q}}{n}} \Rightarrow .660 \pm 1.96\sqrt{\frac{.660(.340)}{1007}} \Rightarrow .660 \pm .029 \Rightarrow (.631, .689)$$

We are 95% confident that the proportion of U.S. workers who take their lunch to work is between .631 and .689.

b. First, we compute $\hat{p}$: $\hat{p} = \dfrac{x}{n} = \dfrac{200}{665} = .301$

To see if the sample size is sufficiently large:

$$\hat{p} \pm 3\sigma_{\hat{p}} \approx \hat{p} \pm 3\sqrt{\frac{\hat{p}\hat{q}}{n}} \Rightarrow .301 \pm 3\sqrt{\frac{.301(.699)}{665}} \Rightarrow .301 \pm .053 \Rightarrow (.248, .354)$$

Since the interval is wholly contained in the interval (0,1), we may conclude that the normal approximation is reasonable.

For confidence coefficient .95, $\alpha = .05$ and $\alpha/2 = .05/2 = .025$. From Table IV, Appendix A, $z_{.025} = 1.96$. The confidence interval is:

$$\hat{p} \pm z_{.025}\sqrt{\frac{pq}{n}} \approx \hat{p} \pm 1.96\sqrt{\frac{\hat{p}\hat{q}}{n}} \Rightarrow .301 \pm 1.96\sqrt{\frac{.301(.699)}{665}} \Rightarrow .301 \pm .035 \Rightarrow (.266, .336)$$

We are 95% confident that the proportion of U.S. workers who take their lunch to work who brown-bag their lunch is between .266 and .336.

7.79 a. For confidence coefficient .99, $\alpha = .01$ and $\alpha/2 = .01/2 = .005$. From Table IV, Appendix A, $z_{.005} = 2.58$. The confidence interval is:

$$\bar{x} \pm z_{.005}\frac{s}{\sqrt{n}} \Rightarrow .044 \pm 2.58\frac{.884}{\sqrt{197}} \Rightarrow .044 \pm .162 \Rightarrow (-.118, .206)$$

We are 99% confident that the mean inbreeding coefficient for this species of wasps is between $-.118$ and .206.

b. A coefficient of 0 indicates that the wasp has no tendency to inbreed. Since 0 is in the 99% confidence interval, it is a likely value for the true mean inbreeding coefficient. Thus, there is no evidence to indicate that this species of wasps has a tendency to inbreed.

7.81 From the printout, the 95% confidence interval is (.39, 1.23). We are 95% confident that the mean solution time for the hybrid algorithm is between .39 and 1.23 seconds.

7.83 For confidence coefficient .95, $\alpha = 1 - .95 = .05$ and $\alpha/2 = .05/2 = .025$. From Table VI, Appendix A, with df $= n - 1 = 20 - 1 = 19$, $t_{.025} = 2.093$. The confidence interval is:

$$\bar{x} \pm t_{.025} \frac{s}{\sqrt{n}} \Rightarrow 49.70 \pm 2.093 \left(\frac{.32}{\sqrt{20}} \right) \Rightarrow 49.70 \pm .1498 \Rightarrow (49.5502, 49.8498)$$

We are 95% confident the mean number of gallons of naphtha per drum is between 49.5502 and 49.8498. We must assume the population of contents of naphtha is normally distributed.

7.85 For all parts to this problem, we will use a 95% confidence interval. For confidence coefficient, .95, $\alpha = .05$ and $\alpha/2 = .05/2 = .025$. From Table IV, Appendix A, $z_{.025} = 1.96$.

a. Some preliminary calculations:

$$\hat{p} = x/n = 14/37 = .378$$

We must check to see if the sample size is sufficiently large:

$$\hat{p} \pm 3\sigma_{\hat{p}} \approx \hat{p} \pm 3\sqrt{\frac{\hat{p}\hat{q}}{n}} \Rightarrow .378 \pm 3\sqrt{\frac{.378(.622)}{37}} \Rightarrow .378 \pm .239 \Rightarrow (.139, .617)$$

Since the interval is wholly contained in the interval (0, 1), we may assume that the normal approximation is reasonable.

The confidence interval is:

$$\hat{p} \pm z_{.025}\sqrt{\frac{\hat{p}\hat{q}}{n}} \Rightarrow .378 \pm 1.96\sqrt{\frac{.378(.622)}{37}} \Rightarrow .378 \pm .156 \Rightarrow (.222, .534)$$

We are 95% confident that the true proportion of suicides at the jail that are committed by inmates charged with murder/manslaughter is between .222 and .534.

b. Some preliminary calculations:

$$\hat{p} = x/n = 26/37 = .703$$

We must check to see if the sample size is sufficiently large:

$$\hat{p} \pm 3\sigma_{\hat{p}} \approx \hat{p} \pm 3\sqrt{\frac{\hat{p}\hat{q}}{n}} \Rightarrow .703 \pm 3\sqrt{\frac{.703(.297)}{37}} \Rightarrow .703 \pm .225 \Rightarrow (.478, .928)$$

Since the interval is wholly contained in the interval (0, 1), we may assume that the normal approximation is reasonable.

The confidence interval is:

$$\hat{p} \pm z_{.025}\sqrt{\frac{\hat{p}\hat{q}}{n}} \Rightarrow .703 \pm 1.96\sqrt{\frac{.703(.297)}{37}} \Rightarrow .703 \pm .147 \Rightarrow (.556, .850)$$

We are 95% confident that the true proportion of suicides at the jail that are committed by inmates charged with murder/manslaughter is between .556 and .850.

c.   Some preliminary calculations are:

$$\bar{x} = \frac{\sum x}{n} = \frac{1532}{37} = 41.405$$

$$s^2 = \frac{\sum x^2 - \frac{\left(\sum x\right)^2}{n}}{n-1} = \frac{223,606 - \frac{(1532)^2}{37}}{37-1} = 4,449.2477$$

$$s = \sqrt{s^2} = \sqrt{4,449.2477} = 66.703$$

The confidence interval is:

$$\bar{x} \pm z_{.05}\frac{s}{\sqrt{n}} \Rightarrow 41.405 \pm 1.96\frac{66.703}{\sqrt{37}} \Rightarrow 41.405 \pm 21.493 \Rightarrow (19.912, 62.898)$$

We are 95% confident that the true average length of time an inmate is in jail before committing suicide is between 19.912 and 62.898 days.

d.   Some preliminary calculations:

$$\hat{p} = x/n = 14/37 = .378$$

We must check to see if the sample size is sufficiently large:

$$\hat{p} \pm 3\sigma_{\hat{p}} \approx \hat{p} \pm 3\sqrt{\frac{\hat{p}\hat{q}}{n}} \Rightarrow .378 \pm 3\sqrt{\frac{.378(.622)}{37}} \Rightarrow .378 \pm .239 \Rightarrow (.139, .617)$$

Since the interval is wholly contained in the interval (0, 1), we may assume that the normal approximation is reasonable.

The confidence interval is:

$$\hat{p} \pm z_{.025}\sqrt{\frac{\hat{p}\hat{q}}{n}} \Rightarrow .378 \pm 1.96\sqrt{\frac{.378(.622)}{37}} \Rightarrow .378 \pm .156 \Rightarrow (.222, .534)$$

We are 95% confident that the true proportion of suicides that are committed by white inmates is between .222 and .534. The percentage would be between 22.2% and 53.4%.

7.87   a.   The point estimate for the fraction of the entire market who refuse to purchase bars is:

$$\hat{p} = \frac{x}{n} = \frac{23}{244} = .094$$

b.   To see if the sample size is sufficient:

$$\hat{p} \pm 3\sqrt{\frac{\hat{p}\hat{q}}{n}} \Rightarrow .094 \pm 3\sqrt{\frac{(.094)(.906)}{244}} \Rightarrow .094 \pm .056 \Rightarrow (.038, .150)$$

Since the interval above is contained in the interval (0, 1), the sample size is sufficiently large.

c.   For confidence coefficient .95, $\alpha = 1 - .95 = .05$ and $\alpha/2 = .05/2 = .025$. From Table IV, Appendix A, $z_{.025} = 1.96$. The confidence interval is:

$$\hat{p} \pm z_{.025}\sqrt{\frac{\hat{p}\hat{q}}{n}} \Rightarrow .094 \pm 1.96\sqrt{\frac{.094(.906)}{244}} \Rightarrow .094 \pm .037 \Rightarrow (.057, .131)$$

d.   The best estimate of the true fraction of the entire market who refuse to purchase bars six months after the poisoning is .094. We are 95% confident the true fraction of the entire market who refuse to purchase bars six months after the poisoning is between .057 and .131.

7.89   a.   For confidence coefficient .99, $\alpha = .01$ and $\alpha/2 = .01/2 = .005$. From Table VI, Appendix A, with df $= n - 1 = 3 - 1 = 2$, $t_{.005} = 9.925$. The confidence interval is:

$$\bar{x} \pm t_{.005}\frac{s}{\sqrt{n}} \Rightarrow 49.3 \pm 9.925\frac{1.5}{\sqrt{3}} \Rightarrow 49.3 \pm 8.60 \Rightarrow (40.70, 57.90)$$

b.   We are 99% confident that the mean percentage of B(a)p removed from all soil specimens using the poison is between 40.70% and 57.90%.

c.   We must assume that the distribution of the percentages of B(a)p removed from all soil specimens using the poison is normal.

d.   For confidence coefficient .99, $\alpha = .01$ and $\alpha/2 = .01/2 = .005$. From Table IV, Appendix A, $z_{.005} = 2.58$.

$$n = \frac{z_{\alpha/2}^2\sigma^2}{B^2} = \frac{1.96^2(1.5)^2}{.5^2} = 34.57 \approx 35$$

Inferences Based on a Single Sample: Tests of Hypotheses

8.1   The null hypothesis is the "status quo" hypothesis, while the alternative hypothesis is the research hypothesis.

8.3   The "level of significance" of a test is α. This is the probability that the test statistic will fall in the rejection region when the null hypothesis is true.

8.5   The four possible results are:

1.   Rejecting the null hypothesis when it is true. This would be a Type I error.
2.   Accepting the null hypothesis when it is true. This would be a correct decision.
3.   Rejecting the null hypothesis when it is false. This would be a correct decision.
4.   Accepting the null hypothesis when it is false. This would be a Type II error.

8.7   When you reject the null hypothesis in favor of the alternative hypothesis, this does not prove the alternative hypothesis is correct. We are $100(1 - \alpha)\%$ confident that there is sufficient evidence to conclude that the alternative hypothesis is correct.

        If we were to repeatedly draw samples from the population and perform the test each time, approximately $100(1 - \alpha)\%$ of the tests performed would yield the correct decision.

8.9   a.   To determine if the percentage of senior women who use herbal therapies to prevent or treat health problems is different from 45%, we test:

                H_0: $p = .45$
                H_a: $p \neq .45$

        b.   To determine if the average number of herbal products used in a year by senior women who use herbal therapies is different from 2.5, we test:

                H_0: $\mu = 2.5$
                H_a: $\mu \neq 2.5$

8.11   To determine if the percentage of all cocaine-exposed babies who suffer no major problems is greater than 75%, we test:

        H_0: $p = .75$
        H_a: $p > .75$

8.13   a.   To determine if the average level of mercury uptake in wading birds in the Everglades in 2000 is less than 15 parts per million, we test:

                H_0: $\mu = 15$
                H_a: $\mu < 15$

b. A Type I error is rejecting H_o when H_o is true. In terms of this problem, we would be concluding that the average level of mercury uptake in wading birds in the Everglades in 2000 is less than 15 parts per million, when in fact, the average level of mercury uptake in wading birds in the Everglades in 2000 is equal to 15 parts per million.

c. A Type II error is accepting H_o when H_o is false. In terms of this problem, we would be concluding that the average level of mercury uptake in wading birds in the Everglades in 2000 is equal to 15 parts per million, when in fact, the average level of mercury uptake in wading birds in the Everglades in 2000 is less than 15 parts per million.

8.15 a. A Type I error is rejecting the null hypothesis when it is true. In this problem, we would be concluding that the individual is a liar when, in fact, the individual is telling the truth.

A Type II error is accepting the null hypothesis when it is false. In this problem, we would be concluding that the individual is telling the truth when, in fact, the individual is a liar.

b. The probability of a Type I error would be the probability of concluding the individual is a liar when he/she is telling the truth. From the problem, it stated that the polygraph would indicate that of 500 individuals who were telling the truth, 185 would be liars. Thus, an estimate of the probability of a Type I error would be 185/500 = .37.

The probability of a Type II error would be the probability of concluding the individual is telling the truth when he/she is a liar. From the problem, it stated that the polygraph would indicate that of 500 individuals who were liars, 120 would be telling the truth. Thus, an estimate of the probability of a Type II error would be 120/500 = .24.

8.17 a. Probability of Type I error
$= P(z > 1.96) = .5 - .4750 = .0250$
(From Table IV, Appendix A)

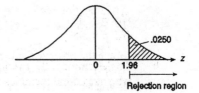

b. Probability of Type I error
$= P(z > 1.645) = .5 - .4500 = .05$
(From Table IV, Appendix A)

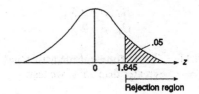

c. Probability of Type I error
$= P(z > 2.58) = .5 - .4951 = .0049$
(From Table IV, Appendix A)

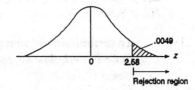

Inferences Based on a Single Sample: Tests of Hypotheses   153

d. Probability of Type I error
$= P(z < -1.28) = .5 - .3997 = .1003$
(From Table IV, Appendix A)

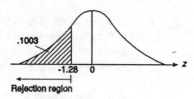

e. Probability of Type I error
$= P(z < -1.645) + P(z > 1.645)$
$= .5 - .4500 + .5 - .4500 = .05 + .05 = .10$
(From Table IV, Appendix A)

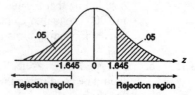

f. Probability of Type I error
$= P(z < -2.58) + P(z > 2.58)$
$= .5 - .4951 + .5 - .4951 = .0049 + .0049$
$= .0098$ (From Table IV, Appendix A)

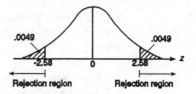

8.19 a. H_0: $\mu = 100$
H_a: $\mu > 100$

The test statistic is $z = \dfrac{\bar{x} - \mu_0}{\sigma_{\bar{x}}} = \dfrac{\bar{x} - \mu_0}{\sigma/\sqrt{n}} = \dfrac{110 - 100}{60/\sqrt{100}} = 1.67$

The rejection region requires $\alpha = .05$ in the upper tail of the z distribution. From Table IV, Appendix A, $z_{.05} = 1.645$. The rejection region is $z > 1.645$.

Since the observed value of the test statistic falls in the rejection region, ($z = 1.67 > 1.645$), H_0 is rejected. There is sufficient evidence to indicate the true population mean is greater than 100 at $\alpha = .05$.

b. H_0: $\mu = 100$
H_a: $\mu \neq 100$

The test statistic is $z = \dfrac{\bar{x} - \mu_0}{\sigma_{\bar{x}}} = \dfrac{110 - 100}{60/\sqrt{100}} = 1.67$

The rejection region requires $\alpha/2 = .05/2 = .025$ in each tail of the z distribution. From Table IV, Appendix A, $z_{.025} = 1.96$. The rejection region is $z < -1.96$ or $z > 1.96$.

Since the observed value of the test statistic does not fall in the rejection region, ($z = 1.67 \not> 1.96$), H_0 is not rejected. There is insufficient evidence to indicate μ does not equal 100 at $\alpha = .05$.

c. In part **a**, we rejected H_0 and concluded the mean was greater than 100. In part **b**, we did not reject H_0. There was insufficient evidence to conclude the mean was different from 100. Because the alternative hypothesis in part **a** is more specific than the one in **b**, it is easier to reject H_0.

8.21 To determine if the mean alkalinity level of water in the tributary exceeds 50 mpl, we test:

H_0: $\mu = 50$
H_a: $\mu > 50$

The test statistic is $z = \dfrac{\bar{x} - \mu_0}{\sigma_{\bar{x}}} = \dfrac{67.8 - 50}{14.4 / \sqrt{100}} = 12.36$

The rejection region requires $\alpha = .01$ in the upper tail of the z distribution. From Table IV, Appendix A, $z_{.01} = 2.33$. The rejection region is $z > 2.33$.

Since the observed value of the test statistic falls in the rejection region ($z = 12.36 > 2.33$), H_0 is rejected. There is sufficient evidence to indicate that the mean alkalinity level of water in the tributary exceeds 50 mpl at $\alpha = .01$.

8.23 a. If we wish to test the research hypothesis that the mean GHQ score for all unemployed men exceeds 10, we test:

H_0: $\mu = 10$
H_a: $\mu > 10$

This is a one-tailed test. We are only interested in rejecting H_0 if the mean GHQ score for all unemployed men is greater than 10.

b. The rejection region requires $\alpha = .05$ in the upper tail of the z distribution. From Table IV, Appendix A, $z_{.05} = 1.645$. The rejection region is $z > 1.645$.

c. The test statistic is $z = \dfrac{\bar{x} - \mu_0}{\sigma_{\bar{x}}} = \dfrac{10.94 - 10.0}{5.10 / \sqrt{49}} = 1.29$

Since the observed value of the test statistic does not fall in the rejection region ($z = 1.29 \ngtr 1.645$), H_0 is not rejected. There is insufficient evidence to indicate the mean GHQ score for all unemployed men is greater than 10 at $\alpha = .05$.

8.25 a. To determine if the true mean PTSD score of all World War II avaiator POWs is less than 16, we test:

H_0: $\mu = 16$
H_a: $\mu < 16$

b. The test statistic is $z = \dfrac{\bar{x} - \mu_0}{\sigma_{\bar{x}}} = \dfrac{9.80 - 16}{9.32 / \sqrt{33}} = -4.31$

The rejection region requires $\alpha = .10$ in the lower tail of the z distribution. From Table IV, Appendix A, $z_{.10} = 1.28$. The rejection region is $z < -1.28$.

Since the observed value of the test statistic falls in the rejection region ($z = -4.31 < -1.28$), H_0 is rejected. There is sufficient evidence to indicate that the true mean PTSD score of all World War II aviator POWs is less than 16 at $\alpha = .10$.

c. Only 33 of the total 239 World War II aviator POWs volunteered to participate in the study. These POWs were self-selected, not random. Usually those who self-select themselves tend to have strong feelings about the subject. They may not be representative of the entire group.

8.27 a. To determine if the population mean ratio of all bones of this particular species differs from 8.5, we test:

H_0: $\mu = 8.5$
H_a: $\mu \neq 8.5$

The test statistic is $z = \dfrac{\bar{x} - \mu_0}{\sigma_{\bar{x}}} = \dfrac{9.258 - 8.5}{1.204/\sqrt{41}} = 4.03$

The rejection region requires $\alpha/2 = .01/2 = .005$ in each tail of the z distribution. From Table IV, Appendix A, $z_{.005} = 2.58$. The rejection region is $z < -2.58$ or $z > 2.58$.

Since the observed value of the test statistic falls in the rejection region ($z = 4.03 > 2.58$), H_0 is rejected. There is sufficient evidence to indicate that the true population mean ratio of all bones of this particular species differs from 8.5 at $\alpha = .01$.

b. The practical implications of the test in part **a** is that the species from which these particular bones came from is probably not species A.

8.29 a. To determine if the mean social interaction score of all Connecticut mental health patients differs from 3, we test:

H_0: $\mu = 3$
H_a: $\mu \neq 3$

The test statistic is $z = \dfrac{\bar{x} - \mu_0}{\sigma_{\bar{x}}} = \dfrac{2.95 - 3}{1.10/\sqrt{6,681}} = -3.72$

The rejection region requires $\alpha/2 = .01/2 = .005$ in each tail of the z distribution. From Table IV, Appendix A, $z_{.005} = 2.58$. The rejection region is $z < -2.58$ or $z > 2.58$.

Since the observed value of the test statistic falls in the rejection region ($z = -3.72 < -2.58$), H_0 is rejected. There is sufficient evidence to indicate that the mean social interaction score of all Connecticut mental health patients differs from 3 at $\alpha = .01$.

b. From the test in part a, we found that the mean social interaction score was statistically different from 3. However, the sample mean score was 2.95. Practically speaking, 2.95 is very similar to 3.0. The very large sample size, $n = 6,681$, makes it very easy to find statistical significance, even when no practical significance exists.

c.	Because the variable of interest is measured on a 5-point scale, it is very unlikely that the population of the ratings will be normal. However, because the sample size was extremely large, ($n = 6,681$), the Central Limit Theorem will apply. Thus, the distribution of $\bar{x}$ will be normal, regardless of the distribution of x. Thus, the analysis used above is appropriate.

8.31	a.	Since the p-value = .10 is greater than $\alpha = .05$, H_0 is not rejected.

b.	Since the p-value = .05 is less than $\alpha = .10$, H_0 is rejected.

c.	Since the p-value = .001 is less than $\alpha = .01$, H_0 is rejected.

d.	Since the p-value = .05 is greater than $\alpha = .025$, H_0 is not rejected.

e.	Since the p-value = .45 is greater than $\alpha = .10$, H_0 is not rejected.

8.33	p-value $= P(z \geq 2.17) = .5 - P(0 < z < 2.17) = .5 - .4850 = .0150$

(using Table IV, Appendix A)

8.35	$z = \dfrac{\bar{x} - \mu_0}{\sigma_{\bar{x}}} = \dfrac{49.4 - 50}{4.1/\sqrt{100}} = -1.46$

p-value $= P(z \geq -1.46) = .5 + .4279 = .9279$

There is no evidence to reject H_0 for $\alpha \leq .10$.

8.37	a.	The p-value reported by SPSS is for a two-tailed test. Thus, $P(z \leq -1.63) + P(z \geq 1.63) = .1032$. For this one-tailed test, the p-value $= P(z \leq -1.63) = .1032/2 = .0516$.

Since the p-value = .0516 $> \alpha = .05$, H_0 is not rejected. There is insufficient evidence to indicate $\mu < 75$ at $\alpha = .05$.

b.	For this one-tailed test, the p-value $= P(z \leq 1.63)$. Since $P(z \leq -1.63) = .1032/2 = .0516$, $P(z \leq 1.63) = 1 - .0516 = .9484$.

Since the p-value = .9484 $> \alpha = .10$, H_0 is not rejected. There is insufficient evidence to indicate $\mu < 75$ at $\alpha = .10$.

c.	For this one-tailed test, the p-value $= P(z \geq 1.63) = .1032/2 = .0516$.

Since the p-value = .0516 $< \alpha = .10$, H_0 is rejected. There is sufficient evidence to indicate $\mu > 75$ at $\alpha = .10$.

d.	For this two-tailed test, the p-value = .1032.

Since the p-value = .1032 $> \alpha = .01$, H_0 is not rejected. There is insufficient evidence to indicate $\mu \neq 75$ at $\alpha = .01$.

8.39 a. To determine if the average age of all MSNBC's viewers is greater than 50 years, we test:

H_o: $\mu = 50$
H_a: $\mu > 50$

The test statistic is $z = \dfrac{\bar{x} - \mu_0}{\sigma_{\bar{x}}} = \dfrac{51.3 - 50}{7.1/\sqrt{50}} = 1.29$.

The p-value for this one-tailed test is $P(z \geq 1.29) = .5 - .4015 = .0985$
(using Table IV, Appendix A).

   b. If $\bar{x}$ had been larger, the value of z would also have been larger. Thus, the p-value would have been smaller.

8.41 From the printout, the p-value = .0002. Since the p-value = .0002 < α = .01, H_0 is rejected. There is sufficient evidence to indicate that the true population mean ratio of all bones of this particular species differs from 8.5 at α = .01.

8.43 a. The test statistic is $z = \dfrac{\bar{x} - \mu_0}{\sigma_{\bar{x}}} = \dfrac{10.2 - 0}{31.3/\sqrt{50}} = 2.3$

   b. To determine if the mean level of feminization differs from 0%, we test:

H_0: $\mu = 0$
H_a: $\mu \neq 0$

Since the alternative hypothesis contains $\neq$, this is a two-tailed test. The p-value is
$p = P(z \leq -2.3) + P(z \geq 2.3) = .5 - P(-2.3 < z < 0) + .5 - P(0 < z < 2.3) =$
$.5 - .4893 + .5 - .4893 = .0214$.

   b. The test statistic is $z = \dfrac{\bar{x} - \mu_0}{\sigma_{\bar{x}}} = \dfrac{15 - 0}{25.1/\sqrt{50}} = 4.22$.

Since the alternative hypothesis contains $\neq$, this is a two-tailed test. The p-value is
$p = P(z \leq -4.22) + P(z \geq 4.22) = .5 - P(-4.22 < z < 0) + .5 - P(0 < z < 4.22) \approx$
$.5 - .5 + .5 - .5 = 0$.

8.45 We should use the t distribution in testing a hypothesis about a population mean if the sample size is small, the population being sampled from is normal, and the variance of the population is unknown.

8.47 a. $P(t > 1.440) = .10$
          (Using Table VI, Appendix A, with df = 6)

   b. $P(t < -1.782) = .05$
          (Using Table VI, Appendix A, with df = 12)

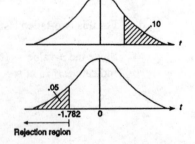

c. $P(t < -2.060) = P(t > 2.060) = .025$
(Using Table VI, Appendix A, with df = 25)

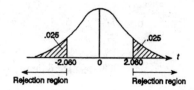

8.49   a.   The rejection region requires $\alpha/2 = .05/2 = .025$ in each tail of the t distribution with df = $n - 1 = 14 - 1 = 13$. From Table VI, Appendix A, $t_{.025} = 2.160$. The rejection region is $t < -2.160$ or $t > 2.160$.

b.   The rejection region requires $\alpha = .01$ in the upper tail of the t distribution with df = $n - 1 = 24 - 1 = 23$. From Table VI, Appendix A, $t_{.01} = 2.500$. The rejection region is $t > 2.500$.

c.   The rejection region requires $\alpha = .10$ in the upper tail of the t distribution with df = $n - 1 = 9 - 1 = 8$. From Table VI, Appendix A, $t_{.10} = 1.397$. The rejection region is $t > 1.397$.

d.   The rejection region requires $\alpha = .01$ in the lower tail of the t distribution with df = $n - 1 = 12 - 1 = 11$. From Table VI, Appendix A, $t_{.01} = 2.718$. The rejection region is $t < -2.718$.

e.   The rejection region requires $\alpha/2 = .10/2 = .05$ in each tail of the t distribution with df = $n - 1 = 20 - 1 = 19$. From Table VI, Appendix A, $t_{.05} = 1.729$. The rejection region is $t < -1.729$ or $t > 1.729$.

f.   The rejection region requires $\alpha = .05$ in the lower tail of the t distribution with df = $n - 1 = 4 - 1 = 3$. From Table VI, Appendix A, $t_{.05} = 2.353$. The rejection region is $t < -2.353$.

8.51   a. H_0: $\mu = 6$
H_a: $\mu < 6$

The test statistic is $t = \dfrac{\bar{x} - \mu_0}{s/\sqrt{n}} = \dfrac{4.8 - 6}{1.3/\sqrt{5}} = -2.064$.

The necessary assumption is that the population is normal.

The rejection region requires $\alpha = .05$ in the lower tail of the t distribution with df = $n - 1 = 5 - 1 = 4$. From Table VI, Appendix A, $t_{.05} = 2.132$. The rejection region is $t < -2.132$.

Since the observed value of the test statistic does not fall in the rejection region ($t = -2.064 \nless -2.132$), H_0 is not rejected. There is insufficient evidence to indicate the mean is less than 6 at $\alpha = .05$.

b. H_0: $\mu = 6$
H_a: $\mu \neq 6$

The test statistic is $t = -2.064$ (from **a**).

The assumption is the same as in **a**.

The rejection region requires $\alpha/2 = .05/2 = .025$ in each tail of the t distribution with df $= n - 1 = 5 - 1 = 4$. From Table VI, Appendix A, $t_{.025} = 2.776$. The rejection region is $t < -2.776$ or $t > 2.776$.

Since the observed value of the test statistic does not fall in the rejection region ($t = -2.064 \not< -2.776$), H_0 is not rejected. There is insufficient evidence to indicate the mean is different from 6 at $\alpha = .05$.

c. For part **a**, the p-value $= P(t \leq -2.064)$.

From Table VI, with df $= 4$, $.05 < P(t \leq -2.064) < .10$ or $.05 < p\text{-value} < .10$.

For part **b**, the p-value $= P(t \leq -2.064) + P(t \geq 2.064)$.

From Table VI, with df $= 4$, $2(.05) < p\text{-value} < 2(.10)$ or $.10 < p\text{-value} < .20$.

8.53 To determine if the mean Dental Anxiety Scale score for college students differs from 11, we test:

H_0: $\mu = 11$
H_a: $\mu \neq 11$

The test statistic is $t = \dfrac{\bar{x} - \mu_0}{s/\sqrt{n}} = \dfrac{107 - 11}{3.6/\sqrt{27}} = -.43$

The rejection region requires $\alpha/2 = .05/2 = .025$ in each tail of the t distribution with df $= n - 1 = 27 - 1 = 26$. From Table VI, Appendix A, $t_{.025} = 2.056$. The rejection region is $t < -2.056$ or $t > 2.056$.

Since the observed value of the test statistic does not fall in the rejection region ($t = -.43 \not< -2.056$), H_0 is not rejected. There is insufficient evidence to indicate the mean Dental Anxiety Scale score for college students differs from 11 at $\alpha = .05$.

8.55 a. From the printout, the alternative hypothesis is mu not $= 1$ or H_a: $\mu \neq 1.0$.

b. From the printout, the value of the test statistic is $T = 1.44$ or $t = 1.44$.

c. From the printout, the p-value is $P = .169$ or $p = .169$. Since the p-value is not small, H_0 would not be rejected for any reasonable value of α. There is insufficient evidence to indicate the mean crop content of the spinifex pigeons is different from 1.0 for $\alpha < 0.169$.

8.57   To determine if the true mean crack intensity of the Mississippi highway exceeds the AASHTO recommended maximum, we test:

H_0: $\mu = .100$
H_a: $\mu > .100$

The test statistic is $t = \dfrac{\bar{x} - \mu_0}{s/\sqrt{n}} = \dfrac{.210 - .100}{\sqrt{.011}/\sqrt{8}} = 2.97$

The rejection region requires $\alpha = .01$ in the upper tail of the t distribution with df $= n - 1$ $= 8 - 1 = 7$. From Table VI, Appendix A, $t_{.01} = 2.998$. The rejection region is $t > 2.998$.

Since the observed value of the test statistic does not fall in the rejection region ($t = 2.97 \not> 2.998$), H_0 is not rejected. There is insufficient evidence to indicate that the true mean crack intensity of the Mississippi highway exceeds the AASHTO recommended maximum at $\alpha = .01$.

8.59   To determine if the true mean PST score of all patients with mild to moderate traumatic brain injury exceeds the "normal" value of 40, we test:

H_0: $\mu = 40$
H_a: $\mu > 40$

The test statistic is $t = \dfrac{\bar{x} - \mu_0}{s/\sqrt{n}} = \dfrac{48.43 - 48}{20.76/\sqrt{23}} = 1.95$

The rejection region requires $\alpha = .05$ in the upper tail of the t distribution with df $= n - 1$ $= 23 - 1 = 22$. From Table VI, Appendix A, $t_{.05} = 1.717$. The rejection region is $t > 1.717$.

Since the observed value of the test statistic falls in the rejection region ($t = 1.95 > 1.717$), H_0 is rejected. There is sufficient evidence to indicate that the true mean PST score of all patients with mild to moderate traumatic brain injury exceeds the "normal" value of 40 at $\alpha = .05$.

8.61   The sample size is large enough if the interval $p_0 \pm 3\sigma_{\hat{p}}$ is contained in the interval $(0, 1)$.

a. $p_0 \pm 3\sqrt{\dfrac{p_0 q_0}{n}} \Rightarrow .05 \pm 3\sqrt{\dfrac{(.05)(.95)}{500}} \Rightarrow .05 \pm .029 \Rightarrow (.021, .079)$

Since the interval is contained in the interval $(0, 1)$, the sample size is large enough.

b. $p_0 \pm 3\sqrt{\dfrac{p_0 q_0}{n}} \Rightarrow .99 \pm 3\sqrt{\dfrac{(.99)(.01)}{100}} \Rightarrow .99 \pm .030 \Rightarrow (.960, 1.020)$

Since the interval is not contained in the interval $(0, 1)$, the sample size is not large enough.

c. $p_0 \pm 3\sqrt{\dfrac{p_0 q_0}{n}} \Rightarrow .2 \pm 3\sqrt{\dfrac{(.2)(.8)}{50}} \Rightarrow .2 \pm .170 \Rightarrow (.030, .370)$

Since the interval is contained in the interval $(0, 1)$, the sample size is large enough.

d. $p_0 \pm 3\sqrt{\dfrac{p_0 q_0}{n}} \Rightarrow .2 \pm 3\sqrt{\dfrac{(.2)(.8)}{20}} \Rightarrow .2 \pm .268 \Rightarrow (-.068, 468)$

Since the interval is not contained in the interval $(0, 1)$, the sample size is not large enough.

e. $p_0 \pm 3\sqrt{\dfrac{p_0 q_0}{n}} \Rightarrow .4 \pm 3\sqrt{\dfrac{(.4)(.6)}{10}} \Rightarrow .4 \pm .465 \Rightarrow (-.065, .865)$

Since the interval is not contained in the interval $(0, 1)$, the sample size is not large enough.

8.63 a. $z = \dfrac{\hat{p} - p_0}{\sqrt{\dfrac{p_0 q_0}{n}}} = \dfrac{.84 - .9}{\sqrt{\dfrac{.9(.1)}{100}}} = -2.00$

b. The denominator in Exercise 8.62 is $\sqrt{\dfrac{.75(.25)}{100}} = .0433$ as compared to $\sqrt{\dfrac{.9(.1)}{100}} = .03$ in part **a**. Since the denominator in this problem is smaller, the absolute value of z is larger.

c. The rejection region requires $\alpha = .05$ in the lower tail of the z distribution. From Table IV, Appendix A, $z_{.05} = 1.645$. The rejection region is $z < -1.645$.

Since the observed value of the test statistic falls in the rejection region $(z = -2.00 < -1.645)$, H_0 is rejected. There is sufficient evidence to indicate the population proportion is less than $.9$ at $\alpha = .05$.

d. The p-value $= P(z \le -2.00) = .5 - .4772 = .0228$ (from Table IV, Appendix A).

8.65 From Exercise 7.40, $n = 50$ and since p is the proportion of consumers who do not like the snack food, will be:

$$\hat{p} = \dfrac{\text{Number of 0's in sample}}{n} = \dfrac{29}{50} = .58$$

First, check to see if the normal approximation will be adequate:

$$p_0 \pm 3\sigma_{\hat{p}} \Rightarrow p_0 \pm 3\sqrt{\dfrac{pq}{n}} \approx p_0 \pm 3\sqrt{\dfrac{p_0 q_0}{n}} \Rightarrow .5 \pm 3\sqrt{\dfrac{.5(1-.5)}{50}} \Rightarrow .5 \pm .2121$$
$$\Rightarrow (.2879, .7121)$$

Since the interval lies completely in the interval $(0, 1)$, the normal approximation will be adequate.

a. H_0: $p = .5$
H_a: $p > .5$

The test statistic is $z = \dfrac{\hat{p} - p_0}{\sigma_{\hat{p}}} = \dfrac{\hat{p} - p_0}{\sqrt{\dfrac{p_0 q_0}{n}}} = \dfrac{.58 - .5}{\sqrt{\dfrac{.5(1-.5)}{50}}} = 1.13$

The rejection region requires $\alpha = .10$ in the upper tail of the z distribution. From Table IV, Appendix A, $z_{.10} = 1.28$. The rejection region is $z > 1.28$.

Since the observed value of the test statistic does not fall in the rejection region ($z = 1.13 \not> 1.28$), H_0 is not rejected. There is insufficient evidence to indicate the proportion of customers who do not like the snack food is greater than .5 at $\alpha = .10$.

b. p-value $= P(z \geq 1.13) = .5 - .3708 = .1292$

8.67 Some preliminary calculations:

$$\hat{p} = \frac{x}{n} = \frac{401}{835} = .48$$

First we check to see if the normal approximation is adequate:

$$p_0 \pm 3\sigma_{\hat{p}} \Rightarrow p_0 \pm 3\sqrt{\frac{pq}{n}} \approx p_0 \pm 3\sqrt{\frac{p_0 q_0}{n}} \Rightarrow .45 \pm 3\sqrt{\frac{.45(.55)}{835}} \Rightarrow .45 \pm .052 \Rightarrow (.398, .502)$$

Since the interval falls completely in the interval (0, 1), the normal distribution will be adequate.

To determine if more than 45% of male youths are raised in a single-parent family, we test:

H_0: $p = .45$
H_a: $p > .45$

The test statistic is $z = \dfrac{\hat{p} - p_0}{\sqrt{\dfrac{p_0 q_0}{n}}} = \dfrac{.48 - .45}{\sqrt{\dfrac{.45(.55)}{835}}} = 1.74$

The rejection region requires $\alpha = .05$ in the upper tail of the z distribution. From Table IV, Appendix A, $z_{.05} = 1.645$. The rejection region is $z > 1.645$.

Since the observed value of the test statistic falls in the rejection region ($z = 1.74 > 1.645$), H_0 is rejected. There is sufficient evidence that more than 45% of male youths are raised in a single-parent family at $\alpha = .05$.

8.69 Some preliminary calculations:

$$\hat{p} = \frac{x}{n} = \frac{15}{60} = .25$$

First we check to see if the normal approximation is adequate:

$$p_0 \pm 3\sigma_{\hat{p}} \Rightarrow p_0 \pm 3\sqrt{\frac{pq}{n}} \approx p_0 \pm 3\sqrt{\frac{p_0 q_0}{n}} \Rightarrow .40 \pm 3\sqrt{\frac{.40(.60)}{60}} \Rightarrow .40 \pm .19 \Rightarrow (.21, .59)$$

Since the interval falls completely in the interval (0, 1), the normal distribution will be adequate.

To determine if the claim made by Creative Good is incorrect, we test:

H_0: $p = .40$
H_a: $p \neq .40$

The test statistic is $z = \dfrac{\hat{p} - p_0}{\sqrt{\dfrac{p_0 q_0}{n}}} = \dfrac{.25 - .40}{\sqrt{\dfrac{.40(.60)}{60}}} = -2.37$.

The rejection region requires $\alpha/2 = .01/2 = .005$ in each tail of the z distribution. From Table IV, Appendix A, $z_{.005} = 2.58$. The rejection region is $z < -2.58$ or $z > 2.58$.

Since the observed value of the test statistic does not fall in the rejection region $(z = -2.37 \nless -2.58)$, H_0 is not rejected. There is insufficient evidence to reject the claim that 40% of shoppers fail in their attempts to purchase merchandise online at $\alpha = .01$.

8.71   a.   Let p = proportion of the subjects who felt that the masculinization of face shape decreased attractiveness of the male face. If there is no preference for the unaltered or morphed male face, then $p = .5$.

        For this problem, $\hat{p} = x/n = 58/67 = .866$.

        First we check to see if the normal approximation is adequate:

$$p_0 \pm 3\sigma_{\hat{p}} \Rightarrow p_0 \pm 3\sqrt{\dfrac{p_0 q_0}{n}} \Rightarrow .5 \pm 3\sqrt{\dfrac{.5(.5)}{67}} \Rightarrow .5 \pm .183 \Rightarrow (.317, .683)$$

        Since the interval falls completely in the interval (0, 1), the normal distribution will be adequate.

        To determine if the subjects showed a preference for either the unaltered or morphed face, we test:

        H_0: $p = .5$
        H_a: $p \neq .5$

   b.   The test statistic is $z = \dfrac{\hat{p} - p_0}{\sqrt{\dfrac{p_0 q_0}{n}}} = \dfrac{.866 - .50}{\sqrt{\dfrac{.50(.50)}{67}}} = 5.99$

   c.   By definition, the p-value is the probability of observing our test statistic or anything more unusual, given H_0 is true. For this problem, $p = P(z \leq -5.99) + P(z \geq 5.99)$ $= .5 - P(-5.99 < z < 0) + .5 - P(0 < z < 5.99) \approx .5 - .5 + .5 - .5 = 0$. This corresponds to what is listed in the Exercise.

   d.   The rejection region requires $\alpha/2 = .01/2 = .005$ in each tail of the z distribution. From Table IV, Appendix A, $z_{.005} = 2.575$. The rejection region is $z < -2.575$ or $z > 2.575$.

        Since the observed value of the test statistic falls in the rejection region $(z = 5.99 > 2.575)$, H_0 is rejected. There is sufficient evidence to indicate that the subjects showed a preference for either the unaltered or morphed face at $\alpha = .01$.

8.73   Let p = proportion of patients taking the pill who reported an improved condition.

First we check to see if the normal approximation is adequate:

$$p_0 \pm 3\sigma_{\hat{p}} \Rightarrow p_0 \pm 3\sqrt{\frac{p_0 q_0}{n}} \Rightarrow .5 \pm 3\sqrt{\frac{.5(.5)}{7000}} \Rightarrow .5 \pm .018 \Rightarrow (.482, .518)$$

Since the interval falls completely in the interval (0, 1), the normal distribution will be adequate.

To determine if there really is a placebo effect at the clinic, we test:

H_0: $p = .5$
H_a: $p > .5$

The test statistic is $z = \dfrac{\hat{p} - p_0}{\sqrt{\dfrac{p_0 q_0}{n}}} = \dfrac{.7 - .5}{\sqrt{\dfrac{.5(.5)}{7000}}} = 33.47$

The rejection region requires $\alpha = .05$ in the upper tail of the z distribution. From Table IV, Appendix A, $z_{.05} = 1.645$. The rejection region is $z > 1.645$.

Since the observed value of the test statistic falls in the rejection region ($z = 33.47 > 1.645$), H_0 is rejected. There is sufficient evidence to indicate that there really is a placebo effect at the clinic at $\alpha = .05$.

8.75   a.   We want to show that more than 17,000 of the 18,200 signatures are valid. Thus, we want to show that the proportion of valid signatures is greater than 17,000/18,200 = .934.

From the Exercise, $\hat{p} = 98/100 = .98$.

First, we check to see if the normal approximation is adequate:

$$p_0 \pm 3\sigma_{\hat{p}} \Rightarrow p_0 \pm 3\sqrt{\frac{p_0 q_0}{n}} \Rightarrow .934 \pm 3\sqrt{\frac{.934(.066)}{100}} \Rightarrow .934 \pm .074 \Rightarrow (.860, 1.008)$$

Since the interval does not fall completely in the interval (0, 1), the normal distribution may not be adequate.

To determine if more than .934 of the signatures are valid, we test:

H_0: $p = .934$
H_a: $p > .934$

The test statistic is $z = \dfrac{\hat{p} - p_0}{\sqrt{\dfrac{p_0 q_0}{n}}} = \dfrac{.98 - .934}{\sqrt{\dfrac{.934(.066)}{100}}} = 1.85$

Since no α is given in the problem, we will choose $\alpha = .05$. The rejection region requires $\alpha = .05$ in the upper tail of the z distribution. From Table IV, Appendix A, $z_{.05} = 1.645$. The rejection region is $z > 1.645$.

Since the observed value of the test statistic falls in the rejection region ($z = 1.85 > 1.645$), H_0 is rejected. There is sufficient evidence to indicate that the true proportion of valid signatures is greater than .934 at $\alpha = .05$. This indicates that more than 17,000 of the 18,200 signatures are valid.

b.   We want to show that more than 16,000 of the 18,200 signatures are valid. Thus, we want to show that the proportion of valid signatures is greater than $16,000/18,200 = .879$.

Again, from the Exercise, $\hat{p} = 98/100 = .98$.

First we check to see if the normal approximation is adequate:

$$p_0 \pm 3\sigma_{\hat{p}} \Rightarrow p_0 \pm 3\sqrt{\frac{p_0 q_0}{n}} \Rightarrow .879 \pm 3\sqrt{\frac{.879(.121)}{100}} \Rightarrow .879 \pm .098 \Rightarrow (.781, .977)$$

Since the interval is completely in the interval $(0, 1)$, the normal distribution will be adequate.

To determine if more than .879 of the signatures are valid, we test:

H_0: $p = .879$
H_a: $p > .879$

The test statistic is $z = \dfrac{\hat{p} - p_0}{\sqrt{\dfrac{p_0 q_0}{n}}} = \dfrac{.98 - .879}{\sqrt{\dfrac{.879(.121)}{100}}} = 3.09$

Since no α is given in the problem, we will choose $\alpha = .05$. The rejection region requires
$\alpha = .05$ in the upper tail of the z distribution. From Table IV, Appendix A, $z_{.05} = 1.645$. The rejection region is $z > 1.645$.

Since the observed value of the test statistic falls in the rejection region ($z = 3.09 > 1.645$), H_0 is rejected. There is sufficient evidence to indicate that the true proportion of valid signatures is greater than .879 at $\alpha = .05$. This indicates that more than 16,000 of the 18,200 signatures are valid.

8.77   a.

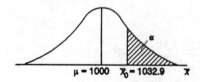

μ = 1000 $\bar{x}_0$ = 1032.9   x

b. $z = \dfrac{\bar{x}_0 - \mu_0}{\sigma_{\bar{x}}} \Rightarrow \bar{x}_0 = \mu_0 + z_\alpha \sigma_{\bar{x}} = \mu_0 + z_\alpha \dfrac{\sigma}{\sqrt{n}}$ where $z_\alpha = z_{.05} = 1.645$ from Table IV, Appendix A.

Thus, $\bar{x}_0 = 1000 + 1.645\dfrac{120}{\sqrt{36}} = 1032.9$.

c.

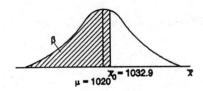

d. $\beta = P(\bar{x}_0 < 1032.9 \text{ when } \mu = 1020) = P\left(z < \dfrac{1032.9 - 1020}{120/\sqrt{36}}\right) = P(z < .65)$

$= .5 + .2422 = .7422$

e. Power $= 1 - \beta = 1 - .7422 = .2578$

8.79 a. The sampling distribution of $\bar{x}$ will be approximately normal (by the Central Limit Theorem) with $\mu_{\bar{x}} = \mu = 50$ and $\sigma_{\bar{x}} = \dfrac{\sigma}{\sqrt{n}} = \dfrac{20}{\sqrt{64}} = 2.5$.

b. The sampling distribution of $\bar{x}$ will be approximately normal (by the Central Limit Theorem) with $\mu_{\bar{x}} = \mu = 45$ and $\sigma_{\bar{x}} = \dfrac{\sigma}{\sqrt{n}} = \dfrac{20}{\sqrt{64}} = 2.5$.

c. First, find $\bar{x}_0$ so that $P(\bar{x} < \bar{x}_0) = .10$.

$$P(\bar{x} < \bar{x}_0) = P\left(z < \dfrac{\bar{x}_0 - 50}{20/\sqrt{64}}\right) = P\left(z < \dfrac{\bar{x}_0 - 50}{2.5}\right) = P(z < z_0) = .10$$

From Table IV, Appendix A, $z_{.10} = -1.28$

$$z_0 = \dfrac{\bar{x}_0 - 50}{2.5} \Rightarrow -1.28 = \dfrac{\bar{x}_0 - 50}{2.5} \Rightarrow \bar{x}_0 = 46.8$$

Now, find $\beta = P(\bar{x} > 46.8 \text{ when } \mu = 45) = P\left(z > \dfrac{46.8 - 45}{20/\sqrt{64}}\right) = P(z > .72)$

$= .5 - .2642 = .2358$

d. Power $= 1 - \beta = 1 - .2358 = .7642$

8.81   a.   The sampling distribution of $\bar{x}$ will be approximately normal (by the Central Limit Theorem) with $\mu_{\bar{x}} = \mu = 10$ and $\sigma_{\bar{x}} = \dfrac{\sigma}{\sqrt{n}} = \dfrac{1.00}{\sqrt{100}} = 0.1$.

        b.   The sampling distribution of $\bar{x}$ will be approximately normal (CLT) with $\mu_{\bar{x}} = \mu = 9.9$ and $\sigma_{\bar{x}} = \dfrac{\sigma}{\sqrt{n}} = \dfrac{1.00}{\sqrt{100}} = 0.1$.

        c.   First, find $\bar{x}_{0,L}$ and $\bar{x}_{0,U}$ such that $P(\bar{x} < x_{0,L}) = P(\bar{x} > x_{0,U}) = .025$

$$P(\bar{x} < x_{0,L}) = P\left(z < \frac{\bar{x}_{0,L} - 10}{.1}\right) = P(z < z_{0,L}) = .025$$

From Table IV, Appendix A, $z_{0,L} = -1.96$

Thus, $z_{0,L} = \dfrac{\bar{x}_{0,L} - 10}{.1} \Rightarrow \bar{x}_{0,L} = -1.96(.1) + 10 = 9.804$

$$P(\bar{x} > x_{0,U}) = P\left(z > \frac{\bar{x}_{0,U} - 10}{.1}\right) = P(z > z_{0,U}) = .025$$

From Table IV, Appendix A, $z_{0,U} = 1.96$

Thus, $z_{0,U} = \dfrac{\bar{x}_{0,U} - 10}{.1} \Rightarrow \bar{x}_{0,U} = 1.96(.1) + 10 = 10.196$

Now, find $\beta = P(9.804 < \bar{x} < 10.196$ when $\mu = 9.9)$
$$= P\left(\frac{9.804 - 9.9}{1/\sqrt{100}} < z < \frac{10.196 - 9.9}{1/\sqrt{100}}\right) = P(-.96 < z < 2.96)$$
$$= .3315 + .4985 = .8300$$

        d. $\beta = P(9.804 < \bar{x} < 10.196$ when $\mu = 10.1)$
$$= P\left(\frac{9.804 - 10.1}{1/\sqrt{100}} < z < \frac{10.196 - 10.1}{1/\sqrt{100}}\right) = P(-2.96 < z < .96)$$
$$= .4985 + .3315 = .8300$$

8.83   a.   First, find $\bar{x}_0$ such that $P(\bar{x} < \bar{x}_0) = .10$

$$P(\bar{x} < \bar{x}_0) = P\left(z < \frac{\bar{x}_0 - 16}{9.32/\sqrt{100}}\right) = P(z < z_0) = .10$$

From Table IV, Appendix A, $z_{.10} = -1.28$

Thus, $z_0 = \dfrac{\bar{x}_0 - 16}{9.32/\sqrt{100}} \Rightarrow \bar{x}_0 = -1.28(1.622) + 16 = 14.81$

For $\mu = 15.5$, the power is:

$$\text{Power} = P(\bar{x} < 14.81) = P\left(z < \dfrac{14.81 - 15.5}{9.32/\sqrt{100}}\right) = P(z < -.74) = .5 - .2704$$
$$= .2296$$

For $\mu = 15.0$, the power is:

$$\text{Power} = P(\bar{x} < 14.81) = P\left(z < \dfrac{14.81 - 15.0}{9.32/\sqrt{100}}\right) = P(z < -.20) = .5 - .0793$$
$$= .4207$$

For $\mu = 14.5$, the power is:

$$\text{Power} = P(\bar{x} < 14.81) = P\left(z < \dfrac{14.81 - 14.5}{9.32/\sqrt{100}}\right) = P(z < .33) = .5 - .1293$$
$$= .6293$$

For $\mu = 14.0$, the power is:

$$\text{Power} = P(\bar{x} < 14.81) = P\left(z < \dfrac{14.81 - 14.0}{9.32/\sqrt{100}}\right) = P(z < .87) = .5 - .3078$$
$$= .8078$$

For $\mu = 13.5$, the power is:

$$\text{Power} = P(\bar{x} < 14.81) = P\left(z < \dfrac{14.81 - 13.5}{9.32/\sqrt{100}}\right) = P(z < 1.41) = .5 + .4207$$
$$= .9207$$

b.   The plot of the power is:

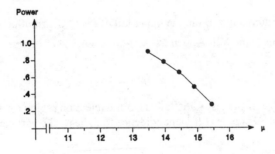

8.85   First, find $\bar{x}_0$ such that $P(\bar{x} < \bar{x}_0) = .05$.

$$P(\bar{x} < \bar{x}_0) = P\left(z < \frac{\bar{x}_0 - 10}{1.2/\sqrt{48}}\right) = P(z < z_0) = .05.$$

From Table IV, Appendix A, $z_0 = -1.645$.

Thus, $z_0 = \dfrac{\bar{x}_0 - 10}{1.2/\sqrt{48}} \Rightarrow \bar{x}_0 = -1.645(.173) + 10 = 9.715$

The probability of a Type II error is:

$$\beta = P(\bar{x} \geq 9.715 \mid \mu = 9.5) = P\left(z \geq \frac{9.715 - 9.5}{1.2/\sqrt{48}}\right) = P(z \geq 1.24) = .5 - .3295$$

$$= .1075$$

8.87   a.   df $= n - 1 = 16 - 1 = 15$; reject H_0 if $\chi^2 < 6.26214$ or $\chi^2 > 27.4884$

b.   df $= n - 1 = 23 - 1 = 22$; reject H_0 if $\chi^2 > 40.2894$

c.   df $= n - 1 = 15 - 1 = 14$; reject H_0 if $\chi^2 > 21.0642$

d.   df $= n - 1 = 13 - 1 = 12$; reject H_0 if $\chi^2 < 3.57056$

e.   df $= n - 1 = 7 - 1 = 6$; reject H_0 if $\chi^2 < 1.63539$ or $\chi^2 > 12.5916$

f.   df $= n - 1 = 25 - 1 = 24$; reject H_0 if $\chi^2 < 13.8484$

8.89   a. H_0: $\sigma^2 = 1$
        H_a: $\sigma^2 > 1$

The test statistic is $\chi^2 = \dfrac{(n-1)s^2}{\sigma_0^2} = \dfrac{(100-1)4.84}{1} = 479.16$

The rejection region requires $\alpha = .05$ in the upper tail of the χ^2 distribution with df $= n - 1 = 100 - 1 = 99$. From Table VII, Appendix A, $\chi_{.05}^2 \approx 124.324$. The rejection region is
$\chi^2 > 124.324$.

Since the observed value of the test statistic falls in the rejection region ($\chi^2 = 479.16 > 124.324$), H_0 is rejected. There is sufficient evidence to indicate the variance is larger than 1 at $\alpha = .05$.

b.   In part **b** of Exercise 8.88, the test statistic was $\chi^2 = 29.04$. The conclusion was to reject H_0 as it was in this problem.

8.91   a.   To determine if the true standard deviation of the point-spread errors exceed 15 (variance exceeds 225), we test:

H_0: $\sigma^2 = 225$
H_a: $\sigma^2 > 225$

b.   The test statistic is $\chi^2 = \dfrac{(n-1)s^2}{\sigma_0^2} = \dfrac{(240-1)13.3^2}{225} = 187.896$

c.   The rejection region requires α in the upper tail of the χ^2 distribution with df $= n - 1 = 240 - 1 = 239$. The maximum value of df in Table VII is 100. Thus, we cannot find the rejection region using Table VII. Using a statistical package, the p-value associated with

$\chi^2 = 187.896$ is .9938.

Since the p-value is so large, there is no evidence to reject H_0. There is insufficient evidence to indicate that the true standard deviation of the point-spread errors exceed 15 for any reasonable value of α.

(Since the observed variance (or standard deviation) is less than the hypothesized value of the variance (or standard deviation) under H_0, there is no way H_0 will be rejected for any reasonable value of α.)

8.93   To determine if the standard deviation of the percent recovery using the new method differs from 15, we test:

H_0: $\sigma^2 = 15^2 = 225$
H_a: $\sigma^2 \neq 225$

The statistic is $\chi^2 = \dfrac{(n-1)s^2}{\sigma_0^2} = \dfrac{(7-1)9^2}{225} = 2.16$

The rejection region requires $\alpha/2 = .10/2 = .05$ in each tail of the χ^2 distribution with df $= n - 1 = 7 - 1 = 6$. From Table VII, Appendix A, $\chi^2_{.05} = 12.5916$ and $\chi^2_{.95} = 1.63539$. The rejection region is $\chi^2 < 1.63539$ or $\chi^2 > 12.5916$.

Since the observed value of the test statistic does not fall in the rejection region ($\chi^2 = 2.16 \not< 1.63539$ and $\chi^2 = 2.16 \not> 12.5916$), H_0 is not rejected. There is insufficient evidence to indicate that the standard deviation of the percent recovery using the new method differs from 15 (or the variance differs from 225) at $\alpha = .10$.

8.95   a.   Some preliminary calculations:

$$s^2 = \dfrac{\sum x^2 - \dfrac{\left(\sum x\right)^2}{n}}{n-1} = \dfrac{29345.78 - \dfrac{(342.6)^2}{4}}{4-1} = .6967$$

To determine if the variation in the CO_2 concentration measurements using the LRM method differs from 1, we test:

H_0: $\sigma^2 = 1$
H_a: $\sigma^2 \neq 1$

The test statistic is $\chi^2 = \dfrac{(n-1)s^2}{\sigma_0^2} = \dfrac{(4-1).6967^2}{1} = 2.09$

Since no α was given we will use $\alpha = .05$. The rejection region requires $\alpha/2 = .05/2 = .025$ in each tail of the χ^2 distribution with df $= n - 1 = 4 - 1 = 3$. From Table VII, Appendix A, $\chi^2_{.025} = 9.34840$ and $\chi^2_{.975} = 0.215795$. The rejection region is $\chi^2 < 0.215795$ or $\chi^2 > 9.34840$.

Since the observed value of the test statistic does not fall in the rejection region ($\chi^2 = 2.09 \nless 0.215795$ and $\chi^2 = 2.09 \ngtr 9.34840$), H_0 is not rejected. There is insufficient evidence to indicate that the variation in the CO_2 concentration measurements using the LRM method differs from 1 at $\alpha = .05$.

b.   We must assume that the population of CO_2 amounts are normally distributed.

8.97   The smaller the p-value associated with a test of hypothesis, the stronger the support for the **alternative** hypothesis. The p-value is the probability of observing your test statistic or anything more unusual, given the null hypothesis is true. If this value is small, it would be very unusual to observe this test statistic if the null hypothesis were true. Thus, it would indicate the alternative hypothesis is true.

8.99   The elements of the test of hypothesis that should be specified prior to analyzing the data are: null hypothesis, alternative hypothesis, and significance level.

8.101   The larger the p-value associated with a test of hypothesis, the stronger the support for the **null** hypothesis. The p-value is the probability of observing your test statistic or anything more unusual, given the null hypothesis is true. If this value is large, it would not be very unusual to observe this test statistic if the null hypothesis were true. Thus, it would lend support that the null hypothesis is true.

8.103   a. H_a: $p = .35$
H_a: $p < .35$

The test statistic is $z = \dfrac{\hat{p} - p_0}{\sqrt{\dfrac{p_0 q_0}{n}}} = \dfrac{.29 - .35}{\sqrt{\dfrac{.35(.65)}{200}}} = -1.78$

The rejection region requires $\alpha = .05$ in the lower tail of the z distribution. From Table IV, Appendix A, $z_{.05} = 1.645$. The rejection region is $z < -1.645$.

Since the observed value of the test statistic falls in the rejection region ($z = -1.78 < -1.645$), H_0 is rejected. There is sufficient evidence to indicate $p < .35$ at $\alpha = .05$.

b. H_0: $p = .35$
H_a: $p \neq .35$

The test statistic is $z = -1.78$ (from **a**).

The rejection region requires $\alpha/2 = .05/2 = .025$ in each tail of the z distribution. From Table IV, Appendix A, $t_{.025} = 1.96$. The rejection region is $z < -1.96$ or $z > 1.96$.

Since the observed value of the test statistic does not fall in the rejection region ($z = -1.78 \nless -1.96$), H_0 is not rejected. There is insufficient evidence to indicate p is different from .35 at $\alpha = .05$.

c. For confidence coefficient .95, $\alpha = .05$ and $\alpha/2 = .025$. From Table IV, Appendix A, $z_{.025} = 1.96$. The confidence interval is:

$$\hat{p} \pm z_{.025}\sqrt{\frac{\hat{p}\hat{q}}{n}} \Rightarrow .29 \pm 1.96\sqrt{\frac{.29(.71)}{200}} \Rightarrow .29 \pm .063 \Rightarrow (.227, .353)$$

d. For confidence coefficient .99, $\alpha = .01$ and $\alpha/2 = .005$. From Table IV, Appendix A, $z_{.005} = 2.58$. The confidence interval is:

$$\hat{p} \pm z_{.005}\sqrt{\frac{\hat{p}\hat{q}}{n}} \Rightarrow .29 \pm 2.58\sqrt{\frac{.29(.71)}{200}} \Rightarrow .29 \pm .083 \Rightarrow (.207, .373)$$

e. $n = \dfrac{(z_{\alpha/2})^2\, pq}{B^2} = \dfrac{2.58^2(.29)(.71)}{.05^2} = 548.2 \approx 549$

(We will use $\hat{p} = .29$ to estimate the value of p.)

8.105 a. H_0: $\sigma^2 = 30$
H_a: $\sigma^2 > 30$

The test statistic is $\chi^2 = \dfrac{(n-1)s^2}{\sigma_0^2} = \dfrac{(41-1)(6.9)^2}{30} = 63.48$

The rejection region requires $\alpha = .05$ in the upper tail of the χ^2 distribution with df $= n - 1 = 40$. From Table VII, Appendix A, $\chi^2_{.05} = 55.7585$. The rejection region is $\chi^2 > 55.7585$.

Since the observed value of the test statistic falls in the rejection region ($\chi^2 = 63.48 > 55.7585$), H_0 is rejected. There is sufficient evidence to indicate the variance is larger than 30 at $\alpha = .05$.

b. H_0: $\sigma^2 = 30$
H_a: $\sigma^2 \neq 30$

The test statistic is $\chi^2 = 63.48$ (from part **a**).

The rejection region requires $\alpha/2 = .05/2 = .025$ in each tail of the χ^2 distribution with df $= n - 1 = 40$. From Table VII, Appendix A, $\chi^2_{.025} = 59.3417$ and $\chi^2_{.975} = 24.4331$. The rejection region is $\chi^2 < 24.4331$ or $\chi^2 > 59.3417$.

Since the observed value of the test statistic falls in the rejection region ($\chi^2 = 63.48 > 59.3417$), H_0 is rejected. There is sufficient evidence to indicate the variance is not 30 at $\alpha = .05$.

8.107 a. Since the company must give proof the drug is safe, the null hypothesis would be the drug is unsafe. The alternative hypothesis would be the drug is safe.

   b. A Type I error would be concluding the drug is safe when it is not safe. A Type II error would be concluding the drug is not safe when it is. α is the probability of concluding the drug is safe when it is not. β is the probability of concluding the drug is not safe when it is.

   c. In this problem, it would be more important for α to be small. We would want the probability of concluding the drug is safe when it is not to be as small as possible.

8.109 a. To determine if the true mean score of all sleep-deprived subjects is less than 80, we test:

$$H_0: \ \mu = 80$$
$$H_a: \ \mu < 80$$

The test statistic is $t = \dfrac{\bar{x}_0 - \mu_0}{s/\sqrt{n}} = \dfrac{63 - 80}{17/\sqrt{12}} = -3.46$

The rejection region requires $\alpha = .05$ in the lower tail of the t distribution with df $= n - 1 = 12 - 1 = 11$. From Table VI, Appendix A, $t_{.05} = 1.796$. The rejection region is $t < -1.796$.

Since the observed value of the test statistic falls in the rejection region ($t = -3.46 < -1.796$), H_0 is rejected. There is sufficient evidence to indicate that the true mean score of all sleep-deprived subjects is less than 80 at $\alpha = .05$.

   b. We must assume that the overall test scores of sleep-deprived students are normally distributed.

8.111 a. First, check to see if the normal approximation is adequate:

$$p_0 \pm 3\sigma_{\hat{p}} \Rightarrow p_0 \pm 3\sqrt{\frac{p_0 q_0}{n}} \Rightarrow .75 \pm 3\sqrt{\frac{(.75)(.25)}{470}} \Rightarrow .75 \pm .060 \Rightarrow (.690, .810)$$

Since the interval falls completely in the interval (0, 1), the normal distribution will be adequate.

Let p = proportion of golf balls sold that are white. $\hat{p} = \dfrac{410}{470} = .872$

To determine if the manufacturer's claim is true, we test:

H_0: $p = .75$
H_a: $p > .75$

The test statistic is $z = \dfrac{\hat{p} - p_0}{\sqrt{\dfrac{p_0 q_0}{n}}} = \dfrac{.872 - .75}{\sqrt{\dfrac{.75(.25)}{470}}} = 6.11$

The rejection region requires $\alpha = .01$ in the upper tail of the z distribution. From Table IV, Appendix A, $z_{.01} = 2.33$. The rejection region is $z > 2.33$.

Since the observed value of the test statistic falls in the rejection region ($z = 6.11 > 2.33$), H_0 is rejected. There is sufficient evidence to indicate more than 75% of all golf balls sold are white at $\alpha = .01$.

b. The p-value = $P(z \geq 6.11) \approx .5 - .5 = 0$. There is evidence to reject H_0 for $\alpha > .0001$.

c. To find the rejection region in terms of $\hat{p}$, we use:

$$z = \dfrac{\hat{p} - p_0}{\sqrt{\dfrac{p_0 q_0}{n}}} \Rightarrow 2.33 = \dfrac{\hat{p} - .75}{\sqrt{\dfrac{.75(.25)}{470}}} \Rightarrow \hat{p} = .7965$$

For $p_0 = .80$, $\beta = P(\hat{p} \leq .7965) = P\left(z \leq \dfrac{.7965 - .80}{\sqrt{\dfrac{.80(.20)}{470}}} \right) = P(z \leq -.19)$

$$= .5 - .0753 = .4247$$

8.113　a. To test the designer's research hypothesis, we test:

H_0: $\mu = 250$
H_a: $\mu > 250$

b. The test statistic is $z = \dfrac{\bar{x} - \mu_0}{\sigma_{\bar{x}}} = \dfrac{256.3 - 250}{43.4/\sqrt{135}} = 1.69$

The p-value = $P(z \geq 1.69) = .5 - .4545 = .0455$

There is evidence to reject H_0 for $\alpha = .05$. There is evidence to indicate the mean driving distance now is greater than 250 yards at $\alpha = .05$.

c. The hypotheses would change for a two-tailed test. The new hypotheses would be:

H_0: $\mu = 250$
H_a: $\mu \neq 250$

The p-value for a two-tailed test would be two times the p-value for a one-tailed test. The p-value = $P(z \leq -1.69) + P(z \geq 1.69) = 2P(z \geq 1.69) = 2(.5 - .4545) = .0910$.

There is no evidence to reject H_0 for $\alpha = .05$. There is no evidence to indicate the mean driving distance now is different than 250 yards at $\alpha = .05$.

There is evidence to reject H_0 for $\alpha = .10$. There is evidence to indicate the mean driving distance now is different than 250 yards at $\alpha = .10$.

8.115   a.   The test statistic is $t = \dfrac{\bar{x} - \mu_0}{s/\sqrt{n}} = \dfrac{1173.6 - 1100}{36.3/\sqrt{3}} = 3.512$

The p-value $= P(t \geq 3.512)$. From Table VI with df $= n - 1 = 3 - 1 = 2$, $.025 < p$-value $< .05$.

b.   The p-value $= .0362 = P(t \geq 3.512)$. Since this p-value is fairly small, there is evidence to reject H_0 for $\alpha > .0362$. There is evidence to indicate the mean length of life of a certain mechanical component is longer than 1100 hours.

c.   We must assume the population of lifetimes is normally distributed.

d.   A Type I error would be of most concern for this test. A Type I error would be concluding the mean lifetime is greater than 1100 hours when in fact the mean lifetime is not greater than 1100.

8.117   a.   To determine if the true mean inbreeding coefficient μ for this species of wasp exceeds 0, we test:

H_0: $\mu = 0$
H_a: $\mu > 0$

The test statistic is $z = \dfrac{\bar{x} - \mu_0}{\sigma_{\bar{x}}} = \dfrac{.044 - 0}{.884/\sqrt{197}} = 0.70$

The rejection region requires $\alpha = .05$ in the upper tail of the z distribution. From Table IV, Appendix A, $z_{.05} = 1.645$. The rejection region is $z > 1.645$.

Since the observed value of the test statistic does not fall in the rejection region ($z = 0.70 \not> 1.645$), H_0 is not rejected. There is insufficient evidence to indicate that the true mean inbreeding coefficient μ for this species of wasp exceeds 0 at $\alpha = .05$.

b.   This result agrees with that of Exercise 7.79. The confidence interval in Exercise 7.79 was $(-.118, .206)$. Since this interval contains 0, there is no evidence to indicate that the mean is different from 0. This is the same conclusion that was reached in part **a**.

8.119   a.   To determine if the production process should be halted, we test:

H_0: $\mu = 3$
H_a: $\mu > 3$

where μ = mean amount of PCB in the effluent.

The test statistic is $z = \dfrac{\bar{x} - \mu_0}{\sigma_{\bar{x}}} = \dfrac{3.1 - 3}{.5/\sqrt{50}} = 1.41$

The rejection region requires $\alpha = .01$ in the upper tail of the z distribution. From Table IV, Appendix A, $z_{.01} = 2.33$. The rejection region is $z > 2.33$.

Since the observed value of the test statistic does not fall in the rejection region, ($z = 1.41 \not> 2.33$), H_0 is not rejected. There is insufficient evidence to indicate the mean amount of PCB in the effluent is more than 3 parts per million at $\alpha = .01$. Do not halt the manufacturing process.

b. As plant manager, I do not want to shut down the plant unnecessarily. Therefore, I want $\alpha = P(\text{shut down plant when } \mu = 3)$ to be small.

8.121 a. No, it increases the risk of falsely rejecting H_0, i.e., closing the plant unnecessarily.

b. First, find $\bar{x}_0$ such that $P(\bar{x} > \bar{x}_0) = P(z > z_0) = .05$.

From Table IV, Appendix A, $z_0 = 1.645$

$$z = \frac{\bar{x} - \mu_0}{\sigma/\sqrt{n}} \Rightarrow 1.645 = \frac{\bar{x}_0 - 3}{.5/\sqrt{50}} \Rightarrow \bar{x}_0 = 3.116$$

Then, compute

$$\beta = P(\bar{x}_0 \le 3.116 \text{ when } \mu = 3.1) = P\left(z \le \frac{3.116 - 3.1}{.5/\sqrt{50}}\right) = P(z \le .23)$$

$$= .5 + .0910 = .5910$$

Power $= 1 - \beta = 1 - .5910 = .4090$

c. The power of the test increases as α increases.

8.123 First, check to see if the normal approximation is adequate:

$$p_0 \pm 3\sigma_{\hat{p}} \Rightarrow p_0 \pm 3\sqrt{\frac{p_0 q_0}{n}} \Rightarrow .5 \pm 3\sqrt{\frac{(.5)(.5)}{2,237}} \Rightarrow .5 \pm .032 \Rightarrow (.468, .532)$$

Since the interval falls completely in the interval $(0, 1)$, the normal distribution will be adequate.

$$\hat{p} = \frac{x}{n} = \frac{1,630}{2,237} = .729$$

To determine if the true percentage of engineering Ph.D. degrees awarded to foreign nationals exceeds 50%, we test:

H_0: $p = .5$
H_a: $p > .5$

The test statistic is $z = \dfrac{\hat{p} - p_0}{\sqrt{\dfrac{p_0 q_0}{n}}} = \dfrac{.729 - .5}{\sqrt{\dfrac{(.5)(.5)}{2{,}237}}} = 21.66$

The rejection region requires $\alpha = .01$ in the upper tail of the z distribution. From Table IV, Appendix A, $z_{.01} = 2.33$. The rejection region is $z > 2.33$.

Since the observed value of the test statistic falls in the rejection region ($z = 21.66 > 2.33$), H_0 is rejected. There is sufficient evidence to indicate that the true percentage of engineering Ph.D. degrees awarded to foreign nationals exceeds 50% at $\alpha = .01$.

8.125   a. $P(\bar{x} < \bar{x}_0) = P\left(z < \dfrac{\bar{x}_0 - 5}{.01/\sqrt{100}} \right) = P(z < z_0) = .05$

The rejection region requires $\alpha = .05$ in the lower tail of the z distribution. From Table IV, Appendix A, $z_0 = -1.645$.

$$z_0 = \dfrac{\bar{x}_0 - 4.9975}{.01/\sqrt{100}} \Rightarrow \bar{x}_0 = -1.645(.001) + 4.9975 = 4.998355$$

For $\mu = 4.9975$, $\beta = P(\bar{x} > 4.998355) = P\left(z > \dfrac{4.998355 - 4.9975}{.01/\sqrt{100}} \right)$

$$= P(z > .86) = .5 - .3051 = .1949$$

Failing to reject H_0 when H_0 is false is a Type II error.

b.   For $\mu = 5.0$, $\alpha = P(\bar{x} < 4.998355) = P\left(z < \dfrac{4.998355 - 5}{.01/\sqrt{100}} \right) = P(z < -1.645)$

$$= .5 - .4500 = .05$$

Rejecting H_0 when H_0 is true is a Type I error.

c.   .0025 mm below $\mu_0 = 5.0$ is $\mu = 5 - .0025 = 4.9975$

For $\mu = 4.9975$, Power $= 1 - \beta = P(\bar{x} < 4.998355) = P\left(z < \dfrac{4.99355 - 4.9975}{.01/\sqrt{100}} \right)$

$$= P(z < .86)$$
$$= (.5 + .3051) = .8051$$

Inferences Based on Two Samples: Confidence Intervals and Tests of Hypotheses

9.1   a.   For confidence coefficient .95, $\alpha = .05$ and $\alpha/2 = .025$. From Table IV, Appendix A, $z_{.025} = 1.96$. The confidence interval is:

$$(\bar{x}_1 - \bar{x}_2) \pm z_{.025} \sqrt{\frac{\sigma_1^2}{n_1} + \frac{\sigma_2^2}{n_2}} \Rightarrow (5{,}275 - 5{,}240) \pm 1.96 \sqrt{\frac{150^2}{400} + \frac{200^2}{400}}$$

$$\Rightarrow 35 \pm 24.5 \Rightarrow (10.5, 59.5)$$

We are 95% confident that the difference between the population means is between 10.5 and 59.5.

b.   The test statistic is $z = \dfrac{(\bar{x}_1 - \bar{x}_2) - (\mu_1 - \mu_2)}{\sqrt{\dfrac{\sigma_1^2}{n_1} + \dfrac{\sigma_2^2}{n_2}}} = \dfrac{(5275 - 5240) - 0}{\sqrt{\dfrac{150^2}{400} + \dfrac{200^2}{400}}} = 2.8$

The p-value of the test is $P(z \le -2.8) + P(z \ge 2.8) = 2P(z \ge 2.8) = 2(.5 - .4974)$
$$= 2(.0026) = .0052$$

Since the p-value is so small, there is evidence to reject H_0. There is evidence to indicate the two population means are different for $\alpha > .0052$.

c.   The p-value would be half of the p-value in part **b**. The p-value $= P(z \ge 2.8) = .5 - .4974 = .0026$. Since the p-value is so small, there is evidence to reject H_0. There is evidence to indicate the mean for population 1 is larger than the mean for population 2 for $\alpha > .0026$.

d.   The test statistic is $z = \dfrac{(\bar{x}_1 - \bar{x}_2) - (\mu_1 - \mu_2)}{\sqrt{\dfrac{\sigma_1^2}{n_1} + \dfrac{\sigma_2^2}{n_2}}} = \dfrac{(5275 - 5240) - 25}{\sqrt{\dfrac{150^2}{400} + \dfrac{200^2}{400}}} = .8$

The p-value of the test is $P(z \le -.8) + P(z \ge .8) = 2P(z \ge .8) = 2(.5 - .2881)$
$$= 2(.2119) = .4238$$

Since the p-value is so large, there is no evidence to reject H_0. There is no evidence to indicate that the difference in the 2 population means is different from 25 for $\alpha \le .10$.

e.   We must assume that we have two independent random samples.

9.3   a. $\mu_{\bar{x}_1} = \mu_1 = 14$

$$\sigma_{\bar{x}_1} = \frac{\sigma_1}{\sqrt{n_1}} = \frac{4}{\sqrt{100}} = .4$$

b. $\mu_{\bar{x}_2} = \mu_2 = 10$

$$\sigma_{\bar{x}_2} = \frac{\sigma_2}{\sqrt{n_2}} = \frac{3}{\sqrt{100}} = .3$$

c. $\mu_{\bar{x}_1 - \bar{x}_2} = \mu_1 - \mu_2 = 14 - 10 = 4$

$$\sigma_{\bar{x}_1 - \bar{x}_2} = \sqrt{\frac{\sigma_1^2}{n_1} + \frac{\sigma_2^2}{n_2}} = \sqrt{\frac{4^2}{100} + \frac{3^2}{100}} = \sqrt{\frac{25}{100}} = .5$$

d.   Since $n_1 \geq 30$ and $n_2 \geq 30$, the sampling distribution of $\bar{x}_1 - \bar{x}_2$ is approximately normal by the Central Limit Theorem.

9.5   a.   No. Both populations must be normal.

b.   No. Both populations variances must be equal.

c.   No. Both populations must be normal.

d.   Yes.

e.   No. Both populations must be normal.

9.7   Some preliminary calculations are:

$$\bar{x}_1 = \frac{\sum x_1}{n_1} = \frac{11.8}{5} = 2.36 \qquad s_1^2 = \frac{\sum x_1^2 - \frac{\left(\sum x_1\right)^2}{n_1}}{n_1 - 1} = \frac{30.78 - \frac{(11.8)^2}{5}}{5 - 1} = .733$$

$$\bar{x}_2 = \frac{\sum x_2}{n_2} = \frac{14.4}{4} = 3.6 \qquad s_2^2 = \frac{\sum x_2^2 - \frac{\left(\sum x_2\right)^2}{n_2}}{n_2 - 1} = \frac{53.1 - \frac{(14.4)^2}{4}}{4 - 1} = .42$$

a. $$s_p^2 = \frac{(n_1 - 1)s_1^2 + (n_2 - 1)s_2^2}{n_1 + n_2 - 2} = \frac{(5-1).773 + (4-1).42}{5 + 4 - 2} = \frac{4.192}{7} = .5989$$

b. H_0: $\mu_1 - \mu_2 = 0$
H_a: $\mu_1 - \mu_2 < 0$

The test statistic is $t = \dfrac{(\bar{x}_1 - \bar{x}_2) - D_0}{\sqrt{s_p^2\left(\dfrac{1}{n_1} + \dfrac{1}{n_2}\right)}} = \dfrac{(2.36 - 3.6) - 0}{\sqrt{.5989\left(\dfrac{1}{5} + \dfrac{1}{4}\right)}} = \dfrac{-1.24}{.5191} = -2.39$

The rejection region requires $\alpha = .10$ in the lower tail of the t distribution with df = $n_1 + n_2 - 2 = 5 + 4 - 2 = 7$. From Table VI, Appendix A, $t_{.10} = 1.415$. The rejection region is $t < -1.415$.

Since the test statistic falls in the rejection region ($t = -2.39 < -1.415$), H_0 is rejected. There is sufficient evidence to indicate that $\mu_2 > \mu_1$ at $\alpha = .10$.

    c.   A small sample confidence interval is needed because $n_1 = 5 < 30$ and $n_2 = 4 < 30$.

For confidence coefficient .90, $\alpha = .10$ and $\alpha/2 = .05$. From Table VI, Appendix A, with df $= n_1 + n_2 - 2 = 5 + 4 - 2 = 7$, $t_{.05} = 1.895$. The 90% confidence interval for $(\mu_1 - \mu_2)$ is:

$$(\bar{x}_1 - \bar{x}_2) \pm t_{.05}\sqrt{s_p^2\left(\frac{1}{n_1} + \frac{1}{n_2}\right)} \Rightarrow (2.36 - 3.6) \pm 1.895\sqrt{.5989\left(\frac{1}{5} + \frac{1}{4}\right)}$$

$$\Rightarrow -1.24 \pm .98 \Rightarrow (-2.22, -0.26)$$

    d.   The confidence interval in part **c** provides more information about $(\mu_1 - \mu_2)$ than the test of hypothesis in part **b**. The test in part **b** only tells us that μ_2 is greater than μ_1. However, the confidence interval estimates what the difference is between μ_1 and μ_2.

9.9   a.   The test statistic is $z = -1.58$ and the p-value = .115.

Since the p-value is not small, there is no evidence to reject H_0. There is no evidence to indicate that the population means are different for $\alpha \le .10$.

    b.   The p-value would be half of the p-value in part **a**. The p-value = $1/2(.1150) = .0575$.

There is no evidence to reject H_0 for $\alpha = .05$. There is no evidence to indicate that the mean for population 1 is less than the mean for population 2 for $\alpha = .05$.

There is evidence to reject H_0 for $\alpha > .0575$. There is evidence to indicate that the mean for population 1 is less than the mean for population 2 for $\alpha > .0575$.

9.11   Let μ_1 = mean height of Australian boys who repeated a grade and μ_2 = mean height of Australian boys who never repeated a grade.

    a.   To determine if the average height of Australian boys who repeated a grade is less than the average height of boys who never repeated, we test:

H_0: $\mu_1 = \mu_2$
H_a: $\mu_1 < \mu_2$

The test statistic is $z = \dfrac{\bar{x}_1 - \bar{x}_2}{\sqrt{\dfrac{s_1^2}{n_1} + \dfrac{s_2^2}{n_2}}} = \dfrac{-.04 - .30}{\sqrt{\dfrac{1.17^2}{86} + \dfrac{.97^2}{1346}}} = -2.64$

The rejection region requires $\alpha = .05$ in the lower tail of the z distribution. From Table IV, Appendix A, $z_{.05} = 1.645$. The rejection region is $z < -1.645$.

Since the observed value of the test statistic falls in the rejection region ($z = -2.64 < -1.645$), H_0 is rejected. There is sufficient evidence to indicate that the average height of Australian boys who repeated a grade is less than the average height of boys who never repeated at $\alpha = .05$.

b.   Let μ_1 = mean height of Australian girls who repeated a grade and μ_2 = mean height of Australian girls who never repeated a grade.

To determine if the average height of Australian girls who repeated a grade is less than the average height of girls who never repeated, we test:

$H_0: \mu_1 = \mu_2$
$H_a: \mu_1 < \mu_2$

The test statistic is $z = \dfrac{\bar{x}_1 - \bar{x}_2}{\sqrt{\dfrac{s_1^2}{n_1} + \dfrac{s_2^2}{n_2}}} = \dfrac{.26 - .22}{\sqrt{\dfrac{.94^2}{43} + \dfrac{1.04^2}{1366}}} = .27$

The rejection region requires $\alpha = .05$ in the lower tail of the z distribution. From Table IV, Appendix A, $z_{.05} = 1.645$. The rejection region is $z < -1.645$.

Since the observed value of the test statistic does not fall in the rejection region ($z = .27 \not< -1.645$), H_0 is not rejected. There is insufficient evidence to indicate that the average height of Australian girls who repeated a grade is less than the average height of girls who never repeated at $\alpha = .05$.

c.   From the data, there is evidence to indicate that the average height of Australian boys who repeated a grade is less than the average height of boys who never repeated a grade. However, there is no evidence that the average height of Australian girls who repeated a grade is less than the average height of girls who never repeated.

9.13   Let μ_1 = mean FNE scores for bulimic students and μ_2 = mean FNE score for normal students.

Some preliminary calculations are:

$$\bar{x}_1 = \frac{\sum x_1}{n_1} = \frac{196}{11} = 17.82$$

$$s_1^2 = \frac{\sum x_1^2 - \dfrac{\left(\sum x_1\right)^2}{n_1}}{n_1 - 1} = \frac{3{,}734 - \dfrac{196^2}{11}}{11 - 1} = 24.1636$$

$$s_1 = \sqrt{s_1^2} = \sqrt{24.1636} = 4.916$$

$$\bar{x}_2 = \frac{\sum x_2}{n_2} = \frac{198}{14} = 14.14$$

$$s_2^2 = \frac{\sum x_2^2 - \frac{\left(\sum x_2\right)^2}{n_2}}{n_2 - 1} = \frac{3{,}164 - \frac{198^2}{14}}{14 - 1} = 27.9780$$

$$s_2 = \sqrt{s_2^2} = \sqrt{27.9780} = 5.289$$

$$s_p^2 = \frac{(n_1 - 1)s_1^2 + (n_2 - 1)s_2^2}{n_1 + n_2 - 2} = \frac{(11-1)24.1636 + (14-1)27.9780}{11 + 14 - 2} = 26.3196$$

a. For confidence coefficient .95, $\alpha = .05$ and $\alpha/2 = .05/2 = .025$. From Table VI, Appendix A, with df $= n_1 + n_2 - 2 = 11 + 14 - 2 = 23$, $t_{.025} = 2.069$. The confidence interval is:

$$\left(\bar{x}_1 - \bar{x}_2\right) \pm t_{.025}\sqrt{s_p^2\left(\frac{1}{n_1} + \frac{1}{n_2}\right)} \Rightarrow (17.82 - 14.14) \pm 2.069\sqrt{26.3196\left(\frac{1}{11} + \frac{1}{14}\right)}$$

$$\Rightarrow 3.68 \pm 4.277 \Rightarrow (-.597,\ 7.957)$$

We are 95% confident that the difference in mean FNE scores for bulimic and normal students is between −.597 and 7.957,

b. We must assume that the distribution of FNE scores for the bulimic students and the distribution of the FNE scores for the normal students are normally distributed. We must also assume that the variances of the two populations are equal.

Both sample distributions look somewhat mound-shaped and the sample variances are fairly close in value. Thus, both assumptions appear to be reasonably satisfied.

9.15 a. Let μ_1 = mean mathematics test score for males and μ_2 = mean mathematics test score for females.

To determine if there is a difference between the true mean mathematics test scores of male and female 8th-graders, we test:

H_0: $\mu_1 - \mu_2 = 0$
H_a: $\mu_1 - \mu_2 \neq 0$

The test statistic is $z = \dfrac{(\bar{x}_1 - \bar{x}_2) - 0}{\sqrt{\left(\dfrac{\sigma_1^2}{n_1} + \dfrac{\sigma_2^2}{n_2}\right)}} = \dfrac{(48.9 - 48.4) - 0}{\sqrt{\left(\dfrac{12.96^2}{1764} + \dfrac{11.85^2}{1739}\right)}} = 1.19$

Since no α was given, we will use $\alpha = .05$. The rejection region requires $\alpha/2 = .05/2 = .025$ in each tail of the z distribution. From Table IV, Appendix A, $z_{.025} = 1.96$. The rejection region is $z < -1.96$ or $z > 1.96$.

Since the observed value of the test statistic does not fall in the rejection region ($z = 1.19 \not> 1.96$), H_0 is not rejected. There is insufficient evidence to indicate there is a difference between the true mean mathematics test scores of male and female 8th-graders at $\alpha = .05$.

b. For confidence coefficient .90, $\alpha = .10$ and $\alpha/2 = .10/2 = .05$. From Table IV, Appendix A, $z_{.05} = 1.645$. The 90% confidence interval is:

$$(\bar{x}_1 - \bar{x}_2) \pm z_{.05} \sqrt{\left(\frac{\sigma_1^2}{n_1} + \frac{\sigma_2^2}{n_2} \right)}$$

$$\Rightarrow (48.9 - 48.4) \pm 1.645 \sqrt{\left(\frac{12.96^2}{1764} + \frac{11.85^2}{1739} \right)}$$

$$\Rightarrow .5 \pm .690 \Rightarrow (-.190, 1.190)$$

We are 90% confident that the true differences in mean mathematics test scores between males and females is between $-.190$ and 1.190.

This agrees with the results in part a, even though different α-levels were used. We are 90% confident that the interval contains the true difference in mean scores. Since the interval contains 0, 0 is a likely candidate for the true value of the difference. Thus, we would not be able to reject H_0.

c. Since both sample sizes are large, the only assumption necessary is that the samples are independent.

d. The observed significance level of the test in part a is the same as the p-value. The p-value is the probability of observing your test statistic or anything more unusual, given H_0 is true. Thus, the p-value $= P(z \le -1.19) + P(z \ge 1.19) = .5 - .3830 + .5 - .3830 = .2340$.

9.17 a. Using MINITAB, the output for comparing the mean level of family involvement in science homework assignments of TIPS and ATIPS students is:

```
Two sample T for GSHWS

COND         N      Mean     StDev    SE Mean
ATIPS       98      1.43     1.06      0.11
TIPS       128      2.55     1.27      0.11

95% CI for mu (ATIPS) - mu (TIPS ): ( -1.43,  -0.82)
T-Test mu (ATIPS) = mu (TIPS ) (vs not =): T = -7.24 P = 0.0000 DF = 222
```

Let μ_1 = mean level of involvement in science homework assignments for TIPS students and μ_2 = mean level of involvement in science homework assignments for ATIPS students.

To compare the mean level of family involvement in science homework assignments of TIPS and ATIPS students, we test:

H_0: $\mu_1 = \mu_2$
H_a: $\mu_1 \neq \mu_2$

From the printout, the test statistic is $t = -7.24$ and the p-value is $p = 0.0000$. Since the p-value is less than α ($p = 0.0000 < .05$), H_0 is rejected. There is sufficient evidence to indicate a difference in the mean level of family involvement in science homework assignments between TIPS and ATIPS students at $\alpha = .05$.

b. Using MINITAB, the output for comparing the mean level of family involvement in mathematics homework assignments of TIPS and ATIPS students is:

```
Two sample T for MTHHWS

COND          N       Mean      StDev    SE Mean
ATIPS         98      1.48      1.22      0.12
TIPS         128      1.56      1.27      0.11

95% CI for mu (ATIPS) - mu (TIPS ): ( -0.41,  0.25)
T-Test mu (ATIPS) = mu (TIPS ) (vs not =): T = -0.50  P = 0.62  DF = 212
```

Let μ_1 = mean level of involvement in mathematics homework assignments for TIPS students and μ_2 = mean level of involvement in mathematics homework assignments for ATIPS students.

To compare the mean level of family involvement in mathematics homework assignments of TIPS and ATIPS students, we test:

H_0: $\mu_1 = \mu_2$
H_a: $\mu_1 \neq \mu_2$

From the printout, the test statistic is $t = -0.50$ and the p-value is $p = 0.62$. Since the p-value is not less than α ($p = 0.62 > .05$), H_0 is not rejected. There is insufficient evidence to indicate a difference in the mean level of family involvement in mathematics homework assignments between TIPS and ATIPS students at $\alpha = .05$.

c. Using MINITAB, the output for comparing the mean level of family involvement in language arts homework assignments of TIPS and ATIPS students is:

```
Two sample T for LAHWS

COND          N       Mean      StDev    SE Mean
ATIPS         98      1.01      1.09      0.11
TIPS         128      1.20      1.12      0.099

95% CI for mu (ATIPS) - mu (TIPS ): ( -0.48,  0.106)
T-Test mu (ATIPS) = mu (TIPS ) (vs not =): T = -1.25  P = 0.21  DF = 211
```

Let μ_1 = mean level of involvement in language arts homework assignments for TIPS students and μ_2 = mean level of involvement in language arts homework assignments for ATIPS students.

Inferences Based on Two Samples: Confidence Intervals and Tests of Hypotheses

To compare the mean level of family involvement in language arts homework assignments of TIPS and ATIPS students, we test:

H_0: $\mu_1 = \mu_2$
H_a: $\mu_1 \neq \mu_2$

From the printout, the test statistic is $t = -1.25$ and the p-value is $p = 0.21$. Since the p-value is not less than α ($p = 0.21 > .05$), H_0 is not rejected. There is insufficient evidence to indicate a difference in the mean level of family involvement in language arts homework assignments between TIPS and ATIPS students at $\alpha = .05$.

d. Since both sample sizes are greater than 30, the only assumption necessary is:

1. The samples are random and independent.

From the information given, there is no reason to dispute this assumption.

9.19 Let μ_1 = mean number of new artists entering the *Billboard* chart per week prior to SoundScan and μ_2 = mean number of new artists entering the *Billboard* chart per week after to SoundScan.

Some preliminary calculations are:

$$\bar{x}_1 = \frac{\sum x_1}{n_1} = \frac{33}{12} = 2.75$$

$$s_1^2 = \frac{\sum x_1^2 - \frac{\left(\sum x_1\right)^2}{n_1}}{n_1 - 1} = \frac{119 - \frac{33^2}{12}}{12 - 1} = 2.568$$

$$s_1 = \sqrt{s_1^2} = \sqrt{2.568} = 1.603$$

$$\bar{x}_2 = \frac{\sum x_2}{n_2} = \frac{20}{12} = 1.67$$

$$s_2^2 = \frac{\sum x_2^2 - \frac{\left(\sum x_2\right)^2}{n_2}}{n_2 - 1} = \frac{50 - \frac{20^2}{12}}{12 - 1} = 1.515$$

$$s_2 = \sqrt{s_2^2} = \sqrt{1.515} = 1.231$$

$$s_p^2 = \frac{(n_1 - 1)s_1^2 + (n_2 - 1)s_2^2}{n_1 + n_2 - 2} = \frac{(12 - 1)2.568 + (12 - 1)1.515}{12 + 12 - 2} = 2.0415$$

a. For confidence coefficient .90, $\alpha = .10$ and $\alpha/2 = .10/2 = .05$. From Table VI, Appendix A, with df $= n_1 + n_2 - 2 = 12 + 12 - 2 = 22$, $t_{.05} = 1.717$. The confidence interval is:

$$(\bar{x}_1 - \bar{x}_2) \pm t_{.05}\sqrt{s_p^2\left(\frac{1}{n_1} + \frac{1}{n_2}\right)} \Rightarrow (2.75 - 1.67) \pm 1.717\sqrt{2.0415\left(\frac{1}{12} + \frac{1}{12}\right)}$$

$$\Rightarrow 1.08 \pm 1.002 \Rightarrow (.078,\ 2.082)$$

We are 95% confident that the difference in mean number of new artists entering the *Billboard* chart per week prior to and after SoundScan was introduced is between .078 and 2.082.

b. The test statistic is $t = \dfrac{\bar{x}_1 - \bar{x}_2}{\sqrt{s_p^2\left(\frac{1}{n_1} + \frac{1}{n_2}\right)}} = \dfrac{2.75 - 1.67}{\sqrt{2.0415\left(\frac{1}{12} + \frac{1}{12}\right)}} = 1.85$

The *p*-value is $P(t \leq -1.85) + P(t \geq 1.85)$. From Table VI, Appendix A, with df $= n_1 + n_2 - 2 = 12 + 12 - 2 = 22$, $.025 < P(t \geq 1.85) < .05$. Thus, the two-tailed *p*-value is $.05 <$ *p*-value $< .10$. (Using MINITAB, the p-value is .077.)

For any value of $\alpha < .077$, we would not reject H_0. We would conclude that there was no evidence to indicate a difference in the mean number of new artists entering the *Billboard* chart per week prior to and after SoundScan was introduced.

For any value of $\alpha \geq .077$, we would reject H_0. We would conclude that there was evidence to indicate a difference in the mean number of new artists entering the *Billboard* chart per week prior to and after SoundScan was introduced.

The reported *p*-value $p < .1$. Our *p*-value agrees with this.

9.21 a. Let $\mu_1 =$ mean degree of swelling for mice treated with bear bile and $\mu_2 =$ mean degree of swelling for mice treated with pig bile.

$$s_p^2 = \frac{(n_1 - 1)s_1^2 + (n_2 - 1)s_2^2}{n_1 + n_2 - 2} = \frac{(10 - 1)4.17^2 + (10 - 1)3.33^2}{10 + 10 - 2} = 14.2389$$

For confidence coefficient .95, $\alpha = .05$ and $\alpha/2 = .05/2 = .025$. From Table VI, Appendix A, with df $= n_1 + n_2 - 2 = 10 + 10 - 2 = 18$, $t_{.025} = 2.101$. The confidence interval is:

$$(\bar{x}_1 - \bar{x}_2) + t_{.025}\sqrt{s_p^2\left(\frac{1}{n_1} + \frac{1}{n_2}\right)} \Rightarrow (9.19 - 9.71) \pm 2.101\sqrt{14.289\left(\frac{1}{10} + \frac{1}{10}\right)}$$

$$\Rightarrow -.52 \pm 3.546 \Rightarrow (-4.066, 3.026)$$

We are 95% confident that the difference in the mean degree of swelling for mice treated with bear bile and mice treated with pig bile is between –4.066 and 3.026.

b. We must assume that the distribution of the degree of swelling for all mice receiving bear bile and the distribution of the degree of swelling for all mice receiving pig bile are normal. Also, we must assume that the variances of the two populations are equal.

c. Let μ_3 = mean degree of swelling for mice treated with saline and μ_2 = mean degree of swelling for mice treated with pig bile.

$$s_p^2 = \frac{(n_3-1)s_3^2 + (n_2-1)s_2^2}{n_3 + n_2 - 2} = \frac{(10-1)3.38^2 + (10-1)3.33^2}{10+10-2} = 11.25665$$

For confidence coefficient .95, $\alpha = .05$ and $\alpha/2 = .05/2 = .025$. From Table VI, Appendix A, with df $= n_1 + n_2 - 2 = 10 + 10 - 2 = 18$, $t_{.025} = 2.101$. The confidence interval is:

$$(\bar{x}_3 - \bar{x}_2) + t_{.025}\sqrt{s_p^2\left(\frac{1}{n_3} + \frac{1}{n_2}\right)} \Rightarrow (18.30 - 9.71) \pm 2.101\sqrt{11.25665\left(\frac{1}{10} + \frac{1}{10}\right)}$$

$$\Rightarrow 8.59 \pm 3.152 \Rightarrow (5.438, 11.742)$$

We are 95% confident that the difference in the mean degree of swelling for mice treated with saline and mice treated with pig bile is between 5.438 and 11.742.

9.23 Let μ_1 = mean DIQ score for SLI children and μ_2 = mean DIQ score for YND children.

Some preliminary calculations are:

$$\bar{x}_1 = \frac{\sum x_1}{n_1} = \frac{936}{10} = 93.6$$

$$s_1^2 = \frac{\sum x_1^2 - \frac{\left(\sum x_1\right)^2}{n_1}}{n_1 - 1} = \frac{88352 - \frac{(936)^2}{10}}{10-1} = 82.4889$$

$$\bar{x}_2 = \frac{\sum x_2}{n_2} = \frac{953}{10} = 95.3$$

$$s_2^2 = \frac{\sum x_2^2 - \frac{\left(\sum x_2\right)^2}{n_2}}{n_1 - 1} = \frac{91329 - \frac{(953)^2}{10}}{10-1} = 56.4556$$

$$s_p^2 = \frac{(n_1-1)s_1^2 + (n_2-1)^2 s_2^2}{n_1 + n_2 - 2} = \frac{(10-1)82.4889 + (10-1)56.4556}{10+10-2} = 69.4723$$

To determine if the mean DIQ scores differ for the two groups, we test:

H_0: $\mu_1 - \mu_2 = 0$
H_a: $\mu_1 - \mu_2 \neq 0$

The test statistic is $t = \dfrac{\bar{x}_1 - \bar{x}_2 - 0}{\sqrt{s_p^2\left(\dfrac{1}{n_1} + \dfrac{1}{n_2}\right)}} = \dfrac{(93.6 - 95.3) - 0}{\sqrt{69.4723\left(\dfrac{1}{10} + \dfrac{1}{10}\right)}} = -0.46$

The rejection region requires $\alpha/2 = .10/2 = .05$ in each tail of the t distribution with df $= n_1 + n_2 - 2 = 10 + 10 - 2 = 18$. From Table VI, Appendix A, $t_{.05} = 1.734$. The rejection region is $t < -1.734$ or $t > 1.734$.

Since the observed value of the test statistic does not fall in the rejection region ($t = -.46 \nless -1.734$), H_0 is not rejected. There is insufficient evidence to indicate that the mean DIQ scores differ for the two groups at $\alpha = .10$.

9.25 a. From the information given, we have no idea what the standard deviations are. Whether the population means are different or not depends on how variable the data are.

b. Let μ_1 = mean pain intensity rating for blacks and μ_2 = mean pain intensity rating for whites.

To determine if blacks, on average, have a higher pain intensity rating than whites, we test:
H_0: $\mu_1 = \mu_2$
H_a: $\mu_1 > \mu_2$

The rejection region requires $\alpha = .05$ in the upper tail of the z distribution. From Table IV, Appendix A, $z_{.05} = 1.645$. Thus, in order to reject H_0, the test statistic would have to be greater than 1.645.

Substituting known values into the test statistic, we can solve for s_p.

$$z = \dfrac{\bar{x}_1 - \bar{x}_2}{\sqrt{s_p^2\left(\dfrac{1}{n_1} + \dfrac{1}{n_2}\right)}} > 1.645$$

$$\Rightarrow \dfrac{8.2 - 6.9}{\sqrt{s_p^2\left(\dfrac{1}{55} + \dfrac{1}{159}\right)}} > 1.645$$

$$\Rightarrow 1.3 > 1.645(s_p)\sqrt{\left(\dfrac{1}{55} + \dfrac{1}{159}\right)}$$

$$\Rightarrow \dfrac{1.3}{1.645(.1564)} > s_p$$

$$\Rightarrow 5.053 > s_p$$

Thus, if both s_1 and s_2 were less than 5.053, then we would conclude that blacks, on average would have a higher pain intensity than whites. (It is possible that either s_1 or s_2 could be greater than 5.053 and we would still reject H_0. As long as $s_p < 5.053$, we would reject H_0.)

c. From part **b**, we know we would reject H_0 if $s_p < 5.053$. Thus, we would not reject H_0 if $s_p \geq 5.053$. This could happen if both s_1 and s_2 are greater than 5.053. (It is possible that either s_1 or s_2 could be less than 5.053 and we would still reject H_0. As long as $s_p \geq 5.053$, we would reject H_0.)

9.27 a. The rejection region requires $\alpha = .05$ in the upper tail of the t distribution with df $= n_D - 1 = 10 - 1 = 9$. From Table VI, Appendix A, $t_{.05} = 1.833$. The rejection region is $t > 1.833$.

b. From Table VI, with df $= n_D - 1 = 20 - 1 = 19$, $t_{.10} = 1.328$. The rejection region is $t > 1.328$.

c. From Table VI, with df $= n_D - 1 = 5 - 1 = 4$, $t_{.025} = 2.776$. The rejection region is $t > 2.776$.

d. From Table VI, with df $= n_D - 1 = 9 - 1 = 8$, $t_{.01} = 2.896$. The rejection region is $t > 2.896$.

9.29 a.

Pair	Difference
1	3
2	2
3	2
4	4
5	0
6	1

$$\bar{x}_D = \frac{\sum x_D}{n} = \frac{12}{6} = 2$$

$$s_D^2 = \frac{\sum x_D^2 - \frac{\left(\sum x_D\right)^2}{n_D}}{n_D - 1} = \frac{\left(34 - \frac{(12)^2}{6}\right)}{5} = 2$$

b. $\mu_D = \mu_1 - \mu_2$

c. For confidence coefficient .95, $\alpha = .05$ and $\alpha/2 = .025$. From Table VI, Appendix A, with df $= n_D - 1 = 6 - 1 = 5$, $t_{.025} = 2.571$. The confidence interval is:

$$\bar{x}_D \pm t_{\alpha/2} \frac{s_D}{\sqrt{n_D}} \Rightarrow 2 \pm 2.571 \frac{\sqrt{2}}{\sqrt{6}} \Rightarrow 2 \pm 1.484 \Rightarrow (.516, 3.484)$$

d. H_0: $\mu_D = 0$

    H_a: $\mu_D \neq 0$

The test statistic is $t = \dfrac{\bar{x}_D}{s_D / \sqrt{n_D}} = \dfrac{2}{\sqrt{2}/\sqrt{6}} = 3.46$

The rejection region requires $\alpha/2 = .05/2 = .025$ in each tail of the t distribution with df $= n_D - 1 = 6 - 1 = 5$. From Table VI, Appendix A, $t_{.025} = 2.571$. The rejection region is $t < -2.571$ or $t > 2.571$.

Since the observed value of the test statistic falls in the rejection region ($t = 3.46 > 2.571$), H_0 is rejected. There is sufficient evidence to indicate that the mean difference is different from 0 at $\alpha = .05$.

9.31   Some preliminary calculations:

Pair	Difference $x - y$
1	$55 - 44 = 11$
2	$68 - 55 = 13$
3	$40 - 25 = 15$
4	$55 - 56 = -1$
5	$75 - 62 = 13$
6	$52 - 38 = 14$
7	$49 - 31 = 18$

$$\bar{x}_D = \frac{\sum x_D}{n_D} = \frac{83}{7} = 11.86$$

$$s_D^2 = \frac{\sum x_D^2 - \dfrac{\left(\sum x_D\right)^2}{n_D}}{n_D - 1} = \frac{1205 - \dfrac{83^2}{7}}{7 - 1} = 36.8095$$

$$s_D = \sqrt{s_D^2} = \sqrt{36.8095} = 6.0671$$

a. H_0: $\mu_D = 10$

    H_a: $\mu_D \neq 10$ where $\mu_D = (\mu_1 - \mu_2)$

The test statistic is $t = \dfrac{\bar{x}_D - D_0}{s_D / \sqrt{n_D}} = \dfrac{11.86 - 10}{6.067/\sqrt{7}} = \dfrac{1.86}{2.2931} = .81$

The rejection region requires $\alpha/2 = .05/2 = .025$ in each tail of the t distribution with df $= n_D - 1 = 7 - 1 = 6$. From Table VI, Appendix A, $t_{.025} = 2.447$. The rejection region is $t < -2.447$ or $t > 2.447$.

Since the observed value of the test statistic does not fall in the rejection region ($t = .81 \not> 2.447$), H_0 is not rejected. There is insufficient evidence to conclude $\mu_D \neq 10$ at $\alpha = .05$.

Inferences Based on Two Samples: Confidence Intervals and Tests of Hypotheses

b. p-value $= P(t \le -.81) + P(t \ge .81) = 2P(t \ge .81)$

Using Table VI, Appendix A, with df $= 6$, $P(t \ge .81)$ is greater than .10.

Thus, $2P(t \ge .81)$ is greater than .20.

The probability of observing a value of t as large as .81 or as small as $-.81$ if, in fact, $\mu_D = 10$ is greater than .20. We would conclude that there is insufficient evidence to suggest $\mu_D \ne 10$.

9.33 a. First, some preliminary calculations:

	Testosterone Levels		
Patient	Pretreatment	After 6 months of Depro-Provera	D = P – A
1	849	96	753
2	903	41	862
3	890	31	859
4	1,092	124	968
5	362	46	316
6	900	53	847
7	1,006	113	893
8	672	174	498

$$\bar{x}_D = \frac{\sum x_D}{n_D} = \frac{5,996}{8} = 749.5$$

$$s_D^2 = \frac{\sum x_D^2 - \frac{\left(\sum x_D\right)^2}{n_D}}{n_D - 1} = \frac{4,847,676 - \frac{5,996^2}{8}}{8-1} = 50,524.857$$

For confidence coefficient .99, $\alpha = .01$ and $\alpha/2 = .01/2 = .005$. From Table VI, Appendix A, with df $= n_D - 1 = 8 - 1 = 7$, $t_{.005} = 3.499$. The confidence interval is:

$$\bar{x}_D \pm t_{.005} \frac{s_D}{\sqrt{n_D}} \Rightarrow 749.5 \pm 3.499 \frac{\sqrt{50,524.857}}{\sqrt{8}} \Rightarrow 749.5 \pm 278.068$$

$$\Rightarrow (471.432, 1027.568)$$

We are 99% confident that the true mean difference between the pretreatment and after-treatment testosterone levels of young male TBI patients is between 471.432 and 1,027.568.

b. Since 0 is not in the confidence interval, there is evidence that there is a difference in the testosterone levels pre and post treatment. Thus, there is evidence that Depo-Provera is effective in reducing testosterone levels in young male TBI patients.

9.35 a. Let μ_1 = mean reading response times for the tongue-twister list and μ_2 = mean reading response times for the control list. Then $\mu_D = \mu_1 - \mu_2$.

To compare the mean reading response times for the tongue-twister and control lists, we test:

H_0: $\mu_D = 0$
H_a: $\mu_D \neq 0$

The test statistic is $z = \dfrac{\bar{d} - 0}{s_D/\sqrt{n}} = \dfrac{.25 - 0}{.78/\sqrt{42}} = 2.08$

The p-value is $P(z \leq -2.08) + P(z \geq 2.08) = (.5000 - .4812) + (.5000 - .4812) = .0188 + .0188 = .0376$.

c. Since the observed p-value is less than α ($p = .0376 < .05$), H_0 is rejected. There is sufficient evidence to conclude that the mean reading response times differ for the tongue-twister and control lists at $\alpha = .05$.

9.37 a. Yes. It appears that the researchers have matched the MS and non-MS subjects successfully.

b. Some preliminary calculations:

Pair	MS Subject Age	Non-MS Subject Age	D = MS – Non-MS
1	48	45	3
2	34	34	0
3	34	34	0
4	38	34	4
5	45	39	6
6	42	42	0
7	32	34	–2
8	35	43	–8
9	33	31	2
10	46	43	3

$$\bar{x}_D = \frac{\sum x_D}{n_D} = \frac{8}{10} = .8$$

$$s_D^2 = \frac{\sum x_D^2 - \dfrac{\left(\sum x_D\right)}{n_D}}{n_D - 1} = \frac{142 - \dfrac{8^2}{10}}{10 - 1} = 15.0667$$

$$s_D = \sqrt{s_D^2} = \sqrt{15.0667} = 3.8816$$

To determine if the mean ages of the MS and Non-MS groups are different, we test:

H_0: $\mu_D = 0$
H_a: $\mu_D \neq 0$

The test statistic is $t = \dfrac{\bar{x}_D - 0}{s_D / \sqrt{n_D}} = \dfrac{.8 - 0}{3.8816 / \sqrt{10}} = .65$

The rejection region requires $\alpha/2 = .05/2 = .025$ in each tail of the t distribution with df $= n_D - 1 = 10 - 1 = 9$. From Table VI, Appendix A, $t_{.025} = 2.262$. The rejection region is $t < -2.262$ or $t > 2.262$.

Since the observed value of the test statistic does not fall in the rejection region ($t = .65$ ≯ 2.262), H_0 is not rejected. There is insufficient evidence to indicate the mean ages of the MS and Non-MS groups are different at $\alpha = .05$.

c. To determine if there is a difference in the mean heights between MS subjects and Non-MS subjects, we test:

H_0: $\mu_D = 0$
H_a: $\mu_D \neq 0$

The test statistic is $t = 1.351$.

The p-value *is* $p = .210$. Since the p-value is not small, there is no evidence to indicate a difference in mean heights between MS subjects and Non-MS subjects for any $\alpha < .210$.

To determine if there is a difference in the mean weights between MS subjects and Non-MS subjects, we test:

H_0: $\mu_D = 0$
H_a: $\mu_D \neq 0$

The test statistic is $t = -1.578$.

The p-value is $p = .149$. Since the p-value is not small, there is no evidence to indicate a difference in mean weights between MS subjects and Non-MS subjects for any $\alpha < .149$.

9.39   a. Let μ_1 = mean trait for senior year in high school and μ_2 = mean trait during sophomore year of college. Also, let $\mu_D = \mu_1 - \mu_2$. To determine if there is a decrease in the mean self-concept of females between the senior year in high school and the sophomore year in college, we test:

H_0: $\mu_0 = 0$
H_a: $\mu_0 > 0$

b. Since the same female students were used in each study, the data should be analyzed as paired differences. The samples are not independent.

c. Since the number of pairs is $n_0 = 133$, no assumptions about the population of differences is necessary. We still assume that a random sample of differences is selected.

d. Leadership: The p-value > .05. There is no evidence to reject H_0 for $\alpha = .05$. There is no evidence to indicate that there is a decrease in mean self-concept of females on leadership between the senior year in high school and the sophomore year in college for $\alpha = .05$.

Popularity: The p-value < .05. There is evidence to reject H_0 for $\alpha = .05$. There is evidence to indicate that there is a decrease in mean self-concept of females on popularity between the senior year in high school and the sophomore year in college for $\alpha = .05$.

Intellectual self-confidence: The p-value < .05. There is evidence to reject H_0 for $\alpha = .05$. There is evidence to indicate that there is a decrease in mean self-concept of females on intellectual self-concept between the senior year in high school and the sophomore year in college for $\alpha = .05$.

9.41 Let μ_1 = mean homophone confusion errors for time 1 and μ_2 = mean homophone confusion errors for time 2. Then $\mu_D = \mu_1 - \mu_2$.

To determine if Alzheimer's patients show a significant increase in mean homophone confusion errors over time, we test:

H_0: $\mu_D = 0$
H_a: $\mu_D \neq 0$

The test statistic is $t = -2.306$.

The p-value is $p = .0163$. Since the p-value is small, there is sufficient evidence to indicate an increase in mean homophone confusion errors from time 1 to time 2

We must assume that the population of differences is normally distributed and that the sample was randomly selected. A stem-and-leaf display of the data indicate that the data are mound-shaped. It appears that these assumptions are valid.

9.43 Remember that $\hat{p}_1$ and $\hat{p}_2$ can be viewed as means of the number of successes per n trials in the respective samples. Therefore, when n_1 and n_2 are large, $\hat{p}_1 - \hat{p}_2$ is approximately normal by the Central Limit Theorem.

9.45 a. The rejection region requires $\alpha = .01$ in the lower tail of the z distribution. From Table IV, Appendix A, $z_{.01} = 2.33$. The rejection region is $z < -2.33$.

b. The rejection region requires $\alpha = .025$ in the lower tail of the z distribution. From Table IV, Appendix A, $z_{.025} = 1.96$. The rejection region is $z < -1.96$.

c. The rejection region requires $\alpha = .05$ in the lower tail of the z distribution. From Table IV, Appendix A, $z_{.05} = 1.645$. The rejection region is $z < -1.645$.

d. The rejection region requires $\alpha = .10$ in the lower tail of the z distribution. From Table IV, Appendix A, $z_{.10} = 1.28$. The rejection region is $z < -1.28$.

9.47　For confidence coefficient .95, $\alpha = 1 - .95 = .05$ and $\alpha/2 = .05/2 = .025$. From Table IV, Appendix A, $z_{.025} = 1.96$. The 95% confidence interval for $p_1 - p_2$ is approximately:

a.　$(\hat{p}_1 - \hat{p}_2) \pm z_{\alpha/2} \sqrt{\dfrac{\hat{p}_1\hat{q}_1}{n_1} + \dfrac{\hat{p}_2\hat{q}_2}{n_2}}$

$\Rightarrow (.65 - .58) \pm 1.96 \sqrt{\dfrac{.65(1-.65)}{400} + \dfrac{.58(1-.58)}{400}} \Rightarrow .07 \pm .067 \Rightarrow (.003, .137)$

b.　$(\hat{p}_1 - \hat{p}_2) \pm z_{\alpha/2} \sqrt{\dfrac{\hat{p}_1\hat{q}_1}{n_1} + \dfrac{\hat{p}_2\hat{q}_2}{n_2}}$

$\Rightarrow (.31 - .25) \pm 1.96 \sqrt{\dfrac{.31(1-.31)}{180} + \dfrac{.25(1-.25)}{250}} \Rightarrow .06 \pm .086 \Rightarrow (-026, .146)$

c.　$(\hat{p}_1 - \hat{p}_2) \pm z_{\alpha/2} \sqrt{\dfrac{\hat{p}_1\hat{q}_1}{n_1} + \dfrac{\hat{p}_2\hat{q}_2}{n_2}}$

$\Rightarrow (.46 - .61) \pm 1.96 \sqrt{\dfrac{.46(1-.46)}{100} + \dfrac{.61(1-.61)}{120}} \Rightarrow -.15 \pm .131 \Rightarrow (-.281, -.019)$

9.49　H_0: $(p_1 - p_2) = .1$
H_a: $(p_1 - p_2) > .1$

Since D_0 is not equal to 0, the test statistic is:

$$z = \frac{(\hat{p}_1 - \hat{p}_2) - D_0}{\sqrt{\dfrac{p_1 q_1}{n_1} + \dfrac{p_2 q_2}{n_2}}} = \frac{(\hat{p}_1 - \hat{p}_2) - D_0}{\sqrt{\dfrac{\hat{p}_1\hat{q}_1}{n_1} + \dfrac{\hat{p}_2\hat{q}_2}{n_2}}} = \frac{(.4 - .2) - .1}{\sqrt{\dfrac{.4(1-.4)}{50} + \dfrac{.2(1-.2)}{60}}} = \frac{.1}{.0864}$$

$$= 1.16$$

The rejection region requires $\alpha = .05$ in the upper tail of the z distribution. From Table IV, Appendix A, $z_{.05} = 1.645$. The rejection region is $z > 1.645$.

Since the observed value of the test statistic does not fall in the rejection region ($z = 1.16 \not> 1.645$), H_0 is not rejected. There is insufficient evidence to show $(p_1 - p_2) > .1$ at $\alpha = .05$.

9.51　a.　The observed proportion of St. John's wort patients who were in remission is:

$\hat{p}_1 = \dfrac{x_1}{n_1} = \dfrac{14}{98} = .143$

b.　The observed proportion of placebo patients who were in remission is:

$\hat{p}_2 = \dfrac{x_2}{n_2} = \dfrac{5}{102} = .049$

c. To determine if the proportion of St. John wort patients in remission exceeds the proportion of placebo patients in remission, we test:

H_0: $p_1 = p_2$
H_a: $p_1 > p_2$

The test statistic is $z = \dfrac{\hat{p}_1 - \hat{p}_2}{\sqrt{\hat{p}\hat{q}\left(\dfrac{1}{n_1} + \dfrac{1}{n_2}\right)}} = \dfrac{.143 - .049}{\sqrt{.095(.905)\left(\dfrac{1}{98} + \dfrac{1}{102}\right)}} = 2.27$

The rejection region requires $\alpha = .01$ in the upper tail of the z distribution. From Table IV, Appendix A, $z_{.01} = 2.33$. The rejection region is $z > 2.33$.

Since the observed value of the test statistic does not fall in the rejection region ($z = 2.27 \not> 2.33$), H_0 is not rejected. There is insufficient evidence to indicate the proportion of St. John wort patients in remission exceeds the proportion of placebo patients in remission at $\alpha = .01$.

d. The hypotheses and test statistic are the same as in part **c**.

The rejection region requires $\alpha = .10$ in the upper tail of the z distribution. From Table IV, Appendix A, $z_{.10} = 1.28$. The rejection region is $z > 1.28$.

Since the observed value of the test statistic falls in the rejection region ($z = 2.27 > 1.28$), H_0 is rejected. There is sufficient evidence to indicate the proportion of St. John wort patients in remission exceeds the proportion of placebo patients in remission $\alpha = .10$.

e. By changing the value of α in this problem, the conclusion changes. When α is small ($\alpha = .01$), we did not reject H_0. When α increases ($\alpha = .10$), we rejected H_0. As α increases, the chance of making a Type I error increase, but the chance of a Type II error decrease.

9.53 Let p_1 = proportion of larvae that died in containers containing high carbon dioxide levels and p_2 = proportion of larvae that died in containers containing normal carbon dioxide levels. The parameter of interest for this problem is $p_1 - p_2$, or the difference in the death rates for the two groups.

Some preliminary calculations are:

$\hat{p}_1 = \dfrac{x_1 + x_2}{n_1 + n_2} = \dfrac{.10(80) + .50(80)}{80 + 80} = .075 \qquad \hat{q} = 1 - \hat{p} = 1 - .075 = .925$

To determine if an increased level of carbon dioxide is effective in killing a higher percentage of leaf-eating larvae, we test:

H_0: $p_1 - p_2 = 0$
H_a: $p_1 - p_2 > 0$

The test statistic is $z = \dfrac{(\hat{p}_1 - \hat{p}_2) - 0}{\sqrt{\hat{p}\hat{q}\left(\dfrac{1}{80} + \dfrac{1}{80}\right)}} = \dfrac{(.10 - .05) - 0}{\sqrt{.075(.925)\left(\dfrac{1}{80} + \dfrac{1}{80}\right)}} = 1.201$

The rejection region requires $\alpha = .01$ in the upper tail of the z distribution. From Table IV, Appendix A, $z_{.01} = 2.33$. The rejection region is $z > 2.33$.

Since the observed value of the test statistic does not fall in the rejection region ($z = 1.201 \ngtr 2.33$), H_0 is not rejected. There is insufficient evidence to indicate that an increased level of carbon dioxide is effective in killing a higher percentage of leaf-eating larvae at $\alpha = .01$.

9.55   Let p_1 = recall rate of the younger group and p_2 = recall rate of the older group.

To determine if one group has a higher recall rate than the other, we test:

H_0: $p_1 = p_2$
H_a: $p_1 \neq p_2$

From the printout, the test statistic is $z = 2.13$.

The p-value is $p = .033$. Since the p-value is so small, there is evidence that one group has a higher recall rate than the other for $\alpha > .033$. Since the sample recall rate of the younger group is .775 and the sample recall rate for the older group is .550, there is evidence that the recall rate for the younger group is higher than the recall rate for the older group.

9.57   Let p_1 = proportion of sun-shaded eggs that hatch and p_2 = proportion of unshaded eggs that hatch.

Some preliminary calculations are:

$$\hat{p}_1 = \frac{x_1}{n_1} = \frac{34}{70} = .486; \qquad \hat{p}_2 = \frac{x_2}{n_2} = \frac{31}{80} = .388; \qquad \hat{p} = \frac{x_1 + x_2}{n_1 + n_2} = \frac{34 + 31}{70 + 80} = .433$$

To compare the hatching rates of the two groups of Pacific tree frog eggs, we test:

H_0: $p_1 - p_2 = 0$
H_a: $p_1 - p_2 \neq 0$

The test statistic is $z = \dfrac{(\hat{p}_1 - \hat{p}_2) - 0}{\sqrt{\hat{p}\hat{q}\left(\dfrac{1}{n_1} + \dfrac{1}{n_2}\right)}} = \dfrac{(.486 - .388) - 0}{\sqrt{.433(.567)\left(\dfrac{1}{70} + \dfrac{1}{80}\right)}} = 1.21$

The rejection region requires $\alpha/2 = .01/2 = .005$ in each tail of the z distribution. From Table IV, Appendix A, $z_{.005} = 2.58$. The rejection region is $z < -2.58$ or $z > 2.58$.

Since the observed value of the test statistic does not fall in the rejection region ($z = 1.21 \not> 2.58$), H_0 is not rejected. There is insufficient evidence to indicate the hatching rates of the two groups of Pacific tree frog eggs differ at $\alpha = .01$.

9.59   a.   Let p_1 = proportion of females who have a food craving and p_2 = proportion of males who have a food craving.

Some preliminary calculations are:

$$\hat{p} = \frac{x_1 + x_2}{n_1 + n_2} = \frac{600(.97) + 400(.67)}{600 + 400} = .85$$

To determine if the proportion of women who have food cravings exceeds the proportion of males who have food cravings, we test:

H_0: $p_1 - p_2 = 0$
H_a: $p_1 - p_2 > 0$

The test statistic is $z = \dfrac{(\hat{p}_1 - \hat{p}_2) - 0}{\sqrt{\hat{p}\hat{q}\left(\dfrac{1}{n_1} + \dfrac{1}{n_2}\right)}} = \dfrac{(.97 - .67) - 0}{\sqrt{(.85)(.15)\left(\dfrac{1}{600} + \dfrac{1}{400}\right)}} = 13.02$

The rejection region requires $\alpha = .01$ in the upper tail of the z distribution. From Table IV, Appendix A, $z_{.01} = 2.33$. The rejection region is $z > 2.33$.

Since the observed value of the test statistic falls in the rejection region ($z = 13.02 > 2.33$), H_0 is rejected. There is sufficient evidence to indicate the proportion of females who have food cravings exceeds the proportion of males who have food cravings at $\alpha = .01$.

   b.   This study involved 1,000 McMaster University students. It is very dangerous to generalize the results of this study to the general adult population of North America. The sample of students used may not be representative of the population of interest.

9.61 $n_1 = n_2 = \dfrac{(z_{\alpha/2})^2 (\sigma_1^2 + \sigma_2^2)}{B^2}$

For confidence coefficient .95, $\alpha = 1 - .95 = .05$ and $\alpha/2 = .05/2 = .025$. From Table IV, Appendix A, $z_{.025} = 1.96$.

$$n_1 = n_2 = \frac{1.96^2 (15 + 15)}{2.2^2} = 23.8 \approx 24$$

9.63   a.   For confidence coefficient .99, $\alpha = 1 - .99 = .01$ and $\alpha/2 = .01/2 = .005$. From Table IV, Appendix A, $z_{.005} = 2.58$.

$$n_1 = n_2 = \frac{(z_{\alpha/2})^2(p_1q_1 + p_2q_2)}{B^2} = \frac{2.58^2(.4(1-.4) + .7(1-.7))}{.01^2}$$

$$= \frac{2.99538}{.0001} = 29{,}953.8 \approx 29{,}954$$

   b.   For confidence coefficient .90, $\alpha = 1 - .90 = .10$ and $\alpha/2 = .10/2 = .05$. From Table IV, Appendix A, $z_{.05} = 1.645$. Since we have no prior information about the proportions, we use $p_1 = p_2 = .5$ to get a conservative estimate. For a width of .05, the bound is .025.

$$n_1 = n_2 = \frac{(z_{\alpha/2})^2(p_1q_1 + p_2q_2)}{B^2} = \frac{(1.645)^2(.5(1-.5) + .5(1-.5))}{.025^2} = 2164.82 \approx 2165$$

   c.   From part b, $z_{.05} = 1.645$.

$$n_1 = n_2 = \frac{(z_{\alpha/2})^2(p_1q_1 + p_2q_2)}{B^2} = \frac{1.645^2(.2(1-.2) + .3(1-.3))}{.03^2}$$

$$= \frac{1.00123}{.0009} = 1112.48 \approx 1113$$

9.65   For confidence coefficient .95, $\alpha = .05$ and $\alpha/2 = .05/2 = .025$. From Table IV, Appendix A, $z_{.025} = 1.96$.

$$n_1 = n_2 = \frac{(z_{\alpha/2})^2(\sigma_B^2 + \sigma_N^2)}{B^2} = \frac{1.96^2(25 + 25)}{2^2} = 48.02 \approx 49$$

9.67   For confidence coefficient .90, $\alpha = .10$ and $\alpha/2 = .10/2 = .05$. From Table IV, Appendix A, $z_{.05} = 1.645$.

$$n_1 = n_2 = \frac{(z_{\alpha/2})^2(p_1q_1 + p_2q_2)}{B^2} = \frac{1.645^2(.4(.6) + .6(.4))}{.1^2} = 129.889 \approx 130$$

9.69   For confidence coefficient .95, $\alpha = .05$ and $\alpha/2 = .025$. From Table IV, Appendix A, $z_{.025} = 1.96$.

$$n_1 = n_2 = \frac{(z_{\alpha/2})^2(\sigma_1^2 + \sigma_2^2)}{B^2} = \frac{1.96^2(15^2 + 15^2)}{1^2} = 1728.72 \approx 1729$$

9.71   For confidence coefficient .9, $\alpha = 1 - .9 = .1$ and $\alpha/2 = .1/2 = .05$. From Table IV, Appendix A, $z_{.05} = 1.645$.

$$n_1 = n_2 = \frac{(z_{\alpha/2})^2(\sigma_1^2 + \sigma_2^2)}{B^2} = \frac{1.645^2(5^2 + 5^2)}{1^2} = 135.3 \Rightarrow 136$$

9.73 a. With $v_1 = 2$ and $v_2 = 30$,
$$P(F \geq 4.18) = .025 \quad \text{(Table X, Appendix A)}$$

b. With $v_1 = 24$ and $v_2 = 10$,
$$P(F \geq 1.94) = .10 \quad \text{(Table VIII, Appendix A)}$$

Thus, $P(F < 1.94) = 1 - P(F \geq 1.94) = 1 - .10 = .90$

c. With $v_1 = 9$ and $v_2 = 1$,
$$P(F \geq 6022) = .01 \quad \text{(Table XI, Appendix A)}$$

Thus, $P(F < 6022) = 1 - P(F \geq 6022) = 1 - .01 = .99$

d. With $v_1 = 30$ and $v_2 = 30$,
$$P(F \geq 1.84) = .05 \quad \text{(Table IX, Appendix A)}$$

9.75 The test statistic for each of these is:

$$F = \frac{\text{Larger sample variance}}{\text{Smaller sample variance}}$$

a. The rejection region requires $\alpha/2 = .20/2 = .10$ in the upper tail of the F distribution. If $s_1^2 > s_2^2$, numerator df $= v_1 = 8$ and denominator df $= v_2 = 40$. From Table VIII, Appendix A, $F_{.10} = 1.83$. The rejection region is $F > 1.83$. If $s_1^2 < s_2^2$, numerator df $= v_2 = 40$ and denominator df $= v_1 = 8$. From Table VIII, Appendix A, $F_{.10} = 2.36$. The rejection region is $F > 2.36$.

b. The rejection region requires $\alpha/2 = .10/2 = .05$ in the upper tail of the F distribution. If $s_1^2 > s_2^2$, numerator df $= v_1 = 8$ and denominator df $= v_2 = 40$. From Table IX, Appendix A, $F_{.05} = 2.18$. The rejection region is $F > 2.18$. If $s_1^2 < s_2^2$, numerator df $= v_2 = 40$ and denominator df $= v_1 = 8$. From Table IX, Appendix A, $F_{.05} = 3.04$. The rejection region is $F > 3.04$.

c. The rejection region requires $\alpha/2 = .05/2 = .025$ in the upper tail of the F distribution. If $s_1^2 > s_2^2$, numerator df $= v_1 = 8$ and denominator df $= v_2 = 40$. From Table X, Appendix A, $F_{.025} = 2.53$. The rejection region is $F > 2.53$. If $s_1^2 < s_2^2$, numerator df $= v_2 = 40$ and denominator df $= v_1 = 8$. From Table X, Appendix A, $F_{.025} = 3.84$. The rejection region is $F > 3.84$.

d. The rejection region requires $\alpha/2 = .02/2 = .01$ in the upper tail of the F distribution. If $s_1^2 > s_2^2$, numerator df $= v_1 = 8$ and denominator df $= v_2 = 40$. From Table XI, Appendix A, $F_{.01} = 2.99$. The rejection region is $F > 2.99$. If $s_1^2 < s_2^2$, numerator df $= v_2 = 40$ and denominator df $= v_1 = 8$. From Table XI, Appendix A, $F_{.01} = 5.12$. The rejection region is $F > 5.12$.

9.77   a. H_0: $\sigma_1^2 = \sigma_2^2$

H_a: $\sigma_1^2 \neq \sigma_2^2$

The test statistic is $F = \dfrac{\text{Larger sample variance}}{\text{Smaller sample variance}} = \dfrac{s_2^2}{s_1^2} = \dfrac{9.85}{2.87} = 3.43$

The rejection region requires $\alpha/2 = .05/2 = .025$ in the upper tail of the F distribution with numerator df $\nu_1 = n_2 - 1 = 25 - 1 = 24$ and denominator df $\nu_2 = n_1 - 1 = 16 - 1 = 15$. From Table X, Appendix A, $F_{.025} = 2.70$. The rejection region is $F > 2.70$.

Since the observed value of the test statistic falls in the rejection region ($F = 3.43 > 2.70$), H_0 is rejected. There is sufficient evidence to indicate $\sigma_1^2 \neq \sigma_2^2$ at $\alpha = .05$.

b. H_0: $\sigma_1^2 = \sigma_2^2$

H_a: $\sigma_1^2 < \sigma_2^2$

The test statistic is $F = \dfrac{\text{Larger sample variance}}{\text{Smaller sample variance}} = \dfrac{s_2^2}{s_1^2} = \dfrac{9.85}{2.87} = 3.43$

The rejection region requires $\alpha = .05$ in the upper tail of the F distribution with numerator df $\nu_1 = n_2 - 1 = 25 - 1 = 24$ and denominator df $\nu_2 = n_1 - 1 = 16 - 1 = 15$. From Table IX, Appendix A, $F_{.05} = 2.29$. The rejection region is $F > 2.29$.

Since the observed value of the test statistic falls in the rejection region ($F = 3.43 > 2.29$), H_0 is rejected. There is sufficient evidence to indicate $\sigma_1^2 < \sigma_2^2$ at $\alpha = .05$.

9.79   Let $\sigma_B^2 =$ variance of the FNE scores for bulimic students and $\sigma_N^2 =$ variance of the FNE scores for normal students.

From Exercise 9.13, $s_B^2 = 24.1636$ and $s_N^2 = 27.9780$

To determine if the variances are equal, we test:

H_0: $\sigma_B^2 = \sigma_N^2$

H_a: $\sigma_B^2 \neq \sigma_N^2$

The test statistic is $F = \dfrac{\text{Larger sample variance}}{\text{Smaller sample variance}} = \dfrac{s_N^2}{s_B^2} = \dfrac{27.9780}{24.1636} = 1.16$

The rejection region requires $\alpha/2 = .05/2 = .025$ in the upper tail of the F distribution with numerator df $\nu_2 = n_2 - 1 = 14 - 1 = 13$ and denominator df $\nu_1 = n_1 - 1 = 11 - 1 = 10$. From Table X, Appendix A, $F_{.025} \approx 3.62$. The rejection region is $F > 3.62$.

Since the observed value of the test statistic does not fall in the rejection region ($F = 1.16 \ngtr 3.62$), H_0 is not rejected. There is insufficient evidence that the two variances are different at $\alpha = .05$. It appears that the assumption of equal variances is valid.

9.81   Let σ_1^2 = variance of the mathematics achievement test scores for males and σ_2^2 = variance of the mathematics achievement test scores for females. To determine if the test scores are more variable for the males than the females, we test:

$$H_0: \sigma_1^2 = \sigma_2^2$$
$$H_a: \sigma_1^2 > \sigma_2^2$$

The test statistic is $F = \dfrac{\text{Larger sample variance}}{\text{Smaller sample variance}} = \dfrac{s_1^2}{s_2^2} = \dfrac{12.96^2}{11.85^2} = 1.196$

The rejection region requires $\alpha = .01$ in the upper tail of the F distribution with numerator df $v_1 = n_1 - 1 = 1764 - 1 = 1763$ and denominator df $v_2 = n_2 - 1 = 1739 - 1 = 1738$. From Table XI, Appendix A, $F_{.01} \approx 1.00$. The rejection region is $F > 1.00$.

Since the observed value of the test statistic falls in the rejection region ($F = 1.196 > 1.00$), H_0 is rejected. There is sufficient evidence to indicate the test scores are more variable for the males than the females at $\alpha = .01$.

9.83   Let σ = variance of the effective population size of the outcrossing snails and σ = variance of the effective population size of the selfing snails. To determine if the variances for the two groups differ, we test:

$$H_0: \sigma_1^2 = \sigma_2^2$$
$$H_a: \sigma_1^2 \neq \sigma_2^2$$

The test statistic is $F = \dfrac{\text{Larger sample variance}}{\text{Smaller sample variance}} = \dfrac{s_1^2}{s_2^2} = \dfrac{1{,}932^2}{1{,}890^2} = 1.045$

Since α is not given, we will use $\alpha = .05$. The rejection region requires $\alpha/2 = .05/2 = .025$ in the upper tail of the F distribution with $v_1 = n_1 - 1 = 17 - 1 = 16$ and $v_2 = n_2 - 1 = 5 - 1 = 4$. From Table X, Appendix A, $F_{.025} \approx 8.66$. The rejection region is $F > 8.66$.

Since the observed value of the test statistic does not fall in the rejection region ($F = 1.045 \not> 8.66$), H_0 is not rejected. There is insufficient evidence to indicate the variances for the two groups differ at $\alpha = .05$.

9.85   a. $s_p^2 = \dfrac{(n_1 - 1)s_1^2 + (n_1 - 1)s_2^2}{n_1 + n_2 - 2} = \dfrac{11(74.2) + 13(60.5)}{12 + 14 - 2} = 66.7792$

$$H_0: \mu_1 - \mu_2 = 0$$
$$H_a: \mu_1 - \mu_2 > 0$$

The test statistic is $t = \dfrac{(\bar{x}_1 - \bar{x}_2) - 0}{\sqrt{s_p^2\left(\dfrac{1}{n_1} + \dfrac{1}{n_2}\right)}} = \dfrac{(17.8 - 15.3) - 0}{\sqrt{66.7792\left(\dfrac{1}{12} + \dfrac{1}{14}\right)}} = .78$

The rejection region requires $\alpha = .05$ in the upper tail of the t distribution with df = $n_1 + n_2 - 2 = 12 + 14 - 2 = 24$. From Table VI, Appendix A, for df = 24, $t_{.05} = 1.711$. The rejection region is $t > 1.711$.

Since the observed value of the test statistic does not fall in the rejection region ($0.78 \ngtr 1.711$), H_0 is not rejected. There is insufficient evidence to indicate that $\mu_1 > \mu_2$ at $\alpha = .05$.

b. For confidence coefficient .99, $\alpha = .01$ and $\alpha/2 = .01/2 = .005$. From Table VI, Appendix A, with df = $n_1 + n_2 - 2 = 12 + 14 - 2 = 24$, $t_{.005} = 2.797$. The confidence interval is:

$$(\bar{x}_1 - \bar{x}_2) \pm t_{.005} \sqrt{s_p^2 \left(\frac{1}{n_1} + \frac{1}{n_2} \right)} \Rightarrow (17.8 - 15.3) \pm 2.797 \sqrt{66.7792 \left(\frac{1}{12} + \frac{1}{14} \right)}$$

$$\Rightarrow 2.50 \pm 8.99 \Rightarrow (-6.49, 11.49)$$

c. For confidence coefficient .99, $\alpha = .01$ and $\alpha/2 = .01/2 = .005$. From Table IV, Appendix A, $z_{.005} = 2.58$.

$$n_1 = n_2 = \frac{(z_{\alpha/2})^2 (\sigma_1^2 + \sigma_2^2)}{B^2} = \frac{(2.58)^2 (74.2 + 60.5)}{2^2} = 224.15 \approx 225$$

9.87 a. For confidence coefficient .90, $\alpha = .10$ and $\alpha/2 = .05$. From Table IV, Appendix A, $z_{.05} = 1.645$. The confidence interval is:

$$(\bar{x}_1 - \bar{x}_2) \pm z_{.05} \sqrt{\frac{s_1^2}{n_1} + \frac{s_2^2}{n_2}} \Rightarrow (12.2 - 8.3) \pm 1.645 \sqrt{\frac{2.1}{135} + \frac{3.0}{148}}$$

$$\Rightarrow 3.90 \pm .31 \Rightarrow (3.59, 4.21)$$

b. H_0: $\mu_1 - \mu_2 = 0$
H_a: $\mu_1 - \mu_2 \neq 0$

The test statistic is $z = \dfrac{(\bar{x}_1 - \bar{x}_2) - 0}{\sqrt{\dfrac{s_1^2}{n_1} + \dfrac{s_2^2}{n_2}}} = \dfrac{(12.2 - 8.3) - 0}{\sqrt{\dfrac{2.1}{135} + \dfrac{3.0}{148}}} = 20.60$

The rejection region requires $\alpha/2 = .01/2 = .005$ in each tail of the z distribution. From Table IV, Appendix A, $z_{.005} = 2.58$. The rejection region is $z < -2.58$ or $z > 2.58$.

Since the observed value of the test statistic falls in the rejection region ($20.60 > 2.58$), H_0 is rejected. There is sufficient evidence to indicate that $\mu_1 \neq \mu_2$ at $\alpha = .01$.

c. For confidence coefficient .90, $\alpha = .10$ and $\alpha/2 = .05$. From Table IV, Appendix A, $z_{.05} = 1.645$.

$$n_1 = n_2 = \frac{(z_{\alpha/2})^2(\sigma_1^2 + \sigma_2^2)}{B^2} = \frac{(1.645)^2(2.1 + 3.0)}{.2^2} = 345.02 \approx 346$$

9.89 a. This is a paired difference experiment.

Pair	Difference (Pop. 1 – Pop. 2)
1	6
2	4
3	4
4	3
5	2

$$\bar{x}_D = \frac{\sum x_D}{n_D} = \frac{19}{5} = 3.8 \qquad s_D^2 = \frac{\sum x_D^2 - \dfrac{\left(\sum x_D\right)^2}{n_D}}{n_D - 1} = \frac{81 - \dfrac{19^2}{5}}{5 - 1} = 2.2$$

$$s_D = \sqrt{2.2} = 1.4832$$

$H_0: \ \mu_D = 0$
$H_a: \ \mu_D \neq 0$

The test statistic is $t = \dfrac{\bar{x}_D - 0}{s_D / \sqrt{n_D}} = \dfrac{3.8 - 0}{1.4832 / \sqrt{5}} = 5.73$

The rejection region requires $\alpha/2 = .05/2 = .025$ in each tail of the t distribution with df $= n - 1 = 5 - 1 = 4$. From Table VI, Appendix A, $t_{.025} = 2.776$. The rejection region is $t < -2.776$ or $t > 2.776$.

Since the observed value of the test statistic falls in the rejection region ($5.73 > 2.776$), H_0 is rejected. There is sufficient evidence to indicate that the population means are different at $\alpha = .05$.

b. For confidence coefficient .95, $\alpha = .05$ and $\alpha/2 = .025$. Therefore, we would use the same t value as above, $t_{.025} = 2.776$. The confidence interval is:

$$\bar{x}_D \pm t_{\alpha/2}\frac{s_D}{\sqrt{n_D}} \Rightarrow 3.8 \pm 2.776\frac{1.4832}{\sqrt{5}} \Rightarrow 3.8 \pm 1.84 \Rightarrow (1.96, 5.64)$$

c. The sample of differences must be randomly selected from a population of differences which has a normal distribution.

9.91   If the *p*-value is less than α, reject H_0. Otherwise, do not reject H_0.

a.   *p*-value = .0429 < .05 $\Rightarrow$ Reject H_0

b.   *p*-value = .1984 $\not<$ .05 $\Rightarrow$ Do not reject H_0

c.   *p*-value = .0001 < .05 $\Rightarrow$ Reject H_0

d.   *p*-value = .0344 < .05 $\Rightarrow$ Reject H_0

e.   *p*-value = .0545 $\not<$ .05 $\Rightarrow$ Do not reject H_0

f.   *p*-value = .9633 $\not<$ .05 $\Rightarrow$ Do not reject H_0

g.   It would be necessary to be able to assume that the two samples were independently and randomly taken from normal populations with the same variance.

9.93   a.   Let μ_1 = mean GPA for traditional students and μ_2 = mean GPA for nontraditional students. To determine whether the mean GPAs of traditional and nontraditional students differ, we test:

H_0: $\mu_1 - \mu_2 = 0$
H_a: $\mu_1 - \mu_2 \neq 0$

b.   The test statistic is $z = \dfrac{(\bar{x}_1 - \bar{x}_2) - D_0}{\sqrt{\dfrac{s_1^2}{n_1} + \dfrac{s_2^2}{n_2}}} = \dfrac{(2.9 - 3.5) - 0}{\sqrt{\dfrac{.5^2}{94} + \dfrac{.5^2}{73}}} = -7.69$

The rejection region requires $\alpha/2 = .01/2 = .005$ in each tail of the z distribution. From Table IV, Appendix A, $z_{.005} = 2.58$. The rejection region is $z < -2.58$ or $z > 2.58$.

Since the observed value of the test statistic falls in the rejection region ($z = -7.69 < -2.58$), H_0 is rejected. There is sufficient evidence to indicate that the mean GPAs of traditional and nontraditional students differ for $\alpha = .01$.

c.   We must assume that the two samples are randomly and independently selected from the populations of GPAs.

9.95   a.   The variable measured for this experiment is the time needed to match the picture with the sentence.

b.   The experimental units are the climbers.

c.   Since the same climbers were timed at the base camp and at a camp 5 miles above sea level, the data should be analyzed as a paired experiment.

9.97   a.   There is no reason to believe that the bacteria counts would not be normally distributed.

   b.   Let μ_1 = mean of the bacteria count for the discharge and μ_2 = mean of the bacteria count upstream. Since we want to test if the mean of the bacteria count for the discharge exceeds the mean of the count upstream, we test:

$$H_0:\ \mu_1 - \mu_2 = 0$$
$$H_a:\ \mu_1 - \mu_2 > 0$$

   c.   From the printout, the test statistic is $t = 1.569$. The two-tailed p-value is .148. Since this is a one-tailed test, $p = .148/2 = .074$.

There is no evidence to indicate that the mean of the bacteria count for the discharge exceeds the mean of the count upstream at $\alpha = .05$.

There is evidence to indicate that the mean of the bacteria count for the discharge exceeds the mean of the count upstream at $\alpha = .10$.

   d.   We must assume:
1.   The mean counts per specimen for each location is normally distributed.
2.   The variances of the 2 distributions are equal.
3.   Independent and random samples were selected from each population.

9.99   a.   Let μ_1 = mean rating of concern about product tampering for males and μ_2 = mean rating of concern about product tampering for females. To determine whether a difference exists in the mean level of concern about product tampering between males and females, we test:

$$H_0:\ \mu_1 - \mu_2 = 0$$
$$H_a:\ \mu_1 - \mu_2 \neq 0$$

   b.   The test statistic is $t = -2.69$ and the p-value $= .007$. Since the p-value is so small, there is evidence to reject H_0. There is sufficient evidence to indicate a difference exists in the mean level of concern about product tampering between males and females for $\alpha > .007$.

   c.   We must assume the sample sizes were sufficiently large so that the Central Limit Theorem applies. We must also assume that we selected two random and independent samples from the two populations.

9.101   a.   For each measure, let μ_1 = mean job satisfaction for day-shift nurses and μ_2 = mean job satisfaction for night-shift nurses. To determine whether a difference in job satisfaction exists between day-shift and night-shift nurses, we test:

$$H_0:\ \mu_1 - \mu_2 = 0$$
$$H_a:\ \mu_1 - \mu_2 \neq 0$$

   b.   Hours of work: The p-value $= .813$. Since the p-value is so large, there is no evidence to reject H_0. There is insufficient evidence to indicate a difference in mean job satisfaction exists between day-shift and night-shift nurses on hours of work for $\alpha \leq .10$.

Inferences Based on Two Samples: Confidence Intervals and Tests of Hypotheses

Free time: The p-value = .047. Since the p-value is so small, there is evidence to reject H_0. There is sufficient evidence to indicate a difference in mean job satisfaction exists between day-shift and night-shift nurses on free time for $\alpha > .047$.

Breaks: The p-value = .0073. Since the p-value is so small, there is evidence to reject H_0. There is sufficient evidence to indicate a difference in mean job satisfaction exists between day-shift and night-shift nurses on breaks for $\alpha > .0073$.

c. We must make the following assumptions for each measure:
1. The job satisfaction scores for both day-shift and night-shift nurses are normally distributed.
2. The variances of job satisfaction scores for both day-shift and night-shift nurses are equal.
3. Random and independent samples were selected from both populations of job satisfaction scores.

9.103 a. Since we want to compare the means of two independent groups, we would use a two sample test of hypothesis for independent samples. Since the sample sizes are small (21 for each), we would use a small sample test.

b. We must assume that both populations being sampled from are approximately normally distributed and that the population variances of the two groups are equal.

c. No. We need to know the common variance of the two groups or an estimate of it.

d. Since the p-value was so large ($p = .79$), H_0 would not be rejected for any reasonable value of α. There is insufficient evidence to indicate a difference in the mean scores for the two groups on the question dealing with clean/sterile gloves at $\alpha \leq .10$.

e. Since the p-value was so small ($p = .02$), H_0 is rejected for any $\alpha > .02$. There is sufficient evidence to indicate a difference in the mean scores for the two groups on the question dealing with a stethoscope at $\alpha > .02$.

9.105 a. Let μ_1 = mean blood pressure before training and μ_2 = mean blood pressure after training. Also, let $\mu_D = \mu_1 - \mu_2$. To determine if the mean blood pressure decreases after training, we test:

H_0: $\mu_D = 0$
H_a: $\mu_D > 0$

b. Some preliminary calculations:

Subject	Before	After	d = Before-After
1	136.9	130.2	6.7
2	201.4	180.7	20.7
3	166.8	149.6	17.2
4	150.0	153.2	-3.2
5	173.2	162.6	10.6
6	169.3	160.1	9.2

$$\bar{x}_D = \frac{\sum x_D}{n} = \frac{61.2}{6} = 10.2$$

$$s_D^2 = \frac{\sum x_D^2 - \frac{\left(\sum x_D\right)^2}{n}}{n-1} = \frac{976.46 - \frac{(61.2)^2}{6}}{6-1} = 70.444$$

$$s_D = \sqrt{s_D^2} = \sqrt{70.444} = 8.393$$

The test statistic is $t = \dfrac{\bar{x}_D - 0}{s_D/\sqrt{n}} = \dfrac{10.2 - 0}{8.393/\sqrt{6}} = 2.98$

The rejection region requires $\alpha = .05$ in the upper tail of the t distribution. From Table VI, Appendix A, with df $= n - 1 = 6 - 1 = 5$, $t_{.05} = 2.015$. The rejection region is $t > 2.015$.

Since the observed value of the test statistic falls in the rejection region ($t = 2.98 > 2.015$), H_0 is rejected. There is sufficient evidence to indicate the mean blood pressure decreases with training at $\alpha = .05$.

c.   For confidence coefficient .95, $\alpha = .05$ and $\alpha/2 = .05/2 = .025$. From Table VI, Appendix A, with df $= n - 1 = 6 - 1 = 5$, $t_{.025} = 2.571$. The 95% confidence interval is:

$$\bar{x}_D \pm t_{.025}\frac{s_D}{\sqrt{n}} \Rightarrow 10.2 \pm 2.571\frac{8.393}{\sqrt{6}} \Rightarrow 10.2 \pm 8.81 \Rightarrow (1.39,\ 19.01)$$

We are 95% confident that the decrease in the mean blood pressure after training is between 1.39 and 19.01.

9.107   Let p_1 = proportion of inositol-fed infants who suffer from retinopathy of prematurity and p_2 = proportion of infants on the standard diet who suffer from retinopathy of prematurity.

Some preliminary calculations are:

$$\hat{p}_1 = \frac{x_1}{n_1} = \frac{14}{110} = .127; \quad \hat{p}_2 = \frac{x_2}{n_2} = \frac{29}{110} = .264; \quad \hat{p} = \frac{x_1 + x_2}{n_1 + n_2} = \frac{14 + 29}{110 + 110} = .195$$

To determine if the proportion of infants suffering from retinopathy is less for the inositol-fed group than for the group fed the standard diet, we test:

$H_0:\ p_1 - p_2 = 0$

$H_a:\ p_1 - p_2 < 0$

The test statistic is $z = \dfrac{(\hat{p}_1 - \hat{p}_2) - 0}{\sqrt{\hat{p}\hat{q}\left(\dfrac{1}{n_1} + \dfrac{1}{n_2}\right)}} = \dfrac{(.127 - .264) - 0}{\sqrt{.195(.805)\left(\dfrac{1}{110} + \dfrac{1}{110}\right)}} = -2.56$

The rejection region requires $\alpha = .01$ in the lower tail of the z distribution. From Table IV, Appendix A, $z_{.01} = 2.33$. The rejection region is $z < -2.33$.

Since the observed value of the test statistic falls in the rejection region ($z = -2.56 < -2.33$), H_0 is rejected. There is sufficient evidence to indicate the proportion of infants suffering from retinopathy is less for the inositol-fed group than for the group fed the standard diet at $\alpha = .01$.

9.109   Let μ_1 = mean rating for music majors and μ_2 = mean rating of nonmusic majors.

For confidence coefficient .95, $\alpha = .05$ and $\alpha/2 = .05/2 = .025$. From Table IV, Appendix A, $z_{.025} = 1.96$. The confidence interval is:

$$(\bar{x}_1 - \bar{x}_2) \pm z_{.025}\sqrt{\frac{s_1^2}{n_1} + \frac{s_2^2}{n_2}} \Rightarrow (4.26 - 4.59) \pm 1.96\sqrt{\frac{.81^2}{100} + \frac{.78^2}{100}}$$

$$\Rightarrow -.33 \pm .220 \Rightarrow (-.550, -.110)$$

We are 95% confident that the difference in the mean rating between music majors and nonmusic majors is between $-.55$ and $-.11$.

9.111   a.   Some preliminary calculations are:

$$s_1^2 = \frac{\sum x_1^2 - \frac{\left(\sum x_1\right)^2}{n_1}}{n_1 - 1} = \frac{4209 - \frac{(145)^2}{5}}{5-1} = 1$$

$$s_2^2 = \frac{\sum x_2^2 - \frac{\left(\sum x_2\right)^2}{n_2}}{n_2 - 1} = \frac{4540 - \frac{(150)^2}{5}}{5-1} = 10$$

Let σ_1^2 = variance of measurements for instrument 1 and σ_2^2 = variance of measurements for instrument 2.

To determine if the variances of the measurements for the two instruments differ, we test:

H_0: $\sigma_1^2 = \sigma_2^2$
H_a: $\sigma_1^2 \neq \sigma_2^2$

The test statistic is $F = \dfrac{\text{Larger sample variance}}{\text{Smaller sample variance}} = \dfrac{s_2^2}{s_1^2} = \dfrac{10}{1} = 10$

Since no α was given, we will use $\alpha = .05$. The rejection region requires $\alpha/2 = .05/2 = .025$ in the upper tail of the F distribution with numerator df $v_1 = n_2 - 1 = 5 - 1 = 4$ and denominator df $v_2 = n_1 - 1 = 5 - 1 = 4$. From Table X, Appendix A, $F_{.025} = 9.60$. The rejection region is $F > 9.60$.

Since the observed value of the test statistic falls in the rejection reigon ($F = 10 > 9.60$), H_0 is rejected. There is sufficient evidence to indicate the variances of the measurements for the two instruments differ at $\alpha = .05$.

c. We must assume that the samples are randomly and independently selected from populations that are normally distributed.

9.113 Let p_1 = proportion of patients who suffered a first stroke who had antibodies and p_2 = proportion of patients hospitalized for other reasons who had antibodies.

Some preliminary calculations are:

$$\hat{p}_1 = \frac{x_1}{n_1} = \frac{25}{255} = .098; \quad \hat{p}_2 = \frac{x_2}{n_2} = \frac{11}{257} = .043; \quad \hat{p} = \frac{x_1 + x_2}{n_1 + n_2} = \frac{25 + 11}{255 + 257} = .070$$

To determine if the proportion of renegade antibodies in the blood is greater for patients who suffered a first stroke than for patients hospitalized for other reasons, we test:

H_0: $p_1 - p_2 = 0$
H_a: $p_1 - p_2 > 0$

The test statistic is $z = \dfrac{(\hat{p}_1 - \hat{p}_2) - 0}{\sqrt{\hat{p}\hat{q}\left(\dfrac{1}{n_1} + \dfrac{1}{n_2}\right)}} = \dfrac{(.098 - .043) - 0}{\sqrt{.07(.93)\left(\dfrac{1}{255} + \dfrac{1}{257}\right)}} = 2.44$

Since no α is given, we will use $\alpha = .05$. The rejection region requires $\alpha = .05$ in the upper tail of the z distribution. From Table IV, Appendix A, $z_{.05} = 1.645$. The rejection region is $z > 1.645$.

Since the observed value of the test statistic falls in the rejection region ($z = 2.44 > 1.645$), H_0 is rejected. There is sufficient evidence to indicate the proportion of first stroke patients who had antibodies is greater than the proportion of patients hospitalized for other reasons who had antibodies at $\alpha = .05$.

Analysis of Variance: Comparing More than Two Means

10.1   Since only one factor is utilized, the treatments are the four levels (A, B, C, D) of the qualitative factor.

10.3   One has no control over the levels of the factors in an observational experiment. One does have control of the levels of the factors in a designed experiment.

10.5   a.   This is an observational experiment. The economist has no control over the factor levels or unemployment rates.

         b.   This is a designed experiment. The psychologist selects the feedback programs of interest and randomly assigns five rats to each program.

         c.   This is a designed experiment. The marketer has control of the selection of the national publications.

         d.   This is an observational experiment. The load on the generators is not controlled by the utility.

         e.   This is an observational experiment. One has no control over the distance of the haul, the goods hauled, or the price of diesel fuel.

         f.   This is an observational experiment. The student does not control which state is assigned to a portion of the country.

10.7   a.   The experimental units are the cockatiels.

         b.   This experiment is a designed experiment. The birds were randomly divided into 3 groups and each group received a different treatment.

         c.   There is one factor in this study. The factor is the group.

         d.   There are three levels of the group variable – Group 1 received purified water, Group 2 received purified water and liquid sucrose, and Group 3 received purified water and liquid sodium chloride.

         e.   There are 3 treatments in the study. Because there is only one factor, the treatments are the same as the factor levels.

         f.   The response variable is the water consumption.

10.9   a.   The ethical behavior of the salesperson.

         b.   There are two factors (with two levels at each factor):
                Type of sales job (high tech versus low tech) and
                Sales task (new account development).

c. The treatments are the $2 \times 2 = 4$ factor-level combinations of type of sales job and sales task.

d. The college students are the experimental units.

10.11 a. Using Table IX, $F_{.05} = 6.59$ with $v_1 = 3$, $v_2 = 4$.

b. Using Table XI, $F_{.01} = 16.69$ with $v_1 = 3$, $v_2 = 4$.

c. Using Table VIII, $F_{.10} = 1.61$ with $v_1 = 20$, $v_2 = 40$.

d. Using Table X, $F_{.025} = 3.87$ with $v_1 = 12$, $v_2 = 9$.

10.13 In the second dot diagram **b**, the difference between the sample means is small relative to the variability within the sample observations. In the first dot diagram **a**, the values in each of the samples are grouped together with a range of 4, while in the second diagram **b**, the range of values is 8.

10.15 For each dot diagram, we want to test:

H_0: $\mu_1 = \mu_2$
H_a: $\mu_1 \neq \mu_2$

From Exercise 10.14,

Diagram a	Diagram b
$\bar{x}_1 = 9$	$\bar{x}_1 = 9$
$\bar{x}_2 = 14$	$\bar{x}_2 = 14$
$s_1^2 = 2$	$s_1^2 = 14.4$
$s_2^2 = 2$	$s_2^2 = 14.4$

a.

Diagram a	Diagram b
$s_p^2 = \dfrac{s_1^2 + s_2^2}{2}$	$s_p^2 = \dfrac{s_1^2 + s_2^2}{2}$
$= \dfrac{2+2}{2} = 2 \quad (n_1 = n_2)$	$= \dfrac{14.4 + 14.4}{2} = 14.4 \quad (n_1 = n_2)$
In Exercise 10.14, MSE = 2	In Exercise 10.14, MSE = 14.4

The pooled variance for the two-sample t test is the same as the MSE for the F test.

b.

Diagram a	Diagram b
$$t = \dfrac{\bar{x}_1 - \bar{x}_2}{\sqrt{s_p^2\left(\dfrac{1}{n_1} + \dfrac{1}{n_2}\right)}} = \dfrac{9-14}{\sqrt{2\left(\dfrac{1}{6} + \dfrac{1}{6}\right)}}$$	$$t = \dfrac{\bar{x}_1 - \bar{x}_2}{\sqrt{s_p^2\left(\dfrac{1}{n_1} + \dfrac{1}{n_2}\right)}} = \dfrac{9-14}{\sqrt{14.4\left(\dfrac{1}{6} + \dfrac{1}{6}\right)}}$$
$= -6.12$	$= -2.28$
In Exercise 10.14, $F = 37.5$	In Exercise 10.14, $F = 5.21$

The test statistic for the F test is the square of the test statistic for the t test.

c.

Diagram a	Diagram b
For the t test, the rejection region requires $\alpha/2 = .05/2 = .025$ in each tail of the t distribution with df $= n_1 + n_2 - 2 = 6 + 6 - 2 = 10$. From Table VI, Appendix A, $t_{.025} = 2.228$.	For the t test, the rejection region is the same as Diagram **a** since we are using the same α, n_1, and n_2 for both tests.

The rejection region is $t < -2.228$ or $t > 2.228$.

In Exercise 10.14, the rejection region for both diagrams using the F test is $F > 4.96$.

The tabled F value equals the square of the tabled t value.

d.

Diagram a	Diagram b
For the t test, since the test statistic falls in the rejection region ($t = 6.12 < -2.228$), we would reject H_0. In Exercise 10.14 using the F test, we rejected H_0.	For the t test, since the test statistic falls in the rejection region ($t = -2.28 < -2.228$), we would reject H_0. In Exercise 10.14 using the F test, we rejected H_0.

e.   Assumptions for the t test:

1.   Both populations have relative frequency distributions that are approximately normal.
2.   The two population variances are equal.
3.   Samples are selected randomly and independently from the populations.

Assumptions for the F test:

1.   Both population probability distributions are normal.
2.   The two population variances are equal.
3.   Samples are selected randomly and independently from the respective populations.

The assumptions are the same for both tests.

10.17 For all parts, the hypotheses are:

$H_0: \mu_1 = \mu_2 = \mu_3 = \mu_4 = \mu_5$
H_a: At least two treatment means differ

The rejection region for all parts is the same.

The rejection region requires $\alpha = .10$ in the upper tail of the F distribution with $v_1 = p - 1 = 5 - 1 = 4$ and $v_2 = n - p = 30 - 5 = 25$. From Table VIII, Appendix A, $F_{.10} = 2.18$. The rejection region is $F > 2.18$.

a. $\text{SST} = .2(500) = 100$ $\qquad$ $\text{SSE} = \text{SS(Total)} - \text{SST} = 500 - 100 = 400$

$$\text{MST} = \frac{\text{SST}}{p-1} = \frac{100}{5-1} = 25 \qquad \text{MSE} = \frac{\text{SSE}}{n-p} = \frac{400}{30-5} = 16$$

$$F = \frac{\text{MST}}{\text{MSE}} = \frac{25}{16} = 1.5625$$

Since the observed value of the test statistic does not fall in the rejection region ($F = 1.5625 \not> 2.18$), H_0 is not rejected. There is insufficient evidence to indicate differences among the treatment means at $\alpha = .10$.

b. $\text{SST} = .5(500) = 250$ $\qquad$ $\text{SSE} = \text{SS(Total)} - \text{SST} = 500 - 250 = 250$

$$\text{MST} = \frac{\text{SST}}{p-1} = \frac{250}{5-1} = 62.5 \qquad \text{MSE} = \frac{\text{SSE}}{n-p} = \frac{250}{30-5} = 10$$

$$F = \frac{\text{MST}}{\text{MSE}} = \frac{62.5}{10} = 6.25$$

Since the observed value of the test statistic falls in the rejection region ($F = 6.25 > 2.18$), H_0 is rejected. There is sufficient evidence to indicate differences among the treatment means at $\alpha = .10$.

c. $\text{SST} = .8(500) = 400$ $\qquad$ $\text{SSE} = \text{SS(Total)} - \text{SST} = 500 - 400 = 100$

$$\text{MST} = \frac{\text{SST}}{p-1} = \frac{400}{5-1} = 100 \qquad \text{MSE} = \frac{\text{SSE}}{n-p} = \frac{100}{30-5} = 4$$

$$F = \frac{\text{MST}}{\text{MSE}} = \frac{100}{4} = 25$$

Since the observed value of the test statistic falls in the rejection region, ($F = 25 > 2.18$), H_0 is rejected. There is sufficient evidence to indicate differences among the treatment means at $\alpha = .10$.

d. The F ratio increases as the treatment sum of squares increases.

10.19 a. The number of treatments is $3 + 1 = 4$. The total sample size is $37 + 1 = 38$.

b. To determine if the treatment means differ, we test:

H_0: $\mu_1 = \mu_2 = \mu_3 = \mu_4$
H_a: At least two treatment means differ

The test statistic is $F = 14.80$.

The rejection region requires $\alpha = .01$ in the upper tail of the F distribution with $v_1 = p - 1 = 4 - 1 = 3$ and $v_2 = n - p = 38 - 4 = 34$. From Table XI, Appendix A, $F_{.01} \approx 4.51$. The rejection region is $F > 4.51$.

Since the observed value of the test statistic falls in the rejection region ($F = 14.80 > 4.51$), H_0 is rejected. There is sufficient evidence to indicate differences among the treatment means at $\alpha = .01$.

c. We need the sample means to compare specific pairs of treatment means.

10.21 a. The experimental design used is a completely randomized design.

b. There are 4 treatments in this experiment. The four treatments are the 4 "colonies" to which the robots were assigned – 3, 6, 9, or 12 robots per colony.

c. To determine if the mean energy expended (per robot) of the four different colony sizes differed, we test:

H_0: $\mu_1 = \mu_2 = \mu_3 = \mu_4$
H_a: At least two treatment means differ

d. The test statistic is $F = 7.70$. The p-value is $p < .001$. Since the p-value is less than $\alpha = .05$, H_0 is rejected. There is sufficient evidence to indicate that mean energy expended (per robot) of the four different colony sizes differed at $\alpha = .05$

10.23 a. To determine if the average heights of male Australian school children differ among the age groups, we test:

H_0: $\mu_1 = \mu_2 = \mu_3$
H_a: At least two treatment means differ

b. The test statistic is $F = 4.57$. The p-value is $p = .01$. Since the p-value is less than α ($p = .01 < .05$), H_0 is rejected. There is sufficient evidence to indicate the average heights of male Australian school children differ among the age groups at $\alpha = .05$.

c. To determine if the average heights of female Australian school children differ among the age groups, we test:

H_0: $\mu_1 = \mu_2 = \mu_3$
H_a: At least two treatment means differ

The test statistic is $F = 0.85$. The p-value is $p = .43$. Since the p-value is not less than α ($p = .43 > .05$), H_0 is not rejected. There is insufficient evidence to indicate the average heights of female Australian school children differ among the age groups at $\alpha = .05$.

d. For the boys, differences were found in the mean standardized heights among the three tertiles. For the girls, no differences were found in the mean standardized heights among the three tertiles.

10.25 a. If the samples represent random and independent selections from the populations associated with the four hair colors, then a completely randomized design has been used.

b. Let μ_1, μ_2, μ_3, and μ_4 represent the mean pain threshold scores for all people with light blond, dark blond, light brunette, and dark brunette hair, respectively.

H_0: $\mu_1 = \mu_2 = \mu_3 = \mu_4$
H_a: At least two means differ

Test statistic: $F = \dfrac{\text{MST}}{\text{MSE}} = 6.79$

The rejection region requires $\alpha = .05$ in the upper tail of the F distribution with $v_1 = p - 1 = 4 - 1 = 3$ and $v_2 = n - p = 19 - 4 = 15$. From Table IX, Appendix A, $F_{.05} = 3.29$. The rejection region is $F > 3.29$.

Since the observed value of the test statistic falls in the rejection region ($F = 6.79 > 3.29$), H_0 is rejected. There is sufficient evidence to indicate that at least one of the mean pain thresholds is different from the others at $\alpha = .05$.

c. The p-value is given on the output under "Pr > F" as 0.0041. Because this value is so small, this is a strong indication that the mean pain thresholds differ.

d. The assumptions necessary to assure the validity of the test are as follows:

1. The probability distributions of the pain threshold scores is normal for each hair color.
2. The variances of the probability distributions of the pain threshold scores for each hair color are equal.
3. The samples are selected randomly and independently.

10.27 Using MINITAB, the ANOVA table is:

Source	df	SS	MS	F	P
Expressions	5	23.09	4.62	3.96	0.007
Error	30	34.99	1.17		
Total	35	58.07			

To determine if differences exist among the mean dominance ratings of the six facial expressions, we test:

H_0: $\mu_1 = \mu_2 = \mu_3 = \mu_4 = \mu_5 = \mu_6$
H_a: At least two treatment means differ

where μ_i represents the mean dominance rating for facial expression i.

The test statistic is $F = 3.96$.

The rejection region requires $\alpha = .10$ in the upper tail of the F distribution with $\nu_1 = p - 1 = 6 - 1 = 5$ and $\nu_2 = n - p = 36 - 6 = 30$. Using Table VIII, Appendix A, $F_{.10} = 2.05$. The rejection region is $F > 2.05$.

Since the observed value of the test statistic falls in the rejection region ($F = 3.96 > 2.05$), H_0 is rejected. There is sufficient evidence to indicate that differences exists among the mean dominance rating for facial expressions at $\alpha = .10$.

10.29 a. $\bar{x} = \left(\sum_{i=1}^{3} n_i \bar{x}_i \right) / 41$

$$= \frac{17(0.2) + 13(1.9) + 11(2.9)}{41}$$

$$= \frac{60}{41}$$

$$\approx 1.46$$

$\text{SST} = \sum_{i=1}^{3} n_i (\bar{x}_i - \bar{x})^2$

$$= 17(0.2 - 1.46)^2 + 13(1.9 - 1.46)^2 + 11(2.9 - 1.46)^2$$

$$\approx 52.32$$

Therefore, SST = 52.32

b. $\text{SSE} = (n_1 - 1) s_1^2 + (n_2 - 1) s_2^2 + (n_3 - 1) s_3^2$

$$= (17 - 1)(0.4)^2 + (13 - 1)(1.7)^2 + (11 - 1)(2.0)^2$$

$$= 2.56 + 34.68 + 40$$

$$= 77.24$$

c.

Source	df	SS	MS	F
Treatment	2	52.32	26.16	12.89
Error	38	77.24	2.03	
Total	40	129.56		

$\text{MST} = \dfrac{\text{SST}}{p-1} = \dfrac{52.32}{3-1} = 26.16$

$\text{MSE} = \dfrac{\text{SSE}}{n-p} = \dfrac{77.24}{41-3} = 2.03$

$F = \dfrac{\text{MST}}{\text{MSE}} = \dfrac{26.16}{2.03} = 12.89$

d. To determine if differences exist among the mean number of binges by the students in the three groups, we test:

H_0: $\mu_1 = \mu_2 = \mu_3$
H_a: At least 1 μ_i differs

The test statistic is $F = 12.89$.

The rejection region requires $\alpha = .01$ in the upper tail of the F distribution with $v_1 = p - 1 = 3 - 1 = 2$ and $v_2 = n - p = 41 - 3 = 38$. Using Table XI, Appendix A, $F \approx 5.21$. The rejection region is $F > 5.21$.

Since the observed value of the test statistic falls in the rejection region ($F = 12.89 > 5.21$), H_0 is rejected. There is sufficient evidence to indicate that differences exist among the mean number of binges in the three groups.

e. 1. The assumption that the probability distributions for the 3 groups are normal seems to be met.

   2. The assumption that the variances for each group are equal may not be met because the CBT - WLT group's variance may be significantly lower than the other two groups.

   3. The assumption that the individuals for each group were selected randomly may not be met because those in the first two groups were based on their response to therapy.

   The results of the study may be invalid if the assumptions are not met.

f. Since the subjects in the first two groups were not randomly and independently assigned to their groups, a completely randomized design may not be valid for this study. The results of the study may be invalid.

10.31 A comparisonwise error rate is the error rate (or the probability of declaring the means different when, in fact, they are not different) for each individual comparison. That is, if each comparison is run using $\alpha = .05$, then the comparisonwise error rate is .05.

The experiment wise error rate is the probability of making a Type I error for at least one of all of the comparisons made. If the experimentwise error rate is $\alpha = .05$, then each individual comparison is made at a value of α which is less than .05.

10.33 a. There are $c = \dfrac{p(p-1)}{2} = \dfrac{4(4-1)}{2} = 6$ pairwise comparisons.

   b. There are no significant differences in the mean energy expended among the colony sizes of 3, 6, and 9. However, the mean energy expended for the colony containing 12 robots was significantly less than that for all other colony sizes.

10.35 a.   Yes, there is a significant difference between the standardized height means for the oldest and youngest boys because their sample means are not connected.

b.   Yes, there is a significant difference between the standardized height means for the oldest and middle-aged boys because their sample means are not connected.

c.   No, there is not a significant difference between the standardized height means for the middle-aged and youngest boys because their sample means are connected.

d.   The experiment-wise error rate for the inferences made in parts **a-c** is $\alpha = .05$.

e.   No differences were found among the means of the three groups of girls. Therefore, we do not need to find where the differences are.

10.37 a.   To determine if there are differences in the mean dental fear scores among the three groups, we test:

H_0: $\mu_1 = \mu_2 = \mu_3$
H_a: At least two treatment means differ

The test statistic is $F = 4.43$ and the p-value is $p < .05$. If we use $\alpha = .05$, then the rejection region will be p-value $< .05$. Since the p-value is less than .05, we reject H_0. There is sufficient evidence to indicate there are differences in the mean dental fear scores among the three groups at $\alpha = .05$.

b.   First, we arrange the means in order from the largest to the smallest. Then we draw a line between the means that are not significantly different. The results are:

Questionnaire   53.8

Slide   43.1 |
              |
Control   41.8 |

10.39   In this experiment, we have $p = 4$ causes to compare. Consequently, the number of relevant comparisons is $c = \dfrac{4(3)}{2} = 6$. These intervals calculated by Bonferroni's method are given below.

$\mu_{collision} - \mu_{ground}$	$(-52.74, 97.81)$
$\mu_{collision} - \mu_{fire}$	$(-75.62, 77.78)$
$\mu_{collision} - \mu_{hull}$	$(-59.12, 94.28)$
$\mu_{ground} - \mu_{fire}$	$(-95.01, 52.10)$
$\mu_{ground} - \mu_{hull}$	$(-78.51, 68.60)$
$\mu_{fire} - \mu_{hull}$	$(-58.51, 91.51)$

Since all of the intervals contain zero, there are no significant differences.

10.41 a.   There are 3 blocks used since Block df $= b - 1 = 2$ and 5 treatments since the treatment df $= p - 1 = 4$.

b.   There were 15 observations since the Total df $= n - 1 = 14$.

c. H_0: $\mu_1 = \mu_2 = \mu_3 = \mu_4 = \mu_5$
   H_a: At least two treatment means differ

d. The test statistic is $F = \dfrac{\text{MST}}{\text{MSE}} = 9.109$

e. The rejection region requires $\alpha = .01$ in the upper tail of the F distribution with $v_1 = p - 1 = 5 - 1 = 4$ and $v_2 = n - p - b + 1 = 15 - 5 - 3 + 1 = 8$. From Table XI, Appendix A, $F_{.01} = 7.01$. The rejection region is $F > 7.01$.

f. Since the observed value of the test statistic falls in the rejection region ($F = 9.109 > 7.01$), H_0 is rejected. There is sufficient evidence to indicate that at least two treatment means differ at $\alpha = .01$.

g. The assumptions necessary to assure the validity of the test are as follows:

   1. The probability distributions of observations corresponding to all the block-treatment combinations are normal.
   2. The variances of all the probability distributions are equal.

10.43 a. The ANOVA Table is as follows:

Source	df	SS	MS	F
Treatment	2	12.032	6.016	50.958
Block	3	71.749	23.916	202.586
Error	6	.708	.118	
Total	11	84.489		

b. To determine if the treatment means differ, we test:

   H_0: $\mu_A = \mu_B = \mu_C$
   H_a: At least two treatment means differ

   The test statistic is $F = \dfrac{\text{MST}}{\text{MSE}} = 50.958$

   The rejection region requires $\alpha = .05$ in the upper tail of the F distribution with $v_1 = p - 1 = 3 - 1 = 2$ and $v_2 = n - p - b + 1 = 12 - 3 - 4 + 1 = 6$. From Table IX, Appendix A, $F_{.05} = 5.14$. The rejection region is $F > 5.14$.

   Since the observed value of the test statistic falls in the rejection region ($F = 50.958 > 5.14$), H_0 is rejected. There is sufficient evidence to indicate that the treatment means differ at $\alpha = .05$.

c. To see if the blocking was effective, we test:

   H_0: $\mu_1 = \mu_2 = \mu_3 = \mu_4$
   H_a: At least two block means differ

   The test statistic is $F = \dfrac{\text{MSB}}{\text{MSE}} = 202.586$

The rejection region requires $\alpha = .05$ in the upper tail of the F distribution with $\nu_1 = b - 1 = 4 - 1 = 3$ and $\nu_2 = n - p - b + 1 = 12 - 3 - 4 + 1 = 6$. From Table IX, Appendix A, $F_{.05} = 4.76$. The rejection region is $F > 4.76$.

Since the observed value of the test statistic falls in the rejection region ($F = 202.586 > 4.76$), H_0 is rejected. There is sufficient evidence to indicate that blocking was effective in reducing the experimental error at $\alpha = .05$.

d. From the printouts, we are given the differences in the sample means. The difference between Treatment B and both Treatments A and C are positive (1.125 and 2.450), so Treatment B has the largest sample mean. The difference between Treatment A and C is positive (1.325), so Treatment A has a larger sample mean than Treatment C. So Treatment B has the largest sample mean, Treatment A has the next largest sample mean and Treatment C has the smallest sample mean.

From the printout, all the means are significantly different from each other.

e. The assumptions necessary to assure the validity of the inferences above are:

1. The probability distributions of observations corresponding to all the block-treatment combinations are normal.
2. The variances of all the probability distributions are equal.

10.45 a. This is a randomized block design. The blocks are the 12 plots of land. The treatments are the three methods used on the shrubs: fire, clipping, and control. The response variable is the mean number of flowers produced. The experimental units are the 36 shrubs.

b.

		Treatment	
	Fire	**Clipping**	**Control**
1	Shrub 2	Shrub 1	Shrub 3
2	Shrub 3	Shrub 2	Shrub 1
Plot ⋮	⋮	⋮	⋮
12	Shrub 1	Shrub 3	Shrub 2

c. To determine if there is a difference in the mean number of flowers produced among the three treatments, we test:

H_0: $\mu_1 = \mu_2 = \mu_3$
H_a: The mean number of flowers produced differ for at least two of the methods.

The test statistic is $F = 5.42$ and $p = .009$. We can reject the null hypothesis at the $\alpha > .009$ level of significance. At least two of the methods differ with respect to mean number of flowers produced by pawpaws.

d. The means of Control and Clipping do not differ significantly. The means of Clipping and Burning do not differ significantly. The mean of treatment Burning exceeds that of the Control.

10.47 a. The experimenters expected there to be much variation in the number of participants from week to week (more participants at the beginning and fewer as time goes on). Thus, by blocking on weeks, this extraneous source of variation can be controlled.

b. The ANOVA table is:

Source	df	SS	MS	F	p
Prompt	4	1185.0	296.25	39.87	0.0001
Week	5	386.4	77.28	10.40	0.0001
Error	20	148.6	7.43		
Total	29	1720.0			

c. To determine if a difference exists in the mean number of walkers per week among the five walker groups, we test:

H_0: $\mu_1 = \mu_2 = \mu_3 = \mu_4 = \mu_5$
H_a: At least two treatment means differ

where μ_i represents the mean number of walkers in group i.

The test statistic is $F = 39.87$.

The rejection region requires $\alpha = .05$ in the upper tail of the F distribution with $v_1 = p - 1 = 5 - 1 = 4$ and $v_2 = n - p - b + 1 = 30 - 4 - 6 + 1 = 20$. From Table IX, Appendix A, $F_{.05} = 2.69$. The rejection region is $F > 2.69$.

Since the observed value of the test statistic falls in the rejection region ($F = 39.87 > 2.69$), H_0 is rejected. There is sufficient evidence to indicate differences exist among the mean number of walkers per week among the 5 walker groups at $\alpha = .05$.

d. Using the STATISTIX printout, the following conclusions are drawn:

The mean number of walkers per week in the "Frequent/High" group is significantly higher than the mean number of walkers per week in the "Infrequent/Low" group, the "Infrequent/High" group, and the "Control" group. The mean number of walkers per week in the "Frequent/Low" group is significantly higher than the mean number of walkers per week in the "Infrequent/Low" group, the "Infrequent/High" group, and the "Control" group. The mean number of walkers per week in the "Infrequent/Low" group is significantly higher than the mean number of walkers per week in the "Control" group. The mean number of walkers per week in the "Infrequent/High" group is significantly higher than the mean number of walkers per week in the "Control group.

e. In order for the above inferences to be valid, the following assumptions must hold:

1) The probability distributions of observations corresponding to all block-treatment conditions are normal.
2) The variances of all the probability distributions are equal.

10.49 Using SAS, the ANOVA Table is:

```
The ANOVA Procedure

Dependent Variable: temp

                             Sum of
Source              DF      Squares      Mean Square   F Value   Pr > F

Model               11   18.53700000    1.68518182      0.52    0.8634

Error               18   58.03800000    3.22433333

Corrected Total     29   76.57500000


            R-Square     Coeff Var     Root MSE     temp Mean

            0.242076     1.885189      1.795643     95.25000


Source              DF     Anova SS     Mean Square   F Value   Pr > F

STUDENT              9   18.41500000    2.04611111      0.63    0.7537
PLANT                2    0.12200000    0.06100000      0.02    0.9813
```

To determine if there are differences among the mean temperatures among the three treatments, we test:

H_0: $\mu_1 = \mu_2 = \mu_3$
H_a: At least two treatment means differ

The test statistic is $F = 0.02$. The associated p-value is $p = .9813$. Since the p-value is very large, there is no evidence of a difference in mean temperature among the three treatments. Since there is no difference, we do not need to compare the means. It appears that the presence of plants or pictures of plants does not reduce stress.

10.51 a. A randomized block design was used to control for the different locations. Evidently, the researchers suspected that mosquitoes from different locations would react to the insecticides differently, so they wanted to control for this difference. So, the blocks in this experiment were the seven locations. The treatments were the 5 different insecticides compared.

b. The output using SAS is:

```
The ANOVA Procedure

Dependent Variable: ratio
```

		Sum of			
Source	DF	Squares	Mean Square	F Value	Pr > F
Model	10	96.0194286	9.6019429	2.79	0.0193
Error	24	82.6760000	3.4448333		
Corrected Total	34	178.6954286			

R-Square	Coeff Var	Root MSE	ratio Mean
0.537336	56.43868	1.856026	3.288571

Source	DF	Anova SS	Mean Square	F Value	Pr > F
INSECT	4	39.28400000	9.82100000	2.85	0.0458
LOC	6	56.73542857	9.45590476	2.74	0.0356

Tukey's Studentized Range (HSD) Test for ratio

NOTE: This test controls the Type I experimentwise error rate, but it generally has a higher Type II error rate than REGWQ.

Alpha	0.05
Error Degrees of Freedom	24
Error Mean Square	3.444833
Critical Value of Studentized Range	4.16632
Minimum Significant Difference	2.9227

Means with the same letter are not significantly different.

Tukey Grouping		Mean	N	insect
A		4.9714	7	1
A				
A		3.8429	7	5
A				
A		3.2714	7	4
A				
A		2.2429	7	2
A				
A		2.1143	7	3

To determine if any of the insecticides are more effective than any others, we test:

H_0: $\mu_1 = \mu_2 = \mu_3 = \mu_4 = \mu_5$
H_a: At least two treatment means differ

From the printout, the test statistic is $F = 2.85$ and the p-value is $p = 0.0458$. Using $\alpha = .05$, the conclusion would be to not reject H_0 because the p-value is less than α ($p = .0458 < .05$). There is sufficient evidence to indicate a difference in the mean ratios among the 5 insecticides.

Since differences were found, we need to find where the differences are. Using Tukey's test at the bottom of the output, there are no differences among the 5 means because all 5 means are connected. This contradicts the analysis of variance test. This contradiction can occur occasionally when the p-value of the analysis of variance is close to α.

10.53 a. There are two factors.

      b. No, we cannot tell whether the factors are qualitative or quantitative.

      c. Yes. There are 3 levels of factor A and 5 levels of factor B.

      d. A treatment would consist of a combination of one level of factor A and one level of factor B. There are a total of $3 \times 5 = 15$ treatments.

      e. One problem with only one replicate is there are no degrees of freedom for error. This is overcome by having at least two replicates.

10.55 The ANOVA table is:

Source	df	SS	MS	F
A	2	.8	24.4000	2.00
B	3	5.3	1.7667	8.83
AB	6	9.6	1.6000	8.00
Error	12	2.4	.2000	
Total	23	18.1		

df for A is $a - 1 = 3 - 1 = 2$
df for B is $b - 1 = 4 - 1 = 3$
df for AB is $(a - 1)(b - 1) = 2(3) = 6$
df for Error is $n - ab = 24 - 3(4) = 12$
df for Total is $n - 1 = 24 - 1 = 23$

$$\text{SSE} = \text{SS(Total)} - \text{SS}A - \text{SS}B - \text{SS}AB = 18.1 - .8 - 5.3 - 9.6 = 2.4$$

$$\text{MS}A = \frac{\text{SS}A}{a - 1} = \frac{.8}{3 - 1} = .40 \qquad \text{MS}B = \frac{\text{SS}B}{b - 1} = \frac{5.3}{4 - 1} = 1.7667$$

$$\text{MS}AB = \frac{\text{SS}AB}{(a - 1)(b - 1)} = \frac{9.6}{(3 - 1)(4 - 1)} = 1.60$$

$$\text{MSE} = \frac{\text{MSE}}{n - ab} = \frac{2.4}{24 - 3(4)} = .2000$$

$$F_A = \frac{\text{MS}A}{\text{MSE}} = \frac{.4000}{.2000} = 2.00 \qquad F_B = \frac{\text{MS}B}{\text{MSE}} = \frac{1.7667}{.2000} = 8.83$$

$$F_{AB} = \frac{\text{MS}AB}{\text{MSE}} = \frac{1.6000}{.2000} = 8.00$$

b.   Sum of Squares for Treatment = $SSA + SSB + SSAB = .8 = 5.3 + 2.6 = 15.7$.

$$MST = \frac{SST}{ab-1} = \frac{15.7}{3(4)-1} = 1.4273$$

$$F_T = \frac{MST}{MSE} = \frac{1.4273}{.2000} = 7.14$$

To determine if the treatment means differ, we test:

H_0: $\mu_1 = \mu_2 = \cdots = \mu_{12}$
H_a: At least two treatment means differ

The test statistic is $F = 7.14$.

The rejection region requires $\alpha = .05$ in the upper tail of the F distribution with $v_1 = ab - 1 = 3(4) - 1 = 11$ and $v_2 = n - ab = 24 - 3(4) = 12$. From Table IX, Appendix A, $F_{.05} \approx 2.75$. The rejection region is $F > 2.75$.

Since the observed value of the test statistic falls in the rejection region ($F = 7.14 > 2.75$), H_0 is rejected. There is sufficient evidence to indicate the treatment means differ at $\alpha = .05$.

c.   Yes. We need to partition the Treatment Sum of Squares into the Main Effects and Interaction Sum of Squares. Then we test whether factors A and B interact. Depending on the conclusion of the test for interaction, we either test for main effects or compare the treatment means.

d.   Two factors are said to interact if the effects of one factor on the dependent variable are not the same at different levels of the second factor. If the factors interact, then tests for main effects are not necessary. We need to compare the treatment means for one factor at each level of the second.

e.   To determine if the factors interact, we test:

H_0: Factors A and B do not interact to affect the response mean
H_a: Factors A and B do interact to affect the response mean

The test statistic is $F = \dfrac{MSAB}{MSE} = 8.00$

The rejection region requires $\alpha = .05$ in the upper tail of the F distribution with $v_1 = (a - 1)(b - 1) = (3 - 1)(4 - 1) = 6$ and $v_2 = n - ab = 24 - 3(4) = 12$. From Table IX, Appendix A, $F_{.05} = 3.00$. The rejection region is $F > 3.00$.

Since the observed value of the test statistic falls in the rejection region ($F = 8.00 > 3.00$), H_0 is rejected. There is sufficient evidence to indicate the two factors interact to affect the response mean at $\alpha = .05$.

f.   No. Testing for main effects is not warranted. Instead, we compare the treatment means of one factor at each level of the second factor.

10.57 a. $SSA = .2(1000) = 200$, $SSB = .1(1000) = 100$, $SSAB = .1(1000) = 100$

$SSE = SS(\text{Total}) - SSA - SSB - SSAB = 1000 - 200 - 100 - 100 = 600$

$SST = SSA + SSB + SSAB = 200 + 100 + 100 = 400$

$$MSA = \frac{SSA}{a-1} = \frac{200}{3-1} = 100 \qquad MSB = \frac{SSB}{b-1} = \frac{100}{3-1} = 50$$

$$MSAB = \frac{SSAB}{(a-1)(b-1)} = \frac{100}{(3-1)(3-1)} = 25$$

$$MSE = \frac{SSE}{n-ab} = \frac{600}{27-3(3)} = 33.333 \qquad MST = \frac{SST}{ab-1} = \frac{400}{3(3)-1} = 50$$

$$F_A = \frac{MSA}{MSE} = \frac{100}{33.333} = 3.00 \qquad F_B = \frac{MSB}{MSE} = \frac{50}{33.333} = 1.50$$

$$F_{AB} = \frac{MSAB}{MSE} = \frac{25}{33.333} = .75 \qquad F_T = \frac{MST}{MSE} = \frac{50}{33.333} = 1.50$$

Source	df	SS	MS	F
A	2	200	100	3.00
B	2	100	50	1.50
AB	4	100	25	.75
Error	18	600	33.333	
Total	26	1000		

To determine whether the treatment means differ, we test:

H_0: $\mu_1 = \mu_2 = \cdots = \mu_9$
H_a: At least two treatment means differ

The test statistic is $F = \dfrac{MST}{MSE} = 1.50$

Suppose $\alpha = .05$. The rejection region requires $\alpha = .05$ in the upper tail of the F distribution with $\nu_1 = ab - 1 = 3(3) - 1 = 8$ and $\nu_2 = n - ab = 27 - 3(3) = 18$. From Table IX, Appendix A, $F_{.05} = 2.51$. The rejection region is $F > 2.51$.

Since the observed value of the test statistic does not fall in the rejection region ($F = 1.50 \not> 2.51$), H_0 is not rejected. There is insufficient evidence to indicate the treatment means differ at $\alpha = .05$. Since there are no treatment mean differences, we have nothing more to do.

b. $SSA = .1(1000) = 100$, $SSB = .1(1000) = 100$, $SSAB = .5(1000) = 500$

$SSE = SS(Total) - SSA - SSB - SSAB = 1000 - 100 - 100 - 500 = 300$

$SST = SSA + SSB + SSAB = 100 + 100 + 500 = 700$

$$MSA = \frac{SSA}{a-1} = \frac{100}{3-1} = 50 \qquad\qquad MSB = \frac{SSB}{b-1} = \frac{100}{3-1} = 50$$

$$MSAB = \frac{SSAB}{(a-1)(b-1)} = \frac{500}{(3-1)(3-1)} = 125$$

$$MSE = \frac{SSE}{n-ab} = \frac{300}{27-3(3)} = 16.667 \qquad MST = \frac{SST}{ab-1} = \frac{700}{9-1} = 87.5$$

$$F_A = \frac{MSA}{MSE} = \frac{50}{16.667} = 3.00 \qquad\qquad F_B = \frac{MSB}{MSE} = \frac{50}{16.667} = 3.00$$

$$F_{AB} = \frac{MSAB}{MSE} = \frac{125}{16.667} = 7.50 \qquad\qquad F_T = \frac{MST}{MSE} = \frac{87.5}{16.667} = 5.25$$

Source	df	SS	MS	F
A	2	100	50	3.00
B	2	100	50	3.00
AB	4	500	125	7.50
Error	18	300	16.667	
Total	26	1000		

To determine if the treatment means differ, we test:

H_0: $\mu_1 = \mu_2 = \cdots = \mu_9$
H_a: At least two treatment means differ

The test statistic is $F = \dfrac{MST}{MSE} = 5.25$

The rejection region requires $\alpha = .05$ in the upper tail of the F distribution with $v_1 = ab - 1 = 3(3) - 1 = 8$ and $v_2 = n - ab = 27 - 3(3) = 18$. From Table IX, Appendix A, $F_{.05} = 2.51$. The rejection region is $F > 2.51$.

Since the observed value of the test statistic falls in the rejection region ($F = 5.25 > 2.51$), H_0 is rejected. There is sufficient evidence to indicate the treatment means differ at $\alpha = .05$.

Since the treatment means differ, we next test for interaction between factors A and B. To determine if factors A and B interact, we test:

H_0: Factors A and B do not interact to affect the mean response
H_a: Factors A and B do interact to affect the mean response

The test statistic is $F = \dfrac{MSAB}{MSE} = 7.50$

The rejection region requires $\alpha = .05$ in the upper tail of the F distribution with $v_1 = (a - 1)(b - 1) = (3 - 1)(3 - 1) = 4$ and $v_2 = n - ab = 27 - 3(3) = 18$. From Table IX, Appendix A, $F_{.05} = 2.93$. The rejection region is $F > 2.93$.

Since the observed value of the test statistic falls in the rejection region ($F = 7.50 > 2.93$), H_0 is rejected. There is sufficient evidence to indicate the factors A and B interact at $\alpha = .05$. Since interaction is present, no tests for main effects are necessary.

c. $SSA = .4(1000) = 400$, $SSB = .1(1000) = 100$, $SSAB = .2(1000) = 200$

$SSE = SS(\text{Total}) - SSA - SSB - SSAB = 1000 - 400 - 100 - 200 = 300$

$SST = SSA + SSB + SSAB = 400 + 100 + 200 = 700$

$$MSA = \frac{SSA}{a-1} = \frac{400}{3-1} = 50 \qquad\qquad MSB = \frac{SSB}{b-1} = \frac{100}{3-1} = 50$$

$$MSAB = \frac{SSAB}{(a-1)(b-1)} = \frac{200}{(3-1)(3-1)} = 50$$

$$MSE = \frac{SSE}{n-ab} = \frac{300}{27-3(3)} = 16.667 \qquad MST = \frac{SST}{ab-1} = \frac{700}{3(3)-1} = 87.5$$

$$F_A = \frac{MSA}{MSE} = \frac{200}{16.667} = 12.00 \qquad\qquad F_B = \frac{MSB}{MSE} = \frac{50}{16.667}\ 3.00$$

$$F_{AB} = \frac{MSAB}{MSE} = \frac{50}{16.667} = 3.00 \qquad\qquad F_T = \frac{MST}{MSE} = \frac{87.5}{16.667} = 5.25$$

Source	df	SS	MS	F
A	2	400	200	12.00
B	2	100	50	3.00
AB	4	200	50	3.00
Error	18	300	16.667	
Total	26	1000		

To determine if the treatment means differ, we test:

H_0: $\mu_1 = \mu_2 = \cdots = \mu_9$
H_a: At least two treatment means differ

The test statistic is $F = \dfrac{MST}{MSE} = 5.25$

The rejection region requires $\alpha = .05$ in the upper tail of the F distribution with $v_1 = ab - 1 = 3(3) - 1 = 8$ and $v_2 = n - ab = 27 - 3(3) = 18$. From Table IX, Appendix A, $F_{.05} = 2.51$. The rejection region is $F > 2.51$.

Since the observed value of the test statistic falls in the rejection region ($F = 5.25 > 2.51$), H_0 is rejected. There is sufficient evidence to indicate the treatment means differ at $\alpha = .05$.

Since the treatment means differ, we next test for interaction between factors A and B. To determine if factors A and B interact, we test:

H_0: Factors A and B do not interact to affect the mean response
H_a: Factors A and B do interact to affect the mean response

The test statistic is $F = \dfrac{MSAB}{MSE} = 3.00$

The rejection region requires $\alpha = .05$ in the upper tail of the F distribution with $v_1 = (a - 1)(b - 1) = (3 - 1)(3 - 1) = 4$ and $v_2 = n - ab = 27 - 3(3) = 18$. From Table IX, Appendix A, $F_{.05} = 2.93$. The rejection region is $F > 2.93$.

Since the observed value of the test statistic falls in the rejection region ($F = 3.00 > 2.93$), H_0 is rejected. There is sufficient evidence to indicate the factors A and B interact at $\alpha = .05$. Since interaction is present, no tests for main effects are necessary.

d. $SSA = .4(1000) = 400$, $SSB = .4(1000) = 400$, $SSAB = .1(1000) = 100$

$SSE = SS(Total) - SSA - SSB - SSAB = 1000 - 400 - 400 - 100 = 100$

$SST = SSA + SSB + SSAB = 400 + 400 + 100 = 900$

$MSA = \dfrac{SSA}{a-1} = \dfrac{400}{3-1} = 200 \qquad\qquad MSB = \dfrac{SSB}{b-1} = \dfrac{400}{3-1} = 200$

$MSAB = \dfrac{SSAB}{(a-1)(b-1)} = \dfrac{100}{(3-1)(3-1)} = 25$

$MSE = \dfrac{SSE}{n-ab} = \dfrac{100}{27-3(3)} = 5.556 \qquad MST = \dfrac{SST}{ab-1} = \dfrac{900}{3(3)-1} = 112.5$

$F_A = \dfrac{MSA}{MSE} = \dfrac{200}{5.556} = 36.00 \qquad\qquad F_B = \dfrac{MSB}{MSE} = \dfrac{200}{5.556} = 36.00$

$F_{AB} = \dfrac{MSAB}{MSE} = \dfrac{25}{5.556} = 4.50 \qquad\qquad F_T = \dfrac{MST}{MSE} = \dfrac{112.5}{5.556} = 20.25$

Source	df	SS	MS	F
A	2	400	200	36.00
B	2	400	200	36.00
AB	4	100	25	4.50
Error	18	100	5.556	
Total	26	1000		

To determine if the treatment means differ, we test:

H_0: $\mu_1 = \mu_2 = \cdots = \mu_9$
H_a: At least two treatment means differ

The test statistic is $F = \dfrac{MST}{MSE} = 20.25$

The rejection region requires $\alpha = .05$ in the upper tail of the F distribution with

$v_1 = ab - 1 = 3(3) - 1 = 8$ and $v_2 = n - ab = 27 - 3(3) = 18$. From Table IX, Appendix A, $F_{.05} = 2.51$. The rejection region is $F > 2.51$.

Since the observed value of the test statistic falls in the rejection region ($F = 20.25 > 2.51$), H_0 is rejected. There is sufficient evidence to indicate the treatment means differ at $\alpha = .05$.

Since the treatment means differ, we next test for interaction between factors A and B. To determine if factors A and B interact, we test:

H_0: Factors A and B do not interact to affect the mean response
H_a: Factors A and B do interact to affect the mean response

The test statistic is $F = \dfrac{MSAB}{MSE} = 4.50$

The rejection region requires $\alpha = .05$ in the upper tail of the F distribution with $v_1 = (a - 1)(b - 1) = (3 - 1)(3 - 1) = 4$ and $v_2 = n - ab = 27 - 3(3) = 18$. From Table IX, Appendix A, $F_{.05} = 2.93$. The rejection region is $F > 2.93$.

Since the observed value of the test statistic falls in the rejection region ($F = 4.50 > 2.93$), H_0 is rejected. There is sufficient evidence to indicate the factors A and B interact at $\alpha = .05$. Since interaction is present, no tests for main effects are necessary.

10.59   a.   This is a 6×6 factorial design.

b.   There are 2 factors – level of coagulant and acidity level. The factor "level of coagulation" has 6 levels – 5, 10, 20, 50, 100, and 200 milligrams per liter. The factor "acidity level" has 6 levels – 4.0, 5.0, 6.0, 7.0, 8.0, and 9.0. There are $6 \times 6 = 36$ treatments for this study. Each treatment is a combination of one level of coagulation level and acidity level. An example of a treatment is coagulation level 5 and acidity level 4.0.

10.61   a.   A 2×2 factorial design was used for this study. The two factors are color and type of questions, each at 2 levels. There are $2 \times 2 = 4$ treatments. The 4 treatments are "blue, difficult", "blue, simple", "red, difficult", and "red, simple".

b.   Since the p-value is small (p-value $< .03$), H_0 is rejected. There is sufficient evidence to indicate that color and type of questions interact to affect mean exam scores. This means that the effect of color on the mean exam scores depends on the difficulty of the questions.

c. Using MINITAB, the graph is:

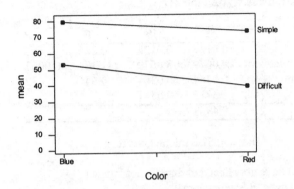

Looking at the graph, the lines are not parallel. The difference between the Blue and Red "simple" questions is not as great as the difference between the Blue and Red "difficult" questions. The effect of color on the exam scores depends on the level of difficulty.

10.63 a.

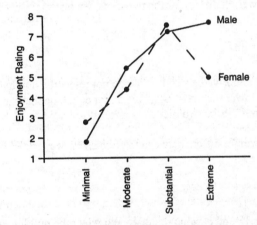

The pattern in the graph suggests interaction between suspense and gender. The difference between the mean ratings of the two genders varies depending on the suspense category. Notice that the lines cross twice.

b. To determine if interaction exists between suspense and gender, we test:

H_0: Gender and Suspense do not interact
H_a: Gender and Suspense interact

The test statistic is $F = 4.42$ and p-value $= .007$. We reject the null hypothesis. For any level of significance greater than .007, we can conclude that Gender and Suspense interact to affect the mean rating.

c.   Because the factors interact, there is evidence to indicate the difference between the mean enjoyment levels of males and females are different for different suspense levels.

10.65   Using MINITAB, the ANOVA table is:
    a.

Source	df	SS	MS	F	P
Diet	1	0.0124	0.0124	0.22	0.645
Size	1	8.0679	8.0679	141.18	<0.0001
Interaction	1	0.0364	0.0364	0.64	0.432
Error	24	1.3715	0.0571		
Total	27	9.4883			

    b.   To determine if Diet and Size interact, we test:

H_0: The factors Diet and Size do not interact
H_a: Diet and Size interact

The test statistic is $F = .64$ and p-value $= .432$. We do not reject the null hypothesis. For any level of significance lower than $\alpha = .432$, we could not conclude that the factors interact. We therefore test the main effects.

To determine if the mean weight differs for the two diets, we test:

H_0: $\mu_{regular} = \mu_{supplement}$
H_a: $\mu_{regular} \neq \mu_{supplement}$

The test statistic is $F = .22$ and p-value $= .645$. We do not reject the null hypothesis. For any significance level lower than $\alpha = .645$, we cannot conclude that the mean weight of the kidney differed for the two diets.

To determine if the mean weight differs between the two sizes, we test:

H_0: $\mu_{lean} = \mu_{obese}$
H_a: $\mu_{lean} \neq \mu_{obese}$

The test statistic is $F = 141.18$ and p-value $< .0001$. We reject the null hypothesis. For any significance level greater than or equal to $\alpha = .0001$, we conclude that the mean weight of the kidney differs for the two sizes.

10.67   Using SAS, the ANOVA results are:

```
                           The ANOVA Procedure

Dependent Variable: RATING

                                    Sum of
Source                  DF          Squares     Mean Square    F Value    Pr > F

Model                    5       84.7045455     16.9409091       4.46     0.0009

Error                  126      478.9545455      3.8012266

Corrected Total        131      563.6590909


              R-Square      Coeff Var       Root MSE      RATING Mean

              0.150276      41.84665        1.949673        4.659091


Source                  DF       Anova SS     Mean Square    F Value    Pr > F

PREP                     1     54.73484848    54.73484848      14.40    0.0002
STANDING                 2     16.50000000     8.25000000       2.17    0.1184
PREP*STANDING            2     13.46969697     6.73484848       1.77    0.1742
```

First, we test for the presence of interaction between preparation and standing. To determine if preparation and class standing interact to affect rating, we test:

H_0: Preparation and standing do not interact
H_a: Preparation and standing do interact

From the printout, the test statistic is $F = 1.77$. The p-value is $p = .1742$. Since the p-value is not small, H_0 is not rejected. There is insufficient evidence to indicate preparation and standing interact to affect rating.

Since there was no evidence of interaction between preparation and standing, we test for the main effects. To determine if there are differences in the mean ratings between the two types of preparation, we test:

H_0: Mean ratings for the two levels of preparation are the same
H_a: Mean ratings for the two levels of preparation are not the same

From the printout, the test statistic is $F = 14.40$. The p-value is $p = .0002$. Since the p-value is very small, H_0 is rejected. There is sufficient evidence to indicate the mean ratings are different for those taking the practice exam and those taking the review session.

To determine if there are differences in the mean ratings among the three class standings, we test:

H_0: Mean ratings for the three levels of class standing are the same
H_a: Mean ratings for the three levels of class standing are not the same

From the printout, the test statistic is $F = 2.17$. The p-value is $p = .1184$. Since the p-value is not small, H_0 is not rejected. There is insufficient evidence to indicate the mean ratings are different among the three levels of class standing.

Since there was a difference between the two levels of preparation, we need to find which level leads to a higher rating. Since there are only two levels, we simply need to find the means for both levels. The mean rating for the practice exam is 5.3030 and the mean rating for the review session is 4.0152. Thus, those taking the practice exam rated the preparation higher than those who attended the review session.

10.69   In a completely randomized design, independent random selection of treatments to be assigned to experimental units is required; in a randomized block design, sets (blocks) of matched experimental units are employed, with an experimental unit in each block randomly assigned a treatment.

10.71   When the overall level of significance of a multiple comparisons procedure is α, the level of significance for each comparison is less than α.

10.73   a. $\text{SS(Treatment)} = \text{SS(Total)} - \text{SS(Block)} - \text{SSE} = 22.308 - 10.688 - .288 = 11.332$

$$\text{MS(Treatment)} = \frac{\text{SS(Treatment)}}{p-1} = \frac{11.332}{4-1} = 3.777, \text{df} = p - 1 = 3$$

$$\text{MS(Block)} = \frac{\text{SS(Block)}}{b-1} = \frac{10.688}{5-1} = 2.672, \text{df} = b - 1 = 4$$

$$\text{MSE} = \frac{\text{SSE}}{n-p-b+1} = \frac{.288}{20-4-5+1} = .024, \text{df} = n - p - b + 1 = 12$$

$$\text{Treatment } F = \frac{\text{MS(Treatment)}}{\text{MSE}} = \frac{3.777}{.024} = 157.375$$

$$\text{Block } F = \frac{\text{MS(Block)}}{\text{MSE}} = \frac{2.672}{.024} = 111.33$$

The ANOVA Table is:

Source	df	SS	MS	F
Treatment	3	11.332	3.777	157.38
Block	4	10.688	2.672	111.33
Error	12	.288	.024	
Total	19	22.308		

b.   To determine if there is a difference among the treatment means, we test:

H_0: $\mu_A = \mu_B = \mu_C = \mu_D$
H_a: At least two treatment means differ

The test statistic is $F = \dfrac{\text{MS(Treatment)}}{\text{MSE}} = 157.38$

The rejection region requires $\alpha = .05$ in the upper tail of the F distribution with $v_1 = p - 1 = 4 - 1 = 3$ and $v_2 = n - 1 - b + 1 = 20 - 4 - 5 + 1 = 12$. From Table IX, Appendix A, $F_{.05} = 3.49$. The rejection region is $F > 3.49$.

Since the observed value of the test statistic falls in the rejection region ($F = 157.38$ > 3.49), H_0 is rejected. There is sufficient evidence to indicate a difference among the treatment means at $\alpha = .05$.

c. Since there is evidence of differences among the treatment means, we need to compare the treatment means. The number of pairwise comparisons is $\dfrac{p(p-1)}{2} = \dfrac{4(4-1)}{2} = 6.$

d. To determine if there are difference among the block means, we test:

> H_0: All the block means are the same
> H_a: At least two block means differ

The test statistic is $F = \dfrac{MS(Block)}{MSE} = 111.33$

The rejection region requires $\alpha = .05$ in the upper tail of the F distribution with $v_1 = b - 1 = 5 - 1 = 4$ and $v_2 = n - p - b + 1 = 20 - 4 - 5 + 1 = 12$. From Table IX, Appendix A, $F_{.05} = 3.26$. The rejection region is $F > 3.26$.

Since the observed value of the test statistic falls in the rejection region ($F = 111.33$ > 3.26), H_0 is rejected. There is sufficient evidence that the block means differ at $\alpha = .05$.

10.75 a. The data are collected as a completely randomized design because five boxes of each size were randomly selected and tested.

b. Yes. The confidence intervals surrounding each of the means do not overlap. This would indicate that there is a difference in the means for the two sizes.

c. No. Several of the confidence intervals overlap. This would indicate that the mean compression strengths of the sizes that have intervals that overlap are not significantly different.

10.77 To determine if differences in the mean percent of pronoun errors exist among the three groups, we test:

> H_0: $\mu_1 = \mu_2 = \mu_3$
> H_a: At least two treatment means differ

where μ_i represents the mean percent of pronoun errors for the ith group.

The test statistic is $F = 8.295$.

Since the p-value is so small ($p = .0016$), H_0 is rejected for any $\alpha > .0016$. There is sufficient evidence to indicate a difference in the mean percent of pronoun errors among the three groups.

A multiple comparison was run on the three group means. A line connects the means of the SLI and the YND groups. The mean percent of pronoun errors is not different for these two groups. The mean percent of pronoun errors for the OND group is significantly smaller than the mean percent of pronoun errors for both the SLI and the YND groups.

10.79 a. To determine whether the mean scores of the four groups differ, we test:

H_0: $\mu_1 = \mu_2 = \mu_3 = \mu_4$
H_a: At least two treatment means differ

where μ_i represents the mean of the ith group.

For the variable "Infrequency":

The test statistic is $F = 155.8$.

The rejection region requires $\alpha = .05$ in the upper tail of the F distribution with $v_1 = p - 1 = 4 - 1 = 3$ and $v_2 = n - p = 278 - 4 = 274$. Using Table IX, Appendix A, $F_{.05} - 2.60$. The rejection region is $F > 2.60$.

Since the observed value of the test statistic falls in the rejection region ($F = 155.5 > 2.60$), H_0 is rejected. There is sufficient evidence to indicate the mean scores on the "Infrequency" variable differ among the four groups at $\alpha = .05$.

For the variable "Obvious":

The test statistic is $F = 49.7$.

The rejection region is $F > 2.60$. (See above.)

Since the observed value of the test statistic falls in the rejection region ($F = 49.7 > 2.60$), H_0 is rejected. There is sufficient evidence to indicate the mean scores on the "Obvious" variable differ among the four groups at $\alpha = .05$.

For the variable "Subtle":

The test statistic is $F = 10.3$.

The rejection region is $F > 2.60$. (See above.)

Since the observed value of the test statistic falls in the rejection region ($F = 10.3 > 2.60$), H_0 is rejected. There is sufficient evidence to indicate the mean scores on the "Subtle" variable differ among the four groups at $\alpha = .05$.

For the variable "Obvious-Subtle":

The test statistic is $F = 45.4$.

The rejection region is $F > 2.60$. (See above.)

Since the observed value of the test statistic falls in the rejection region ($F = 45.4 > 2.60$), H_0 is rejected. There is sufficient evidence to indicate the mean scores on the "Obvious-Subtle" variable differ among the four groups at $\alpha = .05$.

For the variable "Dissimulation":

The test statistic is $F = 39.1$.

The rejection region is $F > 2.60$. (See above.)

Since the observed value of the test statistic falls in the rejection region ($F = 39.1 > 2.60$), H_0 is rejected. There is sufficient evidence to indicate the mean scores on the "Dissimulation" variable differ among the four groups at $\alpha = .05$.

b. No. No information is provided on the sample means. The test of hypotheses performed in part **a** just indicate differences exist, but do not indicate where. Further analysis would be required.

c. For the variable "Infrequency," the mean score for the CSFB group is significantly greater than the mean scores of all the other groups. There is no significant difference in the mean scores for the FP and NFP groups, but both have means significantly greater than the CSH group. Since the mean "Infrequency" score for the CSFB group was significantly greater than all the other groups, the MMPI is effective in detecting distorted responses.

For the variable "Obvious," the mean score for the CSFB group is significantly greater than the mean scores of all the other groups. There is no significant difference in the mean scores for the FP and NFP groups, but both have means significantly greater than the CSH group. Since the mean "Obvious" score for the CSFB group was significantly greater than all the other groups, the MMPI is effective in detecting distorted responses.

For the variable "Subtle," there is no significant differences among any of the groups. Since the mean "Subtle" score for the CSFB group is not significantly greater than all the other groups, the MMPI is not effective in detecting distorted responses.

For the variable "Obvious-subtle," the mean score for the CSFB group is significantly greater than the mean scores of all the other groups. There is no significant difference in the mean scores for the FP, NFP, and CSH groups. Since the mean "Obvious-subtle" score for the CSFB group was significantly greater than all the other groups, the MMPI is effective in detecting distorted responses.

For the variable "Dissimulation," the mean score for the CSFB group is significantly greater than the mean scores of all the other groups. There is no significant difference in the mean scores for the FP, NFP, and CSH groups. Since the mean "Dissimulation" score for the CSFB group was significantly greater than all the other groups, the MMPI is effective in detecting distorted responses.

10.81 a.

Source	df	F-value	p-value
Time Period	3	11.25	.0001
Station	9		
Error	27		
Total	39		

b. To determine if the mean biomass densities differ among the three time periods, we test:

H_0: $\mu_1 = \mu_2 = \mu_3 = \mu_4$
H_a: The mean biomass density differ for at least two of the time periods

The test statistic is $F = 11.25$ and $p = .0001$. We can reject the null hypothesis at the $\alpha > .0001$ level of significance. At least two of the time periods differ with respect to the mean biomass of water boatmen nymphs.

c. The time period with the largest mean biomass is either 7/9-7/23 or 7/24-8/8. There is no significant difference between the two means. The time period with the smallest mean biomass is 8/24-8/31.

10.83 a.
Experimental units: zircaloy components
Blocks: ingots
Response: pressure required to separate the bonded components
Factor: bonding agent or element
Factor type: qualitative
Treatments: nickel, iron, copper

b. This is a designed experiment. The elements and their assignment to the ingots is under the control of the people running the experiment.

c. Using SAS, the output is:

The ANOVA Procedure

Dependent Variable: PRESSURE

Source	DF	Sum of Squares	Mean Square	F Value	Pr > F
Model	8	400.1904762	50.0238095	4.82	0.0076
Error	12	124.4590476	10.3715873		
Corrected Total	20	524.6495238			

R-Square	Coeff Var	Root MSE	PRESSURE Mean
0.762777	4.448490	3.220495	72.39524

Source	DF	Anova SS	Mean Square	F Value	Pr > F
BONDING	2	131.9009524	65.9504762	6.36	0.0131
INGOT	6	268.2895238	44.7149206	4.31	0.0151

```
                         The ANOVA Procedure

                 Tukey's Studentized Range (HSD) Test for PRESSURE

        NOTE: This test controls the Type I experimentwise error rate, but it
              generally has a higher Type II error rate than REGWQ.


                    Alpha                                      0.05
                    Error Degrees of Freedom                    12
                    Error Mean Square                      10.37159
                    Critical Value of Studentized Range    3.77278
                    Minimum Significant Difference          4.5923


              Means with the same letter are not significantly different.


              Tukey Grouping          Mean        N    BONDING

                      A             75.900        7    IRON

                      B             71.100        7    NICKEL
                      B
                      B             70.186        7    COPPER
```

d. Let μ_1, μ_2, and μ_3 represent the mean bonding strength for nickel, iron and copper, respectively.

H_0: $\mu_1 = \mu_2 = \mu_3$
H_a: At least two means differ

The output gives the *p*-value of the analysis of variance F test statistic (which has 2 numerator and 12 denominator degrees of freedom), as 0.0131. Since the observed level of significance is less than α, (0.0131 < 0.05), H_0 is rejected. There is sufficient evidence to indicate that at least one of the elements has a mean bonding strength which is different from the others at $\alpha = .05$.

e. From the printout, the mean pressure for Iron is significantly greater than the mean pressure for the other two agents. No other differences exist.

f. Assumptions:
    1. The probability distributions of the bonding strength for all ingot-element combinations are normal.
    2. The variances for all the probability distributions are equal.
    3. The samples are selected randomly and independently within blocks.

10.85 a. Using SAS, the ANOVA table is:

The ANOVA Procedure

Dependent Variable: SALES

Source	DF	Sum of Squares	Mean Square	F Value	Pr > F
Model	8	5291151.185	661393.898	1336.85	<.0001
Error	18	8905.333	494.741		
Corrected Total	26	5300056.519			

R-Square	Coeff Var	Root MSE	SALES Mean
0.998320	1.434469	22.24277	1550.593

Source	DF	Anova SS	Mean Square	F Value	Pr > F
DISPLAY	2	1691392.519	845696.259	1709.37	<.0001
PRICE	2	3089053.852	1544526.926	3121.89	<.0001
DISPLAY*PRICE	4	510704.815	127676.204	258.07	<.0001

b. To determine whether the treatment means differ, we test:

H_0: $\mu_1 = \mu_2 = \cdots = \mu_9$
H_a: At least two treatment means differ

The test statistic is $F = \dfrac{\text{MST}}{\text{MSE}} = 1336.85$

The rejection region requires $\alpha = .10$ in the upper tail of the F distribution with $v_1 = ab - 1 = 3(3) - 1 = 8$ and $v_2 = n - ab = 27 - 3(3) = 18$. From Table VIII, Appendix A, $F_{.10} = 2.04$. The rejection region is $F > 2.04$.

Since the observed value of the test statistic falls in the rejection region ($F = 1336.85 > 2.04$), H_0 is rejected. There is sufficient evidence to indicate the treatment means differ at $\alpha = .10$.

c. Since there are differences among the treatment means, we next test for the presence of interaction.

H_0: Factors A and B do not interact to affect the response means
H_a: Factors A and B do interact to affect the response means

The test statistic is $F = \dfrac{\text{MS}AB}{\text{MSE}} = 258.07$

The rejection region requires $\alpha = .10$ in the upper tail of the F distribution with $v_1 = (a - 1)(b - 1) = (3 - 1)(3 - 1) = 4$ and $v_2 = n - ab = 17 - 3(3) = 18$. From Table VIII, Appendix A, $F_{.10} = 2.29$. The rejection region is $F > 2.29$.

Since the observed value of the test statistic falls in the rejection region ($F = 258.07 > 2.29$), H_0 is rejected. There is sufficient evidence to indicate the two factors interact at $\alpha = .10$.

d. The main effect tests are not warranted since interaction is present in part **c**.

e. Since the interaction is significant, the nine treatment means need to be compared. To reduce the number of comparisons, we will compare the means of the prices at each level of the displays.

Using SAS, the Tukey's multiple comparisons are:

```
------------------------------ DISPLAY=NORMAL ------------------------------

                          The ANOVA Procedure

                  Tukey's Studentized Range (HSD) Test for SALES

NOTE: This test controls the Type I experimentwise error rate, but it
      generally has a higher Type II error rate than REGWQ.


              Alpha                                    0.05
              Error Degrees of Freedom                    6
              Error Mean Square                     401.8889
              Critical Value of Studentized Range   4.33902
              Minimum Significant Difference          50.221


      Means with the same letter are not significantly different.


      Tukey Grouping         Mean      N    PRICE

                    A      1578.00      3    COST

                    B      1202.67      3    REDUCED

                    C      1014.67      3    REGULAR

------------------------------ DISPLAY=PLUS ------------------------------

                          The ANOVA Procedure

                  Tukey's Studentized Range (HSD) Test for SALES

NOTE: This test controls the Type I experimentwise error rate, but it
      generally has a higher Type II error rate than REGWQ.


              Alpha                                    0.05
              Error Degrees of Freedom                    6
              Error Mean Square                     653.4444
              Critical Value of Studentized Range   4.33902
              Minimum Significant Difference          64.038


      Means with the same letter are not significantly different.


      Tukey Grouping         Mean      N    PRICE

                    A      2510.00      3    COST

                    B      1898.67      3    REDUCED

                    C      1215.00      3    REGULAR
```

```
------------------------------ DISPLAY=TWICE ------------------------------

                         The ANOVA Procedure

               Tukey's Studentized Range (HSD) Test for SALES

       NOTE: This test controls the Type I experimentwise error rate, but it
               generally has a higher Type II error rate than REGWQ.


                  Alpha                                    0.05
                  Error Degrees of Freedom                    6
                  Error Mean Square                      428.8889
                  Critical Value of Studentized Range  4.33902
                  Minimum Significant Difference          51.88


           Means with the same letter are not significantly different.


            Tukey Grouping         Mean     N    PRICE

                         A        1828.67    3    COST

                         B        1505.00    3    REDUCED

                         C        1202.67    3    REGULAR
```

For Display = Normal, there are significant differences among mean sales for all prices. The mean sales for those priced at "cost to supermarket" is significantly greater than the mean sales for those priced at "reduced" price and "regular" price. The mean sales for those priced at "reduced" price is significantly greater than those priced at "regular" price.

For Display = Normal Plus, there are significant differences among mean sales for all prices. The mean sales for those priced at "cost to supermarket" is significantly greater than the mean sales for those priced at "reduced" price and "regular" price. The mean sales for those priced at "reduced" price is significantly greater than those priced at "regular" price.

For Display = Twice Normal, there are significant differences among mean sales for all prices. The mean sales for those priced at "cost to supermarket" is significantly greater than the mean sales for those priced at "reduced" price and "regular" price. The mean sales for those priced at "reduced" price is significantly greater than those priced at "regular" price.

10.87   Using statistical software, the ANOVA Table for this experiment is:

Source	df	SS	MS	F	P
Treatment	4	292.8	73.2	5.29	0.007
Error	15	207.7	13.9		
Total	19	500.6			

b.   To determine if a difference exists among the mean increases in weight for the five inoculins of growth hormone, we test:

H_0: $\mu_1 = \mu_2 = \mu_3 = \mu_4 = \mu_5$
H_a: At least two mean increases in weight differ

where μ_i is the mean increase in weight for the ith inoculin of growth hormone.

The test statistic is $F = \dfrac{\text{MST}}{\text{MSE}} = 5.29$

The rejection region requires $\alpha = .05$ in the upper tail of the F distribution with $v_1 = p - 1 = 5 - 1 = 4$ and $v_2 = n - p = 20 - 5 = 15$. From Table IX, Appendix A, $F_{.05} = 3.06$. The rejection region is $F > 3.06$.

Since the observed value of the test statistic falls in the rejection region ($F = 5.29 > 3.06$), H_0 is rejected. There is evidence of a difference among the mean increases in weight for the five inoculins of growth hormone at $\alpha = .05$.

c. P-value $= P(F \geq 5.29) = .007$

Since $.007 < .01$, there is evidence to reject H_0 for $\alpha > .007$.

d.   The assumptions necessary to assure the validity of the test are as follows and they seem to be satisfied:

1.   The probability distributions for each inoculin are normal.
2.   The variances of the probability distributions for each inoculin are equal.
3.   The samples of shrub cuttings for each inoculin are selected randomly and independently.

10.89 a.   To determine if Herd and Season interact, we test:

H_0: Herd and Season do not interact
H_a: Herd and Season interact

The test statistic is $F = 1.2$.

The p-value is $p > .05$. Since the p-value is greater than $\alpha = .05$, H_0 is not rejected. There is insufficient evidence to indicate that herd and season interact at $\alpha = .05$.

Since the two factors do not interact, the main effect tests are run.

To determine if the mean home range differs among the four herds, we test:

H_0: $\mu_1 = \mu_2 = \mu_3 = \mu_4$
H_a: At least two treatments means differ

where μ_i is the mean home range for herd i.

The test statistic is $F = 17.2$.

The p-value is $p < .001$. Since the p-value is less than $\alpha = .05$, H_0 is rejected. There is sufficient evidence to indicate that the mean home range differs among the four herds at $\alpha = .05$.

To determine if the mean home range differs among the four seasons, we test:

H_0: $\mu_1 = \mu_2 = \mu_3 = \mu_4$
H_a: At least two treatments means differ

where μ_i is the mean home range for season i.

The test statistic is $F = 3.0$.

The p-value is $p > .05$. Since the p-value is greater than $\alpha = .05$, H_0 is not rejected. There is insufficient evidence to indicate that the mean home range differs among the four seasons at $\alpha = .05$.

b.   Yes. Since herd and season do not interact, each main effect factor can be treated separately as if the second factor did not exist.

c.   The mean home range for herd MTZ is significantly greater than the mean home range for the herds PLC and LGN. The mean home range for herd QMD is significantly greater than the mean home range for the herds PLC and LGN. No other differences exist.

Simple Linear Regression

11.1

a.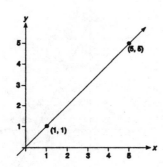

b.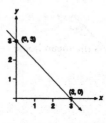

c.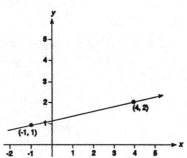

d.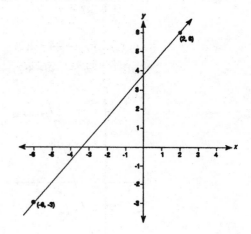

11.3   The two equations are:

$$4 = \beta_0 + \beta_1(-2) \text{ and } 6 = \beta_0 + \beta_1(4)$$

Subtracting the first equation from the second, we get

$$\begin{aligned} 6 &= \beta_0 + 4\beta_1 \\ -(4 &= \beta_0 - 2\beta_1) \end{aligned}$$

$$2 = \quad 6\beta_1 \Rightarrow \beta_1 = \frac{2}{6} = \frac{1}{3}$$

Substituting $\beta_1 = \frac{1}{3}$ into the first equation, we get:

$$4 = \beta_0 + \frac{1}{3}(-2) \Rightarrow \beta_0 = 4 + \frac{2}{3} = \frac{14}{3}$$

The equation for the line is $y = \frac{14}{3} + \frac{1}{3}x$

11.5　To graph a line, we need two points. Pick two values for x, and find the corresponding y values by substituting the values of x into the equation.

a.　Let $x = 0 \Rightarrow y = 4 + (0) = 4$
　　and $x = 2 \Rightarrow y = 4 + (2) = 6$

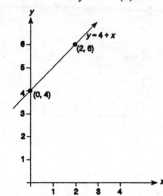

b.　Let $x = 0 \Rightarrow y = 5 \Rightarrow 2(0) = 5$
　　and $x = 2 \Rightarrow y = 5 \Rightarrow 2(2) = 1$

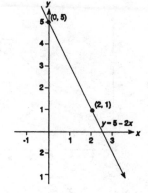

c.　Let $x = 0 \Rightarrow y = -4 + 3(0) = -4$

　　and $x = 2 \Rightarrow y = -4 + 3(2) = 2$

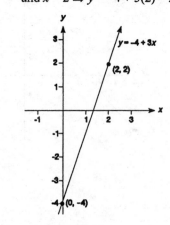

d.　Let $x = 0 \Rightarrow y = -2(0) = 0$
　　and $x = 2 \Rightarrow y = -2(2) = -4$

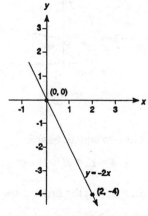

e.　Let $x = 0 \Rightarrow y = 0$
　　and $x = 2 \Rightarrow y = 2$

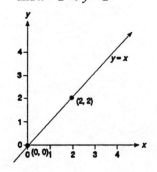

f.　Let $x = 0 \Rightarrow y = .5 + 1.5(0) = .5$
　　and $x = 2 \Rightarrow y = .5 + 1.5(2) = 3.5$

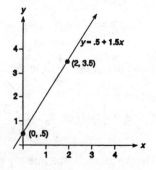

11.7 A deterministic model does not allow for random error or variation, whereas a probabilistic model does. An example where a deterministic model would be appropriate is:

Let y = cost of a 2 × 4 piece of lumber and
  x = length (in feet)

The model would be $y = \beta_1 x$. There should be no variation in price for the same length of wood.

An example where a probabilistic model would be appropriate is:

Let y = sales per month of a commodity and
  x = amount of money spent advertising

The model would be $y = \beta_0 + \beta_1 x + \varepsilon$. The sales per month will probably vary even if the amount of money spent on advertising remains the same.

11.9 No. The random error component, ε, allows the values of the variable to fall above or below the line.

11.11 From Exercise 11.10, $\hat{\beta}_0 = 7.10$ and $\hat{\beta}_1 = -.78$.

The fitted line is $= 7.10 - .78x$. To obtain values for , we substitute values of x into the equation and solve for .

a.

x	y	$\hat{y} = 7.10 - .78x$	$(y - \hat{y})$	$(y - \hat{y})^2$
7	2	1.64	.36	.1296
4	4	3.98	.02	.0004
6	2	2.42	−.42	.1764
2	5	5.54	−.54	.2916
1	7	6.32	.68	.4624
1	6	6.32	−.32	.1024
3	5	4.76	.24	.0576

$$\sum (y - \hat{y}) = 0.02 \qquad \text{SSE} = \sum (y - \hat{y})^2 = 1.2204$$

b.

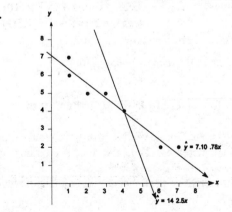

$\hat{y} = 7.10 .78x$

$\hat{y} = 14 2.5x$

c.

x	y	$\hat{y} = 14 - 2.5x$	$(y - \hat{y})$	$(y - \hat{y})^2$
7	2	−3.5	5.5	30.25
4	4	4	0	0
6	2	−1	3	9
2	5	9	−4	16
1	7	11.5	−4.5	20.25
1	6	11.5	−5.5	30.25
3	5	6.5	−1.5	2.25
			= −7	SSE = 108.00

11.13   **a.**

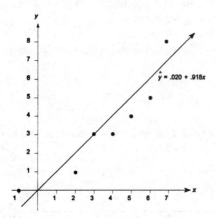

$\hat{y} = .020 + .918x$

b.   As x increases, y tends to increase. Thus, there appears to be a positive, linear relationship between y and x.

c. $\hat{\beta}_1 = \dfrac{SS_{xy}}{SS_{xx}} = \dfrac{39.8571}{43.4286} = .9177616 \approx .918$

$\hat{\beta}_0 = \bar{y} - \hat{\beta}_1\bar{x} = 3.4286 - .9177616(3.7143) = .0197581 \approx .020$

d.   The line appears to fit the data quite well.

e. $\hat{\beta}_0 = .020$ The estimated mean value of y when $x = 0$ is .020.

$\hat{\beta}_1 = .918$ The estimated change in the mean value of y for each unit change in x is .918.

These interpretations are valid only for values of x in the range from −1 to 7.

11.15   **a.**   The straight-line model is $y = \beta_0 + \beta_1 x + \varepsilon$. Based on the theory, we would expect the metal level to decrease as the distance increases. Thus, the slope should be negative.

b.   Yes. As the distance from the plant increases, the concentration of calcium tends to decrease.

c.   No. As the distance from the plant increases, the concentration of arsenic tends to increase.

11.17   a.   The equation for the straight-line model is $y = \beta_0 + \beta_1 x + \varepsilon$.

   b.   Some preliminary calculations are:

$$\bar{y} = \frac{\sum y}{n} = \frac{398}{9} = 44.2222$$

$$\bar{x} = \frac{\sum x}{n} = \frac{1,444}{9} = 160.4444$$

$$SS_{xy} = \sum xy - \frac{\sum x \sum y}{n} = 60,428 - \frac{1,444(398)}{9} = -3,428.88889$$

$$SS_{xx} = \sum x^2 - \frac{\left(\sum x\right)^2}{n} = 235,866 - \frac{1,444^2}{9} = 4,184.2222$$

$$\hat{\beta}_1 = \frac{SS_{xy}}{SS_{xx}} = \frac{-3,428.88889}{4,184.2222} = -.819480593$$

$$\hat{\beta}_o = \bar{y} - \hat{\beta}_1 \bar{x} = \frac{398}{9} - (-.819480593)\left(\frac{1,444}{9}\right) = 175.7033307$$

The fitted model is $\hat{y} = 175.7033 - .8195x$.

   c. $\hat{\beta}_0 = 175.7033$. Since $x = 0$ (age of fish) is not in the observed range, $\hat{\beta}_0$ has no practical meaning.

   d. $\hat{\beta}_1 = -.8195$. For each additional day of age, the mean number of strikes is estimated to decrease by .8195 strikes.

11.19   a.   Some preliminary calculations are:

$$\sum x_i = 6167 \qquad \sum y_i = 135.8 \qquad n = 24$$
$$\sum x_i^2 = 1,641,115 \qquad \sum x_i y_i = 34765$$

$$SS_{xy} = \sum x_i y_i - \frac{\left(\sum x_i\right)\left(\sum y_i\right)}{n}$$

$$= 34765 - \frac{(6167)(135.8)}{24} = -129.94167$$

$$SS_{xx} = \sum x_i^2 - \frac{\left(\sum x_i\right)^2}{n}$$

$$= 1,641,115 - \frac{(6167)^2}{24} = 56452.958$$

$$\hat{\beta}_1 = \frac{SS_{xy}}{SS_{xx}} = \frac{-129.94167}{56452.958} = -.002301769 \approx -.0023$$

$$\hat{\beta}_0 = \bar{y} - \hat{\beta}_1 \bar{x} = \frac{135.8}{24} - (-.002301769)\left(\frac{6167}{24}\right) = 6.249792065 \approx 6.25$$

The least squares line is $\hat{y} = 6.25 - .0023x$

b. $\hat{\beta}_0 = 6.25$. Since $x = 0$ is not in the observed range, $_0$ has no interpretation other than being the y-intercept.

   $\hat{\beta}_1 = -.0023$. For each additional increase of 1 part per million of pectin, the mean sweetness index is estimated to decrease by .0023.

c. $\hat{y} = 6.25 - .0023(300) = 5.56$

11.21 a. A scattergram of the data is:

   It appears that there is a positive linear relationship between total number of words with unfamiliar partner and with familiar partner.

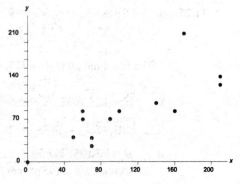

b. A straight line model would be $y = \beta_0 + \beta_1 x + \varepsilon$.

c. $\hat{\beta}_0 = 20.1275$ and $\hat{\beta}_1 = .62442$.

d. $\hat{\beta}_0 = 20.1275$. Since $x = 0$ is in the observed range, the mean number of words in conversation with an unfamiliar partner is estimated to be 20.1275 when the number of words in conversation with a familiar partner is 0.

   $\hat{\beta}_1 = .62442$. For each additional word in conversation with a familiar partner, the mean number of words in conversation with an unfamiliar partner is estimated to increase by .62442.

11.23 Some preliminary calculations are:

$$\bar{y} = \frac{\sum x}{n} = \frac{103.07}{144} = .71576$$

$$\bar{x} = \frac{\sum y}{n} = \frac{792}{144} = 5.5$$

$$SS_{xy} = \sum xy - \frac{\sum x \sum y}{n} = 586.86 - \frac{792(103.07)}{144} = 19.975$$

$$SS_{xx} = \sum x^2 - \frac{\left(\sum x\right)^2}{n} = 5{,}112 - \frac{792^2}{144} = 756$$

$$\hat{\beta}_1 = \frac{SS_{xy}}{SS_{xx}} = \frac{19.975}{756} = .026421957$$

$$\hat{\beta}_o = \bar{y} - \hat{\beta}_1\bar{x} = \frac{103.07}{144} - (.026421957)\left(\frac{792}{144}\right) = .570443121$$

The estimated regression line is $\hat{y} = .5704 + .0264x$.

11.25 a. $SSE = SS_{yy} - \hat{\beta}_1 SS_{xy} = 95 - .75(50) = 57.5$

         $s^2 = \frac{SSE}{n-2} = \frac{57.5}{20-2} = 3.19444$

     b. $SS_{yy} = \sum y^2 - \frac{\left(\sum y\right)^2}{n} = 860 - \frac{50^2}{40} = 797.5$

         $SSE = SS_{yy} - \hat{\beta}_1 SS_{xy} = 797.5 - .2(2700) = 257.5$

         $s^2 = \frac{SSE}{n-2} = \frac{257.5}{40-2} = 6.776315789 \approx 6.7763$

     c. $SS_{yy} = \sum(y_i - \bar{y})^2 = 58$ $\hat{\beta}_1 = \frac{SS_{xy}}{SS_{xx}} = \frac{91}{170} = .535294117$

         $SSE = SS_{yy} - \hat{\beta}_1 SS_{xy} = 58 - .535294117(91) = 9.2882353 \approx 9.288$

         $s^2 = \frac{SSE}{n-2} = \frac{9.2882353}{10-2} = 1.161029413 \approx 1.1610$

11.27   The graph in **b** would have the smallest s^2 because the width of the data points is the smallest.

11.29   From the solution to Exercise 11.17, $SS_{xy} = -3,428.88889$ and $\hat{\beta}_1 = -.76794828$.

    a.   Some preliminary calculations are:

$$SS_{yy} = \sum y^2 - \frac{\left(\sum y\right)^2}{n} = 22,078 - \frac{398^2}{9} = 4,477.55556$$

$$SSE = SS_{yy} - \hat{\beta}_1\left(SS_{xy}\right) = 4,477.55556 - \left(-.819480593\right)\left(-3,428.88889\right) = 1,667.647659$$

$$s^2 = \frac{SSE}{n-2} = \frac{1,667.647659}{9-2} = 238.2353799$$

$$s = \sqrt{s^2} = \sqrt{238.2353799} = 15.43487544$$

    b.   We would expect most of the observed number of strikes to fall within 2s or 2(15.435) or 30.87 units of their predicted values.

11.31   a. $SSE = 14,357.5$, $s^2 = MSE = 1,305.22$, and $s = 36.1279$.

    b.   We would expect most of the observed numbers of words in conversation with an unfamiliar partner to fall within 2s or 2(36.1279) or 72.26 words of their predicted values.

11.33   a.

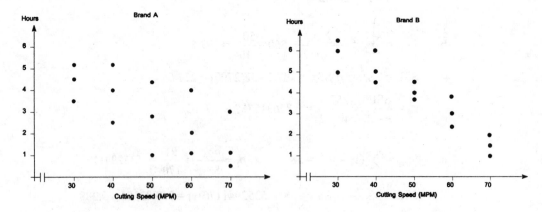

b. For Brand A,

$$\sum x_i = 750 \qquad \sum y_i = 44.8 \qquad \sum x_i y_i = 2022 \qquad \sum x_i^2 = 40{,}500 \qquad \sum y_i^2 = 168.7$$

$$SS_{xx} = \sum x_i^2 - \frac{\left(\sum x_i\right)^2}{n} = 40{,}500 - \frac{750^2}{15} = 40{,}500 - 37{,}500 = 3000$$

$$SS_{xy} = \sum x_i y_i - \frac{\left(\sum x_i\right)\left(\sum y_i\right)}{n} = 2022 - \frac{(750)(44.8)}{15} = 2022 - 2240 = -218$$

$$\hat{\beta}_1 = \frac{SS_{xy}}{SS_{xx}} = \frac{-218}{3000} = -.07266667 \approx -.0727$$

$$\hat{\beta}_0 = \bar{y} - \hat{\beta}_1 \bar{x} = \frac{44.8}{15} - (-.07266667)\left(\frac{750}{15}\right) = 2.9866667 + 3.633333 = 6.62$$

$$\hat{y} = 6.62 - .0727x$$

For Brand B,

$$\sum x_i = 750 \qquad \sum y_i = 58.9 \qquad \sum x_i y_i = 2622 \qquad \sum x_i^2 = 40{,}500 \qquad \sum y_i^2 = 270.89$$

$$SS_{xx} = \sum x_i^2 - \frac{\left(\sum x_i\right)^2}{n} = 40{,}500 - \frac{(750)^2}{15} = 40{,}500 - 37{,}500 = 3000$$

$$SS_{xy} = \sum xy - \frac{\left(\sum x\right)\left(\sum y\right)}{n} = 2622 - \frac{(750)(58.9)}{15} = 2622 - 2945 = -323$$

$$\hat{\beta}_1 = \frac{SS_{xy}}{SS_{xx}} = \frac{-323}{3000} = -.10766667 \approx -.1077$$

$$\hat{\beta}_0 = \bar{y} - \hat{\beta}_1 \bar{x} = \left(\frac{59.9}{15}\right) - (-.10766667)\left(\frac{750}{15}\right) = 3.92667 + 5.38333 = 9.31$$

$$= 9.31 - .1077x$$

c. For Brand A,

$$SS_{yy} = \sum y_i^2 - \frac{\left(\sum y_i\right)^2}{n} = 168.7 - \frac{(44.8)^2}{15} = 168.7 - 133.802667 = 34.8973333$$

$$SSE = SS_{yy} - \hat{\beta}_1 SS_{xy} = 34.8973333 - (-.07266667)(-218)$$

$$= 34.8973333 - 15.8413333 = 19.056$$

$$s^2 = \frac{SSE}{n-2} = \frac{19.056}{13} = 1.465846154 \qquad s = \sqrt{1.465846154} = 1.211$$

For Brand B,

$$SS_{yy} = \sum y_i^2 - \frac{\left(\sum y_i\right)^2}{n} = 270.89 - \frac{(58.9)^2}{15} = 270.89 - 231.2806667$$

$$= 39.6093333$$

$$SSE = SS_{yy} - \hat{\beta}_1 SS_{xy} = 39.6093333 - (-.10766667)(-323)$$

$$= 39.6093333 - 34.7763333 = 4.83299999 \approx 4.833$$

$$s^2 = \frac{\text{SSE}}{n-2} = \frac{4.833}{13} = .37176923 \qquad s = \sqrt{.37176923} = .610$$

d. For Brand A,

$\hat{y} = 6.62 - .0727x$. For $x = 70$, $\hat{y} = 6.62 - .0727(70) = 1.531$

$2s = 2(1.211) = 2.422$

Therefore, $\hat{y} \pm 2s \Rightarrow 1.531 \pm 2.422 \Rightarrow (-.891, 3.593)$

For Brand B, $\hat{y} = 9.31 - .1077x$. For $x = 70$, $\hat{y} = 9.31 - .1077(70) = 1.771$

$2s = 2(.61) = 1.22$

Therefore, $\hat{y} \pm 2s \Rightarrow 1.771 \pm 1.22 \Rightarrow (.551, 2.991)$

e. More confident with Brand B since there is less variation (s is smaller).

11.35 a.

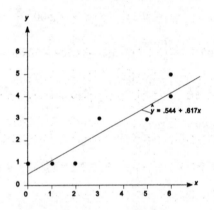

b. Some preliminary calculations are:

$$\sum x = 23 \qquad \sum x^2 = 111 \qquad \sum xy = 81$$
$$\sum y = 18 \qquad \sum y^2 = 62$$

$$SS_{xy} = \sum xy - \frac{\sum x \sum y}{n} = 81 - \frac{23(18)}{7} = 21.85714286$$

$$SS_{xx} = \sum x^2 - \frac{\left(\sum x\right)^2}{n} = 111 - \frac{23^2}{7} = 35.42857143$$

$$SS_{yy} = \sum y^2 - \frac{\left(\sum y\right)^2}{n} = 62 - \frac{18^2}{7} = 15.71428571$$

$$\hat{\beta}_1 = \frac{SS_{xy}}{SS_{xx}} = \frac{21.85714286}{35.42857143} = .616935483 \approx .617$$

$$\hat{\beta}_0 = \bar{y} - \hat{\beta}_1\bar{x} = \frac{18}{7} - .616935483\frac{23}{7} = .544354838 \approx .544$$

The least squares line is $\hat{y} = .544 + .617x$

c. The line is plotted on the graph in **a**.

d. To determine if x contributes information for the linear prediction of y, we test:

$$H_0: \ \beta_1 = 0$$
$$H_a: \ \beta_1 \neq 0$$

e. The test statistic is $t = \dfrac{\hat{\beta}_1 - 0}{\dfrac{s}{\sqrt{SS_{xx}}}} = \dfrac{.617 - 0}{\dfrac{.6678}{\sqrt{35.42857143}}} = 5.50$

where $SSE = SS_{yy} - \hat{\beta}_1 SS_{xy} = 15.71428571 - .616935483(21.85714286) = 2.22983872$

$$s^2 = \frac{SSE}{n-2} = \frac{2.22983872}{7-2} = .44596774 \qquad s = \sqrt{.44596774} = .6678$$

The degrees of freedom are $n - 2 = 7 - 2 = 5$.

f. The rejection region requires $\alpha/2 = .05/2 = .025$ in each tail of the t distribution with df $= 5$. From Table VI, Appendix A, $t_{.025} = 2.571$. The rejection region is $t < -2.571$ or $t > 2.571$.

Since the observed value of the test statistic falls in the rejection region ($t = 5.50 > 2.571$), H_0 is rejected. There is sufficient evidence to indicate x contributes information for the linear prediction of y at $\alpha = .05$.

11.37 Some preliminary calculations are:

$$\sum x = 19 \qquad \sum x^2 = 71 \qquad \sum xy = 66$$
$$\sum y = 20 \qquad \sum y^4 = 74$$

$$SS_{xy} = \sum xy - \frac{\sum x \sum y}{n} = 66 - \frac{19(20)}{6} = 2.66666667$$

$$SS_{xx} = \sum x^2 - \frac{(\sum x)^2}{n} = 71 - \frac{19^2}{6} = 10.83333333$$

$$SS_{yy} = \sum y^2 - \frac{(\sum y)^2}{n} = 74 - \frac{20^2}{6} = 7.33333333$$

$$\hat{\beta}_1 = \frac{SS_{xy}}{SS_{xx}} = \frac{2.66666667}{10.83333333} = .246153846 \approx .246$$

$$SSE = SS_{yy} - \hat{\beta}_1 SS_{xy} = 7.33333333 - (.246153846)(2.66666667) = 6.676923073$$

$$s^2 = \frac{SSE}{n-2} = \frac{6.676923073}{6-2} = 1.669230768 \qquad s = \sqrt{1.669230768} = 1.2920$$

To determine if a straight line is useful for characterizing the relationship between x and y, we test:

$$H_0: \ \beta_1 = 0$$
$$H_a: \ \beta_1 \neq 0$$

The test statistic is $t = \dfrac{\hat{\beta}_1 - 0}{s_{\hat{\beta}_1}} = \dfrac{.246}{\dfrac{1.292}{\sqrt{10.83333333}}} = .627$

The rejection region requires $\alpha/2 = .05/2 = .025$ in each tail of the t distribution with df $= n - 2 = 6 - 2 = 4$. From Table VI, Appendix A, $t_{.025} = 2.776$. The rejection region is $t < -2.776$ or $t > 2.776$.

Since the observed value of the test statistic does not fall in the rejection region ($t = .627 \ngtr 2.776$), H_0 is not rejected. There is insufficient evidence to indicate a straight line is useful for characterizing the relationship between x and y at $\alpha = .05$.

11.39   Some preliminary calculations are:

$$SS_{yy} = \sum y_i^2 - \frac{\left(\sum y_i\right)^2}{n} = 769.72 - \frac{(135.8)^2}{24} = 1.3183333$$

$$SSE = SS_{yy} - \hat{\beta}_1 SS_{xy} = 1.3183333 - (-.002301769)(-129.94167) = 1.019237592$$

$$s^2 = \frac{SSE}{n-2} = \frac{1.019237592}{22} = .046329$$

$$s_{\hat{\beta}_1} = \sqrt{\frac{s^2}{SS_{xx}}} = \sqrt{\frac{.046329}{56452.958}} = .000906$$

For confidence level .90, $\alpha = .10$ and $\alpha/2 = .10/2 = .05$. From Table VI, Appendix A with df $= n - 2 = 24 - 2 = 22$, $t_{.05} = 1.717$.

The confidence interval is:

$$\hat{\beta}_1 \pm t_{.05}\, s_{\hat{\beta}_1} \Rightarrow .0023 \pm 1.717(.000906)$$
$$\Rightarrow (-.0039, -.0007)$$

We are 90% confident that the change in the mean sweetness index for each one unit change in the pectin is between $-.0039$ and $-.0007$.

11.41   a. $\hat{\beta}_0 = .031$. Since $x = 0$ is not in the range (i.e., $x = 0$ means 0% of land covered), $_0$ has no interpretation other than being the y-intercept.

b. $\hat{\beta}_1 = .089$. For additional increase of 1% of land covered by the plants, the mean standing crop increases by an estimated .089 gram per meter squared.

c.   For any significance level $\alpha > .042$, there is sufficient evidence of a linear relationship. Any α-level $> .042$ is greater than the observed p-value of .042. Therefore, the null hypothesis can be rejected.

11.43   a.   To determine if body plus head rotation and active head movement are positively linearly related, we test:

H_0: $\beta_1 = 0$
H_a: $\beta_1 > 0$

The test statistic is $t = \dfrac{\hat{\beta}_1 - 0}{s_{\hat{\beta}_1}} = \dfrac{.88 - 0}{.14} = 6.286$

The rejection region requires $\alpha = .05$ in each tail of the t distribution with df $= n - 2 = 39 - 2 = 37$. From Table VI, Appendix A, $t_{.05} \approx 1.687$. The rejection region is $t > 1.687$.

Since the observed value of the test statistic falls in the rejection region ($t = 6.286 > 1.687$), H_0 is rejected. There is sufficient evidence to indicate that the two variables are positively linearly related at $\alpha = .05$.

b.   For confidence level .90, $\alpha = .10$ and $\alpha/2 = .10/2 = .05$. From Table VI, Appendix A, with df $= n - 2 = 39 - 2 = 37$, $t_{.05} \approx 1.687$. The confidence interval is:

$$\hat{\beta}_1 \pm t_{.05} s_{\hat{\beta}_1} \Rightarrow .88 \pm 1.687(.14)$$
$$\Rightarrow .88 \pm .23618$$
$$\Rightarrow (.6438, 1.1162)$$

We are 90% confident that the true value of β_1 is between .6438 and 1.1162.

c.   Because the interval in part b contains the value 1, there is no evidence that the true slope of the line differs from 1.

11.45   Using the calculations from Exercise 11.23 and these calculations:

$$SS_{yy} = \sum y^2 - \frac{\left(\sum y\right)^2}{n} = 83.474 - \frac{103.07^2}{144} = 9.70021597$$

$$SSE = SS_{yy} - \hat{\beta}_1\left(SS_{xy}\right) = 9.70021597 - (.026421957)(19.975) = 9.172437366$$

$$s^2 = \frac{SSE}{n-2} = \frac{9.172437366}{144-2} = .064594629$$

$$s = \sqrt{s^2} = \sqrt{.064594629} = .254154735$$

To determine if there is a linear trend between the proportion of names recalled and position, we test:

H_0: $\beta_1 = 0$
H_a: $\beta_1 \neq 0$

The test statistic is $t = \dfrac{\hat{\beta}_1 - 0}{s_{\hat{\beta}}} = \dfrac{\hat{\beta}_1 - 0}{s \Big/ \sqrt{SS_{xx}}} = \dfrac{.02642 - 0}{.25415 \Big/ \sqrt{756}} = 2.858$

The rejection region requires $\alpha/2 = .01/2 = .005$ in each tail of the t distribution. From Table VI, Appendix A, with df $= n - 2 = 144 - 2 = 142$, $t_{.005} = 2.576$. The rejection region is $t < -2.576$ or $t > 2.576$.

Since the observed test statistic falls in the rejection region ($t = 2.858 > 2.576$), H_0 is rejected. There is sufficient evidence to indicate the proportion of names recalled is linearly related to position at $\alpha = .01$.

11.47   Answers may vary. Possible answer:

The scaffold-drop survey provides the most accurate estimate of spall rate in a given wall segment. However, the drop areas were not selected at random from the entire complex; rather, drops were made at areas with high spall concentrations. Therefore, if the photo spall rates could be shown to be related to drop spall rates, then the 83 photo spall rates could be used to predict what the drop spall rates would be.

a.   Construct a scattergram for the data.

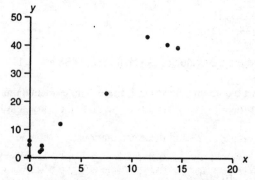

The scattergram shows a positive relationship between the photo spall rate (x) and the drop spall rate (y).

b.   Find the prediction equation for drop spall rate. The MINITAB output shows the results of the analysis.

```
The regression equation is
drop = 2.55 + 2.76 photo

Predictor          Coef       StDev          T          P
Constant          2.548       1.637       1.56      0.154
photo            2.7599      0.2180      12.66      0.000

S = 4.164       R-Sq = 94.7%      R-Sq(adj) = 94.1%

Analysis of Variance

Source              DF          SS          MS          F          P
Regression           1      2777.5      2777.5     160.23      0.000
Residual Error       9       156.0        17.3
Total               10      2933.5

Unusual Observations
Obs       photo        drop         Fit    StDev Fit    Residual    St Resid
11        11.8       43.00       35.11        1.97        7.89       2.15R

R denotes an observation with a large standardized residual
```

$$\hat{y} = 2.55 + 2.76x$$

c.   Conduct a formal statistical hypthesis test to determine if the photo spall rates contribute information for the prediction of drop spall rates.

H_0: $\beta_1 = 0$
H_a: $\beta_1 \neq 0$

The test statistic is $t = 12.66$, with p-value z < .0001.

Reject H_0 for any level of significance $\geq$ .0001. There is sufficient evidence to indicate that photo spall rates contribute information for the prediction of drop spall rates at $\alpha \geq$ .0001.

d.   One could now use the 83 photos spall rates to predict values for 83 drop spall rates. Then use this information to estimate the true spall rate at a given wall segment and estimate to total spall damage.

11.49 a. $r = 1$ implies x and y are perfectly, positively related.

b. $r = -1$ implies x and y are perfectly, negatively related.

c. $r = 0$ implies x and y are not related.

d. $r = .90$ implies x and y are positively related. Since r is close to 1, the strength of the relationship is very high.

e. $r = .10$ implies x and y are positively related. Since r is close to 0, the relationship is fairly weak.

f. $r = -.88$ implies x and y are negatively related. Since r is close to -1, the relationship is fairly strong.

11.51   a.   Some preliminary calculations are:

$$\sum x = 0 \qquad \sum x^2 = 10 \qquad \sum xy = 20$$
$$\sum y = 12 \qquad \sum y^2 = 70$$

$$SS_{xy} = \sum xy - \frac{\sum x \sum y}{n} = 20 - \frac{0(12)}{5} = 20$$

$$SS_{xx} = \sum x^2 - \frac{\left(\sum x\right)^2}{n} = 10 - \frac{0^2}{5} = 10$$

$$SS_{yy} = \sum y^2 - \frac{\left(\sum y\right)^2}{n} = 70 - \frac{12^2}{5} = 41.2$$

$$r = \frac{SS_{xy}}{\sqrt{SS_{xx}SS_{yy}}} = \frac{20}{\sqrt{10(41.2)}} = .9853$$
$$r^2 = .9853^2 = .9709$$

Since $r = .9853$, there is a very strong positive linear relationship between x and y.

Since $r^2 = .9709$, 97.09% of the total sample variability around is explained by the linear relationship between x and y.

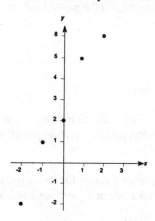

b.   Some preliminary calculations are:

$$\sum x = 0 \qquad \sum x^2 = 10 \qquad \sum xy = -15$$
$$\sum y = 16 \qquad \sum y^2 = 74$$

$$SS_{xy} = \sum xy - \frac{\sum x \sum y}{n} = -15 - \frac{0(16)}{5} = -15$$

$$SS_{xx} = \sum x^2 - \frac{\left(\sum x\right)^2}{n} = 10 - \frac{0^2}{5} = 10$$

$$SS_{yy} = \sum y^2 - \frac{\left(\sum y\right)^2}{n} = 74 - \frac{16^2}{5} = 22.8$$

$$r = \frac{SS_{xy}}{\sqrt{SS_{xx}SS_{yy}}} = \frac{-15}{\sqrt{10(22.8)}} = -.9934$$

$$r^2 = (-.9934)^2 = .9868$$

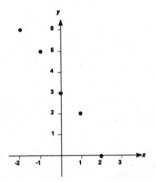

Since $r = -.9934$, there is a very strong negative linear relationship between x and y.

Since $r^2 = .9868$, 98.68% of the total sample variability around is explained by the linear relationship between x and y.

c. Some preliminary calculations are:

$$\sum x = 18 \qquad \sum x^2 = 52 \qquad \sum xy = 36$$

$$\sum y = 14 \qquad \sum y^2 = 32$$

$$SS_{xy} = \sum xy - \frac{\sum x \sum y}{n} = 36 - \frac{18(14)}{7} = 0$$

$$SS_{xx} = \sum x^2 - \frac{\left(\sum x\right)^2}{n} = 52 - \frac{18^2}{7} = 5.71428571$$

$$SS_{yy} = \sum y^2 - \frac{\left(\sum y\right)^2}{n} = 32 - \frac{14^2}{7} = 4$$

$$r = \frac{SS_{xy}}{\sqrt{SS_{xx}SS_{yy}}} = \frac{0}{\sqrt{5.71428571(4)}} = 0$$

$$r^2 = 0^2 = 0$$

Since $r = 0$, this implies that x and y are not related.

Since $r^2 = 0$, 0% of the total sample variability around is explained by the linear relationship between x and y.

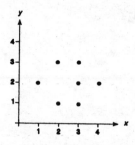

d. Some preliminary calculations are:

$$\sum x = 15 \qquad \sum x^2 = 71 \qquad \sum xy = 12$$
$$\sum y = 4 \qquad \sum y^2 = 6$$

$$SS_{xy} = \sum xy - \frac{\sum x \sum y}{n} = 12 - \frac{15(4)}{5} = 0$$

$$SS_{xx} = \sum x^2 - \frac{\left(\sum x\right)^2}{n} = 71 - \frac{15^2}{5} = 26$$

$$SS_{yy} = \sum y^2 - \frac{\left(\sum y\right)^2}{n} = 6 - \frac{4^2}{5} = 2.8$$

$$r = \frac{SS_{xy}}{\sqrt{SS_{xx}SS_{yy}}} = \frac{0}{\sqrt{26(2.8)}} = 0$$
$$r^2 = 0^2 = 0$$

Since $r = 0$, this implies that x and y are not related.

Since $r^2 = 0$, 0% of the total sample variability around is explained by the linear relationship between x and y.

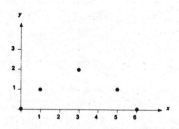

11.53 a. To determine whether the number of leisure activities and GPA are linearly related, we test:

H_0: $\rho = 0$
H_a: $\rho \neq 0$

b.   The p-value = .0512. This implies that H_0 will not be rejected for $\alpha = .05$. There is insufficient evidence to indicate that the number of leisure activities and GPA are linearly related for $\alpha = .05$. For $\alpha > .0512$, H_0 will be rejected. There is sufficient evidence to indicate that the number of leisure activities and GPA are linearly related for $\alpha > .0512$.

11.55   a.   Piano: $r = .447$
Because this value if near .5, there is s slight positive linear relationship between recognition exposure time and goodness of view for piano.

Bench: $r = -.057$
Because this value is extremely close to 0, there is an extremely weak negative linear relationship between recognition exposure time and goodness of view for bench.

Motorbike: $r = .619$
Because this value is near .5, there is a moderate positive linear relationship between recognition exposure time and goodness of view for motorbike.

Armchair: $r = .294$
Because this value is fairly close to 0, there is a weak positive linear relationship between recognition exposure time and goodness of view for armchair.

Teapot: $r = .949$
Because this value is very close to 1, there is a strong positive linear relationship between recognition exposure time and goodness of view for teapot.

b.   Piano: $r^2 = (.447)^2 = .1998$
19.98% of the total sample variability around the sample mean recognition exposure time is explained by the linear relationship between the recognition exposure time and the goodness of view for piano.

Bench: $r^2 = (-.057)^2 = .0032$
.32% of the total sample variability around the sample mean recognition exposure time is explained by the linear relationship between the recognition exposure time and the goodness of view for bench.

Motorbike: $r^2 = (.619)^2 = .3832$
38.32% of the total sample variability around the sample mean recognition exposure time is explained by the linear relationship between the recognition exposure time and the goodness of view for motorbike.

Armchair: $r^2 = (.294)^2 = .0864$
8.64% of the total sample variability around the sample mean recognition exposure time is explained by the linear relationship between the recognition exposure time and the goodness of view for armchair.

Teapot: $r^2 = (.949)^2 = .9006$
90.06% of the total sample variability around the sample mean recognition exposure time is explained by the linear relationship between the recognition exposure time and the goodness of view for teapot.

c.   The test is:

$$H_0: \beta_1 = 0$$
$$H_a: \beta_1 \neq 0$$

Following are the values of α and $t_{\alpha/2}$ that correspond to df $= n - 2 = 25 - 2 = 23$.

α	.20	.10	.05	.02	.01	.002	.001
$t_{\alpha/2}$	1.319	1.714	2.069	2.500	2.807	3.485	3.767

Piano: $t = 2.40$
$2.069 < 2.40 < 2,500, p \approx .025$
For levels of significance greater than $\alpha = .025$, H_0 can be rejected. There is sufficient evidence to indicate that there is a linear relationship between goodness of view and recognition exposure time for piano for $\alpha > .025$.

Bench: $t = .27$
$.27 < 1.319, p > .2$
H_0 is not rejected. There is insufficient evidence to indicate that there is a linear relationship between goodness of view and recognition exposure time for bench for $\alpha \leq .2$.

Motorbike: $t = 3.78$
$3.78 > 3.767, p < .001$
H_0 can be rejected for $\alpha \geq .001$. There is sufficient evidence to indicate that there is a linear relationship between goodness of view and recognition exposure time for motorbike for $\alpha \geq .001$.

Armchair: $t = 1.47$
$1.319 < 1.47 < 1.714, p \approx .15$
H_0 cannot be rejected for levels of significance $\alpha < .15$. There is insufficient evidence to indicate that there is a linear relationship between goodness of view and recognition exposure time for armchair for $\alpha < .15$.

Teapot: $t = 14.50$
$14.50 > 3.767, p < .001$
H_0 can be rejected for $\alpha \geq .001$. There is sufficient evidence to indicate that there is a linear relationship between goodness of view and recognition exposure time for teapot for $\alpha \geq .001$.

11.57 a.   A straight line model is $y = \beta_0 + \beta_1 x + \varepsilon$.

b.   The researcher hypothesized that therapists with more years of formal dance training will report a higher perceived success rate in cotherapy relationships. This indicates that $\beta_1 > 0$.

c. $r = -.26$. Because this value is fairly close to 0, there is a weak negative linear relationship between years of formal training and reported success rate.

d. To determine if there is a positive linear relationship between years of formal training and reported success rate, we test:

H_0: $\beta_1 = 0$
H_0: $\beta_1 > 0$

The test statistic is $t = \dfrac{r}{\sqrt{(1-r^2)/(n-2)}} = \dfrac{-.26}{\sqrt{(1-(-.26^2))/(136-2)}} = -3.12$

The rejection region requires $\alpha = .05$ in the upper tail of the t distribution with df $= n - 2 = 136 - 2 = 134$. From Table VI, Appendix A, $t_{.05} \approx 1.645$. The rejection region is $t > 1.645$.

Since the observed value of the test statistic does not fall in the rejection region ($t = -8.66 \not> 1.645$), H_0 is not rejected. There is insufficient evidence to indicate that there is a positive linear relationship between years of formal training and perceived success rates at $\alpha = .05$.

11.59 Using the values computed in Exercise 11.42:

$$r = \frac{SS_{xy}}{\sqrt{SS_{xx}SS_{yy}}} = \frac{48.125}{\sqrt{379.9375(18.75)}} = .5702$$

Because r is moderately small, there is a rather weak positive linear relationship between blood lactate concentration and perceived recovery.

$r^2 = .5702^2 = .3251$.

32.51% of the sample variance of blood lactate concentration around the sample mean is explained by the linear relationship between blood lactate concentration and perceived recovery.

11.61 a. Since there was an inverse relationship, the value of r must be negative.

b. If the result was significant, then the test statistic must fall in the rejection region. For a one tailed test, $\alpha = .05$ must fall in the lower tail of the t distribution with df $= n - 1 = 337 - 2 = 335$. From Table VI, Appendix A, $t_{.05} \approx 1.645$. The rejection region is $t < -1.645$.

Using the equation given, then:

$$t = \frac{r\sqrt{(n-2)}}{\sqrt{(1-r^2)}} < -1.645$$

$$\Rightarrow \frac{r^2(n-2)}{1-r^2} > (-1.645)^2$$

$$\Rightarrow r^2(337-2) > 2.706025(1-r^2)$$

$$\Rightarrow r^2(335) + 2.706025r^2 > 2.706025$$

$$\Rightarrow r^2(337.706025) > 2.706025$$

$$\Rightarrow r^2 > \frac{2.706025}{337.706025} = .00801296$$

$$\Rightarrow r < -\sqrt{.00801296} = -.0895$$

11.63   a,b.

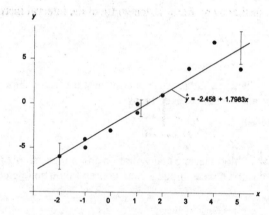

c.   Some preliminary calculations are:

$$SSE = SS_{yy} - \hat{\beta}_1 SS_{xy} = 168.1 - 1.798319328(85.6) = 14.1638655$$

$$s^2 = \frac{SSE}{n-2} = \frac{14.1638655}{10-2} = 1.770483188 \qquad s = \sqrt{1.770483188} = 1.3306$$

$$\bar{x} = \frac{\sum x}{n} = \frac{12}{10} = 1.2$$

For $x_p = 5$, $\hat{y} = -2.458 + 1.7983(5) = 6.5335$

For confidence coefficient .95, $\alpha = 1 - .95 = .05$ and $\alpha = .05/2 = .025$. From Table VI, Appendix A, $t_{.025} = 2.306$ with df $= n - 2 = 10 - 2 = 8$. The 95% confidence interval is:

$$\hat{y} \pm t_{\alpha/2}s\sqrt{\frac{1}{n} + \frac{(x_p - \bar{x})^2}{SS_{xx}}} \Rightarrow 6.5335 \pm 2.306(1.3306)\sqrt{\frac{1}{10} + \frac{(5 - 1.2)^2}{47.6}}$$

$$\Rightarrow 6.5335 \pm 1.9487 \Rightarrow (4.5848, 8.4822)$$

d. For $x_p = 1.2$, $\hat{y} = -2.458 + 1.7983(1.2) = -.3$. The 95% confidence interval is:

$$-.3 \pm 2.306(1.3306)\sqrt{\frac{1}{10} \pm \frac{(1.2 - 1.2)^2}{47.6}} \Rightarrow -.3 \pm .9703 \Rightarrow (-1.2703, .6703)$$

For $x_p = -2$, $\hat{y} = -2.458 + 1.7983(-2) = -6.0546$. The 95% confidence interval is:

$$-6.0546 \pm 2.306(1.3306)\sqrt{\frac{1}{10} + \frac{(-2 - 1.2)^2}{47.6}} \Rightarrow -6.0546 \pm 1.7225$$

$$\Rightarrow (-7.7771, -4.3321)$$

e. The width of the confidence interval when $x = 1.2$ is smaller than when $x = -2$ or $x = 5$. The widths of the intervals change because of the term $\dfrac{(x_p - \bar{x})^2}{SS_{xx}}$. If $x_p = \bar{x}$, the term is 0. As x_p moves further away from $\bar{x}$, the width of the interval increases.

11.65 a, b. The scattergram is:

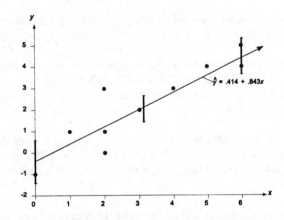

c. $SSE = SS_{yy} - \hat{\beta}_1 SS_{xy} = 33.6 - .84318766(32.8) = 5.94344473$

$$s^2 = \frac{SSE}{n - 2} = \frac{5.94344473}{10 - 2} = .742930591 \qquad\qquad s = \sqrt{.742930591} = .8619$$

$$\bar{x} = \frac{31}{10} = 3.1$$

The form of the confidence interval is $\hat{y} \pm t_{\alpha/2}\sqrt{\dfrac{1}{n} + \dfrac{(x_p - \bar{x})^2}{SS_{xx}}}$

For $x_p = 6$, $\hat{y} = -.414 + .843(6) = 4.644$

For confidence coefficient .95, $\alpha = .05$ and $\alpha/2 = .025$. From Table VI, Appendix A, with df $= n - 2 = 10 - 2 = 8$, $t_{.025} = 2.306$. The confidence interval is:

$$4.644 \pm 2.306(.8619)\sqrt{\frac{1}{10} + \frac{(6-3.1)^2}{38.9}} \Rightarrow 4.644 \pm 1.118 \Rightarrow (3.526, 5.762)$$

d. For $x_p = 3.2$, $\hat{y} = -.414 + .843(3.2) = 2.284$

The confidence interval is:

$$2.284 \pm 2.306(.8619)\sqrt{\frac{1}{10} + \frac{(3.2-3.1)^2}{38.9}} \Rightarrow 2.284 \pm .629 \Rightarrow (1.655, 2.913)$$

For $x_p = 0$, $\hat{y} = -.414 + .843(0) = -.414$

The confidence interval is:

$$-.414 \pm 2.306(.8619)\sqrt{\frac{1}{10} + \frac{(0-3.1)^2}{38.9}} \Rightarrow -.414 \pm 1.717 \Rightarrow (-1.585, .757)$$

e. The width of the confidence interval for the mean value of y depends on the distance x_p is from . The width of the interval for $x_p = 3.2$ is the smallest because 3.2 is the closest to = 3.1. The width of the interval for $x_p = 0$ is the widest because 0 is the farthest from = 3.1.

11.67 a. No. We know there is a significant linear relationship between sale price and appraised value. However, the actual sale prices maybe scattered quite far from the predicted line.

b. From the printout, the 95% prediction interval for the actual sale price when the appraised value is $300,000 is (275.86, 407.26) or ($275,860, $407,260). We are 95% confident that the actual sale price for a home appraised at $300,000 is between $275,860 and $407,260.

c. From the printout, the 95% confidence interval for the mean sale price when the appraised value is $300,000 is (332.95, 350.17) or ($332,950, $350,170). We are 95% confident that the mean sale price for a home appraised at $300,000 is between $332,950 and $350,170.

11.69 a. For $x = 64$, the 95% prediction interval is (1.0516, 6.6347). We are 95% confident that the actual number of flycatchers killed is between 1.0516 and 6.6347 when the nest box nit occupancy is 64%.

b. The width of the 95% confidence interval for the mean number of flycatchers killed when the nest box nit occupancy is 64% would be smaller than the 95% prediction interval.

c. We would not recommend that this model be used to predict the number of flycatchers killed at a site with a nest box nit occupancy rate of 15% because 15% is outside the observed occupancy rates (the occupancy rates ranged from 24% to 64%). We have no idea if the relationship between the number of flycatchers killed and the nest box nit occupancy rate is the same outside the observed range.

11.71 a. The equation for the straight-line model relating duration to frequency is $y = \beta_0 + \beta_1 x + \varepsilon$.

b. Some preliminary calculations are:

$$\bar{y} = \frac{\sum y}{n} = \frac{1,287}{11} = 117$$

$$\bar{x} = \frac{\sum x}{n} = \frac{394}{11} = 35.818$$

$$SS_{xy} = \sum xy - \frac{\sum x \sum y}{n} = 30,535 - \frac{394(1,287)}{11} = -15,563$$

$$SS_{xx} = \sum x^2 - \frac{\left(\sum x\right)^2}{n} = 28,438 - \frac{394^2}{11} = 14,325.63636$$

$$\hat{\beta}_1 = \frac{SS_{xy}}{SS_{xx}} = \frac{-15,563}{14,325.63636} = -1.086374079$$

$$\hat{\beta}_o = \bar{y} - \hat{\beta}_1 \bar{x} = \frac{1,287}{11} - (-1.086374079)\left(\frac{394}{11}\right) = 155.9119443$$

The least squares prediction equation is $\hat{y} = 155.912 - 1.086x$.

c. Some preliminary calculations are:

$$SS_{yy} = \sum y^2 - \frac{\left(\sum y\right)^2}{n} = 203,651 - \frac{1,287^2}{11} = 53,072$$

$$SSE = SS_{yy} - \hat{\beta}_1\left(SS_{xy}\right) = 53,072 - (-1.086374079)(-15,563) = 36,164.76021$$

$$s^2 = \frac{SSE}{n-2} = \frac{36,164.76021}{11-2} = 4,018.30669$$

$$s = \sqrt{s^2} = \sqrt{4,018.30669} = 63.39011508$$

To determine if there is a linear relationship between duration and frequency, we test:

H_0: $\beta_1 = 0$
H_a: $\beta_1 \neq 0$

The test statistic is $t = \dfrac{\hat{\beta}_1 - 0}{s_{\hat{\beta}}} = \dfrac{\hat{\beta}_1 - 0}{s / \sqrt{SS_{xx}}} = \dfrac{-1.086 - 0}{63.3901 / \sqrt{14,325.63636}} = -2.051$

The rejection region requires $\alpha/2 = .05/2 = .025$ in each tail of the t distribution. From Table VI, Appendix A, with df $= n - 2 = 11 - 2 = 9$, $t_{.025} = 2.262$. The rejection region is $t < -2.262$ or $t > 2.262$.

Since the observed test statistic does not fall in the rejection region ($t = -2.051 \nless -2.262$), H_0 is not rejected. There is insufficient evidence to indicate that duration and frequency are linearly related at $\alpha = .05$.

d. For $x = 25$, the predicted duration is $\hat{y} = 155.912 - 1.086(25) = 128.762$.

For confidence coefficient .95, $\alpha = .05$ and $\alpha/2 = .05/2 = .025$. From Table VI, Appendix A, with df $= n - 2 = 11 - 2 = 9$, $t_{.025} = 2.262$. The 95% prediction interval is:

$$\hat{y} \pm t_{\alpha/2}s\sqrt{1 + \frac{1}{n} + \frac{(x_p - \bar{x})^2}{SS_{xx}}} \Rightarrow 128.762 \pm 2.262(63.3901)\sqrt{1 + \frac{1}{11} + \frac{(25 - 35.818)^2}{14,325.63636}}$$

$$\Rightarrow 128.762 \pm 150.324 \Rightarrow (-21.562, \ 279.086)$$

We are 95% confident that the actual duration of a person who participates 25 times a year is between -21.562 and 279.086 days. Since the duration cannot be negative, the actual duration will be between 0 and 279.086.

11.73 a. From Exercise 11.33, $SS_{xx} = 3000$ and $= 50$.

Also, for Brand A, $s = 1.211$; for Brand B, $s = .610$.

For Brand A, $\hat{y} = 6.62 - .0727(45) = 3.349$, while for Brand B, $\hat{y} = 9.31 - .1077(45) = 4.464$.

The degrees of freedom for both brands is $n - 2 = 15 - 2 = 13$. For confidence coefficient .90, (i.e., for all parts of this question), $\alpha = .10$ and $\alpha/2 = .05$. From Table VI, Appendix A, with df $= 13$, $t_{.05} = 1.771$.

The form of both confidence intervals is $\hat{y} \pm t_{\alpha/2}s\sqrt{\frac{1}{n} + \frac{(x_p - \bar{x})^2}{SS_{xx}}}$

For Brand A, we obtain:

$$3.349 \pm 1.771(1.211)\sqrt{\frac{1}{15} + \frac{(45 - 50)^2}{3000}} \Rightarrow 3.349 \pm .587 \Rightarrow (2.762, 3.936)$$

For Brand B, we obtain:

$$4.464 \pm 1.771(.610)\sqrt{\frac{1}{15} + \frac{(45 - 50)^2}{3000}} \Rightarrow 4.464 \pm .296 \Rightarrow (4.168, 4.760)$$

The first interval is wider, caused by the larger value of s.

b. The form of both prediction intervals is $t_{\alpha/2}s\sqrt{1+\dfrac{1}{n}+\dfrac{(x_p-\bar{x})^2}{SS_{xx}}}$

For Brand A, we obtain:

$$3.349 \pm 1.771(1.211)\sqrt{1+\frac{1}{15}+\frac{(45-50)^2}{3000}} \Rightarrow 3.349 \pm 2.224 \Rightarrow (1.125, 5.573)$$

For Brand B, we obtain:

$$4.464 \pm 1.771(.610)\sqrt{1+\frac{1}{15}+\frac{(40+50)^2}{3000}} \Rightarrow 4.464 \pm 1.120 \Rightarrow (3.344, 5.584)$$

Again, the first interval is wider, caused by the larger value of s. Each of these intervals is wider than its counterpart from part **a**, since, for the same x, a prediction interval for an individual y is always wider than a confidence interval for the mean of y. This is due to an individual observation having a greater variance than the variance of the mean of a set of observations.

c. To obtain a confidence interval for the life of a brand A cutting tool that is operated at 100 meters per minute, we use:

$$\hat{y} \pm t_{\alpha/1}s\sqrt{1+\frac{1}{n}+\frac{(x_p-\bar{x})}{SS_{xx}}}$$

For $x = 100$, $\hat{y} = 6.62 - .0727(100) = -.65$.

The degrees of freedom are $n - 2 = 15 - 2 = 13$. For confidence coefficient .95, $\alpha = .05$ and $\alpha/2 = .025$. From Table VI, Appendix A, with df = 13, $t_{.025} = 2.160$.

Here, we obtain:

$$-.65 \pm 2.160(1.211)\sqrt{1+\frac{1}{15}+\frac{(100-50)^2}{3000}} \Rightarrow -.65 \pm 3.606 \Rightarrow (-4.256, 2.956)$$

The additional assumption would be that the straight line model fits the data well for the x's actually observed all the way up to the value under consideration, 100. Clearly from the estimated value of $-.65$, this is not true (usually, negative "useful lives" are not found).

11.75 a.

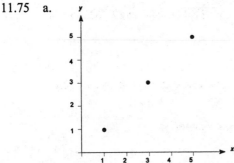

b. One possible line is $\hat{y} = x$.

x	y	$\hat{y}$	$y - \hat{y}$
1	1	1	0
3	3	3	0
5	5	5	0
			0

For this example $\sum(y - \hat{y}) = 0$

A second possible line is $\hat{y} = 3$.

x	y	$\hat{y}$	$y - \hat{y}$
1	1	3	-2
3	3	3	0
5	5	3	2
			0

For this example $\sum(y - \hat{y}) = 0$

c. Some preliminary calculations are:

$$\sum x_i = 9 \qquad \sum x_i^2 = 35 \qquad \sum x_i y_i = 35$$

$$\sum y_i = 9 \qquad \sum y_i^2 = 35$$

$$SS_{xy} = \sum x_i \sum y_i - \frac{\sum x_i \sum y_i}{n} = 35 - \frac{9(9)}{3} = 8$$

$$SS_{xx} = \sum x_i^2 - \frac{\left(\sum x_i\right)^2}{n} = 35 - \frac{9^2}{3} = 8$$

$$SS_{yy} = \sum y_i^2 - \frac{\left(\sum y_i\right)^2}{n} = 35 - \frac{9^2}{3} = 8$$

$$\hat{\beta}_1 = \frac{SS_{xy}}{SS_{xx}} = \frac{8}{8} = 1 \qquad\qquad \hat{\beta}_0 = \bar{y} - \hat{\beta}_1 \bar{x} = \frac{9}{3} - 1\left(\frac{9}{3}\right) = 0$$

The least squares line is $\hat{y} = 0 + 1x = x$.

d. For $\hat{y} = x$, SSE = $SS_{yy} - \hat{\beta}_1 SS_{xy} = 8 - 1(8) = 0$

For $\hat{y} = 3$, SSE = $\sum(y_i - \hat{y}_i)^2 = (1-3)^2 + (3-3)^2 + (5-3)^2 = 8$

The least squares line has the smallest SSE of all possible lines.

11.77 a. From the printout, the least squares equation is $\hat{y} = 40.7842 + .7656x$.

b.

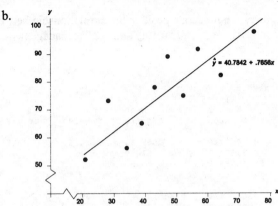

c. To determine if there is sufficient evidence to indicate that a positive correlation exists between verbal score and final grade, we test:

H_0: $\beta_1 = 0$
H_a: $\beta_1 > 0$

The test statistic is $t = \dfrac{\hat{\beta}_1 - 0}{s_{\hat{\beta}_1}} = \dfrac{.76556}{.71498} = 4.38$

The rejection region requires $\alpha = .01$ in the upper tail of the t distribution with df $= n - 2 = 10 - 2 = 8$. From Table VI, Appendix A, $t_{.01} = 2.896$. The rejection region is $t > 2.896$.

Since the observed value of the test statistic falls in the rejection region ($t = 4.38 > 2.896$), H_0 is rejected. There is sufficient evidence to indicate that a positive correlation exists between verbal score and final grade at $\alpha = .01$.

d. The form of the confidence interval for β_1 is $\hat{\beta}_1 \pm t_{\alpha/2} s_{\hat{\beta}_1}$

For confidence coefficient .95, $\alpha = .05$ and $\alpha/2 = .025$. From Table VI, Appendix A, with df $= n - 2 = 10 - 2 = 8$, $t_{.025} = 2.306$.

The confidence interval is $.7656 \pm 2.306(.17498) \Rightarrow .7656 \pm .4035 \Rightarrow (.3621, 1.1691)$.

e. From the printout, the 95% prediction interval for a student's final grade when his/her verbal score is 50 is (57.950, 100.17). We are 95% confident that a student's final grade in the introductory course will be between 57.950 and 100.17 if his/her verbal score was 50.

f. From the printout, the 95% confidence interval for the mean final grade for all students' whose verbal score is 50 is (72.513, 85.611). We are 95% confident that the mean final grade in the introductory course will be between 72.513 and 85.611 for all students whose verbal score was 50.

Simple Linear Regression

11.79   a. $r = .570$

   b.   Because $r = .570$ in near .5, there is a moderate positive linear relationship between managers attitudes for U.S. and Asia regarding quality and quality management.

11.81   Dependent Variable: VALUE

Analysis of Variance

Source	DF	Sum of Squares	Mean Square	F Value	Pr > F
Model	1	865746	865746	10.55	0.0021
Error	48	3939797	82079		
Corrected Total	49	4805543			

Root MSE	286.49451	R-Square	0.1802
Dependent Mean	128.90000	Adj R-Sq	0.1631
Coeff Var	222.26106		

Parameter Estimates

Variable	DF	Parameter Estimate	Standard Error	t Value	Pr > |t|
Intercept	1	-92.45768	79.29106	-1.17	0.2494
AGE	1	8.34682	2.57005	3.25	0.0021

Answers may vary. One possible answer may include:

The least squares line is $\hat{y} = -92.45768 + 8.34682x$. To determine if age can be used to predict market value, we test:

H_0: $\beta_1 = 0$
H_a: $\beta_1 \neq 0$

The test statistic is $t = 3.25$ with p-value = .0021. Reject the null hypothesis for levels of significance $\alpha > .0021$. There is sufficient evidence to indicate that age contributes information for the prediction of market value (y) at $\alpha > .0021$.

$r = .42$; Since this value is near .5, there is a moderate positive linear relationship between the value and age of the Beanie Baby.

11.83   a. $\hat{\beta}_0 = -13.490347$ –   has no meaning since $x = 0$ is not in the observed range. It is the y-intercept.

   $\hat{\beta}_1 = -.052829$ –   is the estimated change in the mean proportion of impurity passing through helium for each additional degree.

b. For confidence coefficient .95, $\alpha = .05$ and $\alpha/2 = .025$. From Table VI, Appendix A, with df $= n - 2 = 10 - 2 = 8$, $t_{.025} = 2.306$. The confidence interval is:

$$\hat{\beta}_1 \pm t_{\alpha/2} s_{\hat{\beta}_1} \Rightarrow -.0528 \pm 2.306(.00772828) \Rightarrow -.0528 \pm .0178 \Rightarrow (-.0706, \ -.0350)$$

We are 95% confident that the change in mean proportion of impurity passing through helium for each additional degree is between $-.0706$ and $-.0350$. Since 0 is not in the interval, there is evidence to indicate that temperature contributes information about the proportion of impurity passing through helium.

c. $r^2 = R$-square $= .8538$. 85.38% of the total sample variation around the mean proportion of impurity is explained by the linear relationship between proportion of impurity and temperature.

d. From the printout, the 95% prediction interval is $(.5987, 1.2653)$.

e. We cannot be sure that the relationship between the proportion of impurity passing through helium and temperature is the same outside the observed range.

11.85 a. The scattergram of the data is:

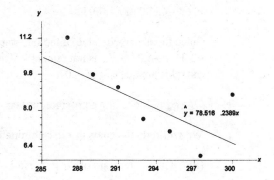

b. Some preliminary calculations are:

$$\sum x_i = 2{,}053 \qquad\qquad \sum y_i = 59.2 \qquad\qquad n = 7$$
$$\sum x_i^2 = 602{,}265.5 \qquad \sum y_i^2 = 517.22 \qquad\qquad \sum x_i y_i = 17{,}326.7$$

$$SS_{xy} = \sum x_i y_i - \frac{\left(\sum x_i\right)\left(\sum y_i\right)}{n} = 17{,}326.7 - \frac{2053(59.2)}{7} = -35.81429$$

$$SS_{xx} = \sum x_i^2 - \frac{\left(\sum x_i\right)^2}{n} = 602{,}265.5 - \frac{2053^2}{7} = 149.9286$$

$$SS_{yy} = \sum y_i^2 - \frac{\left(\sum y_i\right)^2}{n} = 517.22 - \frac{59.2^2}{7} = 16.5571429$$

$$\hat{\beta}_1 = \frac{SS_{xy}}{SS_{xx}} = \frac{-35.81429}{149.9286} = -.238875638 \approx -.2389$$

$$\hat{\beta}_0 = \bar{y} - \hat{\beta}_1 \bar{x} = \frac{59.2}{7} - (-.238875638)\frac{2053}{7} = 78.515955 \approx 78.516$$

The least squares line is $\hat{y} = 78.516 - .2389x$.

c.   The least squares line is plotted on the graph in part **a**.

d.   Some additional calculations:

$$SSE = SS_{yy} - \hat{\beta}_1 SS_{xy} = 16.5571429 - (-.238875638)(-35.81429) = 8.001980951$$

$$s^2 = \frac{SSE}{n-2} = \frac{8.001980951}{7-2} = 1.60039619 \quad s = \sqrt{1.60039619} = 1.2651$$

To determine if the model is adequate, we test:

H_0: $\beta_1 = 0$
H_a: $\beta_1 \neq 0$

The test statistic is $t = \dfrac{\hat{\beta}_1 - 0}{s_{\hat{\beta}_1}} = \dfrac{-.2389 - 0}{\dfrac{1.2651}{\sqrt{149.9286}}} = -2.31$

The rejection region requires $\alpha/2 = .01/2 = .005$ in each tail of the t distribution with df $= n - 2 = 7 - 2 = 5$. From Table VI, Appendix A, $t_{.005} = 4.032$. The rejection region is $t < -4.032$ or $t > 4.032$.

Since the observed value of the test statistic does not fall in the rejection region ($t = -2.31 \nleq -4.032$), H_0 is not rejected. There is insufficient evidence to indicate that the model is adequate at $\alpha = .01$.

e.   One point looks like a possible outlier: (301, 8.5).

f.   We will redo the analysis after eliminating the above point:

(a)   The scattergram of the data is:

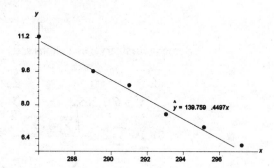

(b)   Some preliminary calculations are:

$$\sum x_i = 1{,}752 \qquad \sum y_i = 50.7 \qquad n = 6$$

$$\sum x_i^2 = 511{,}664.5 \qquad \sum y_i^2 = 444.97 \qquad \sum x_i y_i = 14{,}768.2$$

$$SS_{xy} = \sum x_i y_i - \frac{\left(\sum x_i\right)\left(\sum y_i\right)}{n} = 14{,}768.2 - \frac{1752(50.7)}{6} = -36.2$$

$$SS_{xx} = \sum x_i^2 - \frac{\left(\sum x_i\right)^2}{n} = 511{,}664.5 - \frac{1752^2}{6} = 80.5$$

$$SS_{yy} = \sum y_i^2 - \frac{\left(\sum y_i\right)^2}{n} = 444.97 - \frac{50.7^2}{6} = 16.555$$

$$\hat{\beta}_1 = \frac{SS_{xy}}{SS_{xx}} = \frac{-36.2}{80.5} = -.449689441 \approx -.4497$$

$$\hat{\beta}_0 = \bar{y} - \hat{\beta}_1\bar{x} = \frac{50.7}{6} - (-.449689441)\frac{1752}{6} = 139.7593168 \approx 139.759$$

The least squares line is $\hat{y} = 139.759 - .4497x$.

(c) The least squares line is plotted on the graph in part **f(a)**.

(d) Some additional calculations:

$$SSE = SS_{yy} - \hat{\beta}_1 SS_{xy} = 16.555 - (-.449689441)(-36.2) = .27624224$$

$$s^2 = \frac{SSE}{n-2} = \frac{.27624224}{6-2} = 0.06906056 \qquad s = \sqrt{0.06906056} = 0.2628$$

To determine if the model is adequate, we test:

H_0: $\beta_1 = 0$
H_a: $\beta_1 \neq 0$

The test statistic is $t = \dfrac{\hat{\beta}_1 - 0}{s_{\hat{\beta}_1}} = \dfrac{-.4497 - 0}{\dfrac{0.2628}{\sqrt{80.5}}} = -15.35$

The rejection region requires $\alpha/2 = .01/2 = .005$ in each tail of the t distribution with df $= n - 2 = 6 - 2 = 4$. From Table VI, Appendix A, $t_{.005} = 4.604$. The rejection region is $t < -4.604$ or $t > 4.604$.

Since the observed value of the test statistic falls in the rejection region ($t = -15.35 < -4.604$), H_0 is rejected. There is sufficient evidence to indicate that the model is adequate at $\alpha = .01$.

11.87 a. $r = .26$. Because this value is fairly close to 0, there is a weak positive linear relationship between a daughter's loneliness score and a mother's loneliness score.

b. $r = .19$. Because this value is fairly close to 0, there is a weak positive linear relationship between a daughter's loneliness score and a father's loneliness score.

c. $r = .14$. Because this value is fairly close to 0, there is a weak positive linear relationship between a daughter's loneliness score and a mother's self-esteem score.

d. $r = .01$. Because this value is extremely close to 0, there is an extremely weak positive linear relationship between a daughter's loneliness score and a father's assertiveness score. Because this value is so close to 0, these two variables are probably not linearly related.

e. (a) $r^2 = .26^2 = .0676$. 6.76% of the total sample variability around the sample mean daughter's loneliness score is explained by the linear relationship between a daughter's loneliness score and a mother's loneliness score.

(b) $r^2 = .19^2 = .0361$. 3.61% of the total sample variability around the sample mean daughter's loneliness score is explained by the linear relationship between a daughter's loneliness score and a father's loneliness score.

(c) $r^2 = .14^2 = .0196$. 1.96% of the total sample variability around the sample mean daughter's loneliness score is explained by the linear relationship between a daughter's loneliness score and a mother's self-esteem score.

(d) $r^2 = .01^2 = .0001$. 0.01% of the total sample variability around the sample mean daughter's loneliness score is explained by the linear relationship between a daughter's loneliness score and a father's assertiveness score.

Multiple Regression

12.1   a. $E(y) = \beta_0 + \beta_1 x_1 + \beta_2 x_2$

b. $E(y) = \beta_0 + \beta_1 x_1 + \beta_2 x_2 + \beta_3 x_3 + \beta_4 x_4$

c. $E(y) = \beta_0 + \beta_1 x_1 + \beta_2 x_2 + \beta_3 x_3 + \beta_4 x_4 + \beta_5 x_5$

12.3   a.   We are given $\hat{\beta}_2 = 2.7$, $s_{\hat{\beta}_2} = 1.86$, and $n = 30$.

$$H_0:\ \beta_2 = 0$$
$$H_a:\ \beta_2 \neq 0$$

The test statistic is $t = \dfrac{\hat{\beta}_2 - 0}{s_{\hat{\beta}_2}} = \dfrac{2.7}{1.86} = 1.45$

The rejection region requires $\alpha/2 = .05/2 = .025$ in each tail of the t distribution with df $= n - (k + 1) = 30 - (3 + 1) = 26$. From Table VI, Appendix A, $t_{.025} = 2.056$. The rejection region is $t < -2.056$ or $t > 2.056$.

Since the observed value of the test statistic does not fall in the rejection region ($t = 1.45$ $\not> 2.056$), H_0 is not rejected. There is insufficient evidence to indicate $\beta_2 \neq 0$ at $\alpha = .05$.

b.   We are given $\beta_3 = .93$, $s_{\hat{\beta}_3} = .29$, and $n = 30$.

Test $H_0:\ \beta_3 = 0$
          $H_a:\ \beta_3 \neq 0$

The test statistic is $t = \dfrac{\hat{\beta}_3 - 0}{s_{\hat{\beta}_3}} = \dfrac{.93}{.29} = 3.21$

The rejection region is the same as part **a**, $t < -2.056$ or $t > 2.056$.

Since the observed value of the test statistic falls in the rejection region ($t = 3.21 > 2.056$), H_0 is rejected. There is sufficient evidence to indicate $\beta_3 \neq 0$ at $\alpha = .05$.

c. $\hat{\beta}_3$ has a smaller estimated standard error than $\hat{\beta}_2$. Therefore, the test statistic is larger for $\hat{\beta}_3$ even though $\hat{\beta}_3$ is smaller than $\hat{\beta}_2$.

12.5   The number of degrees of freedom available for estimating σ^2 is $n - (k + 1)$ where k is the number of variables in the regression model. Each additional independent variable placed in the model causes a corresponding decrease in the degrees of freedom.

Multiple Regression   281

12.7   a.   From the output, the least squares prediction equation is
$$\hat{y} = 3.70 + .34x_1 + .49x_2 + .72x_3 + 1.14x_4 + 1.51x_5 + .26x_6 - .14x_7 - .10x_8 - .10x_9.$$

b. H_0: $\beta_7 = 0$
H_a: $\beta_7 < 0$

The test statistic is $t = \dfrac{\hat{\beta}_7 - 0}{s_{\hat{\beta}_7}} = \dfrac{-.14 - 0}{.14} = -1.00$

The rejection region requires $\alpha = .05$ in the lower tail of the t distribution. From Table VI, Appendix A, with df $= n - (k + 1) = 234 - (9 + 1) = 224$, $t_{.05} \approx 1.645$. The rejection region is $t < -1.645$.

Since the observed value of the test statistic does not fall in the rejection region ($t = -1.00 \nless -1.645$), H_0 is not rejected. There is insufficient evidence to indicate that there is a negative linear relationship between the number of runs scored and the number of times caught stealing at $\alpha = .05$.

c.   For confidence coefficient .95, $\alpha = .05$ and $\alpha/2 = .05/2 = .025$. From Table VI, Appendix A, with df $= n - (k + 1) = 234 - (9 + 1) = 224$, $t_{.025} \approx 1.96$. The 95% confidence interval is:

$$\hat{\beta}_5 \pm t_{.025} s_{\hat{\beta}_5} \Rightarrow 1.51 \pm 1.96(.05) \Rightarrow 1.51 \pm .098 \Rightarrow (1.412, 1.608)$$

We are 95% confident that for each additional home run, the mean number of runs scored will increase by anywhere from 1.412 to 1.608, holding all the other variables constant.

12.9   a.   The SAS printout for the model is:

DEPENDENT VARIABLE: Y

SOURCE	DF	SUM OF SQUARES	MEAN SQUARE	F VALUE
MODEL	5	1052894700508.240	210578940101.648	190.75
ERROR	19	20975246806.001	1103960358.211	PR > F
CORRECTED TOTAL	24	1073869947314.240		0.0001

R-SQUARE	C.V.	ROOT MSE	Y MEAN
0.980468	11.4346	33225.899	290573.52000000

PARAMETER	ESTIMATE	T FOR HO: PARAMETER=0	PR > T	STD ERROR OF ESTIMATE
INTERCEPT	93073.85223495	3.24	0.0043	28720.89686205
X1	4152.20700875	2.78	0.0118	1491.62587008
X2	-854.94161450	-2.86	0.0099	298.44765134
X3	0.92424393	0.32	0.7515	2.87673442
X4	2692.46175182	1.71	0.1041	1577.28622584
X5	15.54276851	10.62	0.0001	1.46287006

The least squares prediction equation is:

$$\hat{y} = 93{,}074 + 4152x_1 - 855x_2 + .924x_3 + 2692x_4 + 15.5x_5$$

b. $s = $ ROOT MSE $= 33{,}225.9$. We would expect about 95% of the observations to fall within $\pm 2s$ or $\pm 2(33{,}225.9)$ or $\pm 66{,}452$ units of the regression line.

c. To determine if the value increases with the number of units, we test:

H_0: $\beta_1 = 0$
H_a: $\beta_1 > 0$

The test statistic is $t = \dfrac{\hat{\beta}_1 - 0}{s_{\hat{\beta}_1}} = \dfrac{4152 - 0}{1491.626} = 2.78$

The observed significance level or p-value is $.0118/2 = .0059$. Since this value is less than $\alpha = .05$, H_0 is rejected. There is sufficient evidence to indicate that the value increases as the number of units increases at $\alpha = .05$.

d. $\hat{\beta}_1$: We estimate the mean value will increase by \$4,152 for each additional apartment unit, all other variables held constant.

e. Using SAS, the plot is:

```
PLOT IF Y*X2      LEGEND:    A = 1 OBS, B = 2 OBS, ETC.
          y                             A
900,000  ┤

                                  A
600,000  ┤   A                       A
             A  A
           A  A
300,000  ┤     A               A
               A          A      A  A
           BA           A AA   A A
                      A            D
      0  ┼──────────────────────────────── x₂
         0  5 10 15 20 25 30 35 40 45 50 55 60 65 70 75 80 85   1
```

It appears from the graph that there is not much of a relationship between value (y) and age (x_2).

f. H_0: $\beta_2 = 0$
H_a: $\beta_2 < 0$

The test statistic is $t = \dfrac{\hat{\beta}_2 - 0}{s_{\hat{\beta}_2}} = \dfrac{-855 - 0}{298.447} = -2.86$

The observed significance level or p-value is $.0099/2 = .00495$. Since this value is less than $\alpha = .01$, H_0 is rejected. There is sufficient evidence to indicate that the value and age are negatively related, all other variables in the model held constant at $\alpha = .01$.

A one-tailed rather than two-tailed test is performed because $\hat{\beta}_2$ is negative. If $\hat{\beta}_2 < 0$, we could never reject H_0: $\beta_2 > 0$.

g.   The *p*-value is .0099/2 = .00495 (because we had a one-tailed test).

12.11 a. $\hat{y} = 12.1804 - .02654x_1 - .45783x_2$.

b. $\hat{\beta}_0$ = the estimate of the *y*-intercept

$\hat{\beta}_1 = -.0265$. We estimate that the mean weight change will decrease by .0265% for each additional increase of 1% in digestion efficiency, with acid-detergent fibre held constant.

$\hat{\beta}_2 = -.458$. We estimate that the mean weight change will decrease by .458% for each additional increase of 1% in acid-detergent fibre, with digestion efficiency held constant.

c.   To determine if digestion efficiency is a useful predictor of weight change, we test:

$H_0: \ \beta_1 = 0$
$H_a: \ \beta_1 \neq 0$

The test statistic is $t = -.50$. The *p*-value is $p = .6226$ Since the *p*-value is greater than α ($p = .6226 > .01$), H_0 is not rejected. There is insufficient evidence to indicate that digestion efficiency is a useful linear predictor of weight change at $\alpha = .01$.

d.   For confidence coefficient .99, $\alpha = 1 - .99 = .01$ and $\alpha/2 = .01/2 = .005$. From Table VI, Appendix A, with df $= n - (k + 1) = 42 - (2 + 1) = 39$, $t_{.005} \approx 2.704$. The 99% confidence interval is:

$$\hat{\beta}_2 \pm t_{.005} s_{\hat{\beta}} \Rightarrow -.4578 \pm 2.704 \,(.1283) \Rightarrow -.4578 \pm .3469 \Rightarrow (-.8047, -.1109)$$

We are 99% confident that the change in mean weight change for each unit change in acid-detergent fiber, holding digestion efficiency constant is between $-.8047$ and $-.1109$.

12.13 a. $E(y) = \beta_0 + \beta_1 x_1 + \beta_2 x_2 + \beta_3 x_3 + \beta_4 x_4 + \beta_5 x_5 + \beta_6 x_6 + \beta_7 x_7$

b.   Using MINITAB, the output is:

```
The regression equation is y = 0.998 - 0.0224 x1 + 0.156x2 - 0.0172x3 - 0.00953x4
+ 0.421x5 + 0.417x6 - 0.155x7

Predictor        Coef        StDev                      P
Constant        0.9981       0.2475          4.0       0.002
x1             -0.022429    0.005039         -4.4       0.001
x2              0.15571     0.07429           2.1       0.060
x3             -0.01719     0.01186          -1.4       0.175
x4             -0.009527    0.009619         -0.9       0.343
x5              0.4214      0.1008            4.1       0.002
x6              0.4171      0.4377            0.9       0.361
x7             -0.1552      0.1486           -1.0       0.319

S = 0.4365      R-Sq = 77.1%      R-Sq(adj) = 62.5%
```

```
Analysis of Variance

Source              DF          SS          MS          F          P
Regression          7        7.9578      1.1368       5.29      0.007
Residual           11        2.3632      0.2148
Error
Total              18       10.3210

Sourc       DF          Seq S
e
x1           1          1.401
x2           1          1.926
x3           1          0.117
x4           1          0.044
x5           1          4.077
x6           1          0.156
x7           1          0.234

Unusual Observations

Obs        x1           y       Fit StDev Fit   Residual  St Resid
14        80.0       0.120     0.628     0.328      0.748     2.28R

R denotes an observation with a large standardized residual.
```

The least squares model is $= .9981 - .0224x_1 + .1557x_2 - .0172x_3 - .0095x_4$
$+ .4214x_5 + .4171x_6 - .1552x_7$

c. $\hat{\beta}_0 = .9981 =$ the estimate of the y-intercept.

$\hat{\beta}_1 = -.0224$; We estimate that the mean voltage will decrease by .0224 kw/cm, for each additional increase of 1% of x_1, the disperse phase volume (with all other variables held constant).

$\hat{\beta}_2 = .1557$; We estimate that the mean voltage will increase by .1557 kw/cm for each additional increase of 1% of x_2, the salinity (with all other variables held constant).

$\hat{\beta}_3 = -.0172$; We estimate the the mean voltage will decrease by .0172 kw/cm for each additional increase of 1 degree of x_3, the temperature in Celcius (with all other variables held constant).

$\hat{\beta}_4 = -.0095$; We estimate that the mean voltage will decrease by .0095 kw/cm for each additional increase of 1 hour of x_4, the time delay (with all other variables held constant).

$\hat{\beta}_5 = .4214$; We estimate that the mean voltage will increase by .4124 kw/cm for each additional increase of 1% of x_5, sufficient concentration (with all other variables held constant).

$\hat{\beta}_6 = .4171$; We estimate that the mean voltage will increase by .4171 kw/cm for each additional increase of 1 unit of x_6, span: Triton (with all other variables held constant).

$\hat{\beta}_7 = -.1552$; We estimate that the mean voltage will decrease by .1552 kw/cm for each additional increase of 1% of x_7, the solid particles (with all other variables held constant).

12.15   a.   **Model 1:**

H_0: $\beta_1 = 0$
H_a: $\beta_1 \neq 0$

The test statistic is $t = \dfrac{\hat{\beta}_1 - 0}{s_{\hat{\beta}_1}} = \dfrac{.0354}{.0137} = 2.58$.

Since no α was given, we will use $\alpha = .05$. The rejection region requires $\alpha/2 = .05/2 = .025$ in each tail of the t distribution. From Table VI, Appendix A, with df $= n - (k + 1) = 12 - (1 + 1) = 10$, $t_{.025} = 2.228$. The rejection region is $t < -2.228$ or $t > 2.228$.

Since the observed value of the test statistic falls in the rejection region ($t = 2.58 > 2.228$), H_0 is rejected. There is sufficient evidence to indicate that there is a linear relationship between vintage and the logarithm of price.

Model 2:

H_0: $\beta_1 = 0$
H_a: $\beta_1 \neq 0$

The test statistic is $t = \dfrac{\hat{\beta}_1 - 0}{s_{\hat{\beta}_1}} = \dfrac{.0238}{.00717} = 3.32$

Since no α was given, we will use $\alpha = .05$. The rejection region requires $\alpha/2 = .05/2 = .025$ in each tail of the t distribution. From Table VI, Appendix A, with df $= n - (k + 1) = 12 - (4 + 1) = 7$, $t_{.025} = 2.365$. The rejection region is $t < -2.365$ or $t > 2.365$.

Since the observed value of the test statistic falls in the rejection region ($t = 3.32 > 2.365$), H_0 is rejected. There is sufficient evidence to indicate that there is a linear relationship between vintage and the logarithm of price, adjusting for all other variables.

H_0: $\beta_2 = 0$
H_a: $\beta_2 \neq 0$

The test statistic is $t = \dfrac{\hat{\beta}_2 - 0}{s_{\hat{\beta}_2}} = \dfrac{.616}{.0952} = 6.47$

The rejection region is $t < -2.365$ or $t > 2.365$.

Since the observed value of the test statistic falls in the rejection region ($t = 6.47 > 2.365$), H_0 is rejected. There is sufficient evidence to indicate that there is a linear relationship between average growing season temperature and the logarithm of price, adjusting for all other variables.

H_0: $\beta_3 = 0$
H_a: $\beta_3 \neq 0$

The test statistic is $t = \dfrac{\hat{\beta}_3 - 0}{s_{\hat{\beta}_3}} = \dfrac{-.00386}{.00081} = -4.77$

The rejection region is $t < -2.365$ or $t > 2.365$.

Since the observed value of the test statistic falls in the rejection region ($t = -4.77 < -2.365$), H_0 is rejected. There is sufficient evidence to indicate that there is a linear relationship between Sept./ Aug. rainfall and the logarithm of price, adjusting for all other variables.

H_0: $\beta_4 = 0$
H_a: $\beta_4 \neq 0$

The test statistic is $t = \dfrac{\hat{\beta}_4 - 0}{s_{\hat{\beta}_4}} = \dfrac{.0001173}{.000482} = 0.24.$

The rejection region is $t < -2.365$ or $t > 2.365$.

Since the observed value of the test statistic does not fall in the rejection region ($t = 0.24 - 2.365$), H_0 is not rejected. There is insufficient evidence to indicate that there is a linear relationship between rainfall in months preceding vintage and the logarithm of price, adjusting for all other variables.

Model 3:

H_0: $\beta_1 = 0$
H_a: $\beta_1 \neq 0$

The test statistic is $t = \dfrac{\hat{\beta}_1 - 0}{s_{\hat{\beta}_1}} = \dfrac{.0240}{.00747} = 3.21$

Since no α was given, we will use $\alpha = .05$. The rejection region requires $\alpha/2 = .05/2 = .025$ in each tail of the t distribution. From Table VI, Appendix A, with df $= n - (k + 1) = 12 - (5 + 1) + 7$, $t_{.025} = 2.447$. The rejection region is $t < -2.447$ or $t > 2.447$.

Since the observed value of the test statistic falls in the rejection region ($t = 3.21 > 2.447$), H_0 is rejected. There is sufficient evidence to indicate that there is a linear relationship between vintage and the logarithm of price, adjusting for all other variables.

H_0: $\beta_2 = 0$
H_a: $\beta_2 \neq 0$

The test statistic is $t = \dfrac{\hat{\beta}_2 - 0}{s_{\hat{\beta}_2}} = \dfrac{.608}{.116} = 5.24$.

The rejection region is $t < -2.447$ or $t > 2.447$.

Since the observed value of the test statistic falls in the rejection region ($t = 5.24 > 2.447$), H_0 is rejected. There is sufficient evidence to indicate that there is a linear relationship between average growing season temperature and the logarithm of price, adjusting for all other variables.

H_0: $\beta_3 = 0$
H_a: $\beta_3 \neq 0$

The test statistic is $t = \dfrac{\hat{\beta}_3 - 0}{s_{\hat{\beta}_3}} = \dfrac{-.00380}{.00095} = -4.00$

The rejection region is $t < -2.447$ or $t > -2.447$.

Since the observed value of the test statistic falls in the rejection region ($t = -4.00 > -2.447$), H_0 is rejected. There is sufficient evidence to indicate that there is a linear relationship between Sept./Aug. rainfall and the logarithm of price, adjusting for all other variables.

H_0: $\beta_4 = 0$
H_a: $\beta_4 \neq 0$

The test statistic is $t = \dfrac{\hat{\beta}_4 - 0}{s_{\hat{\beta}_4}} = \dfrac{.00115}{.000505} = 2.28$

The rejection region is $t < -2.447$ or $t > 2.447$.

Since the observed value of the test statistic does not fall in the rejection region ($t = 2.28 \not> 2.365$), H_0 is not rejected. There is insufficient evidence to indicate that there is a linear relationship between rainfall in months preceding vintage and the logarithm of price, adjusting for all other variables.

H_0: $\beta_5 = 0$
H_a: $\beta_5 \neq 0$

The test statistic is $t = \dfrac{\hat{\beta}_5 - 0}{s_{\hat{\beta}_5}} = \dfrac{.00765}{.0565} = 0.14$.

The rejection region is $t < -2.447$ or $t > 2.447$.

Since the observed value of the test statistic does not fall in the rejection region ($t = 0.14 \not> 2.365$), H_0 is not rejected. There is insufficient evidence to indicate that there is a linear relationship between average September temperature and the logarithm of price, adjusting for all other variables.

b. **Model 1:**

$\hat{\beta}_1 = .0354$, $e^{.0354} - 1 = .036$

We estimate that the mean price will increase by 3.6% for each additional increase of unit of x_1, vintage year.

Model 2:

$\hat{\beta}_1 = .0238$, $e^{.0238} - 1 = .024$

We estimate that the mean price will increase by 2.4% for each additional increase of 1 unit of x_1, vintage year (with all other variables held constant).

$\hat{\beta}_2 = .616$, $e^{.616} - 1 = .852$

We estimate that the mean price will increase by 85.2% for each additional increase of 1 unit of x_2, average growing season temperature   C (with all other variables held constant).

$\hat{\beta}_3 = -.00386$, $e^{-.00386} - 1 = -.004$

We estimate that the mean price will decrease by .4% for each additional increase of 1 unit of x_3, Sept./Aug. rainfall in cm (with all other variables held constant).

$\hat{\beta}_4 = .0001173$, $e^{.0001173} - 1 = .0001$

We estimate that the mean price will increase by .01% for each additional increase of 1 unit of x_4, rainfall in months prededing vintage in cm (with all other variables held constant).

Model 3:

$\hat{\beta}_1 = .0240$, $e^{.0240} - 1 = .024$

We estimate that the mean price will increase by 2.4% for each additional increase of 1 unit of x_1, vintage year (with all other variables held constant).

$\hat{\beta}_2 = .608$, $e^{.608} - 1 = .837$

We estimate that the mean price will increase by 83.7% for each additional increase of 1 unit of x_2, average growing season temperatures in °C (with all other variables held constant).

$\hat{\beta}_3 = -.00380$, $e^{-.00380} - 1 = -.004$

We estimate that the mean price will decrease by .4% for each additional increase of 1 unit of x_3, Sept./Aug. rainfall in cm, (with all other variables held constant).

$$\hat{\beta}_4 = .00115, \; e^{.00115} - 1 = .001$$

We estimate that the average mean price will increase by .1% for each additional increase of 1 unit of x_4, rainfall in months preceding vintage in cm (with all other variables held constant).

$$\hat{\beta}_5 = .00765, \; e^{.00765} - 1 = .008$$

We estimate that the average mean price will increase by .8% for each additional increase of 1 unit of x_5, average Sept. temperature in °C (with all other variables held constant).

12.17 a. From the printout, R-Sq = 45.9%. 45.9% of the total sample variability of y around its mean is explained by the linear relationship between y and the two independent variables, x_1 and x_2.

b. From the printout, R-Sq(adj) = 39.6%. 39.6% of the total variability of y around its mean is explained by the linear relationship between y and the two independent variables, x_1 and x_2, adjusting for the sample size and the number of independent variables.

c. H_0: $\beta_1 = \beta_2 = 0$
H_a: At least one $\beta_i \neq 0$, for $i = 1, 2$

The test statistic is $F = \dfrac{R^2/k}{(1-R^2)/[n-(k+1)]} = \dfrac{.459/2}{(1-.459)/[20-(2+1)]} = 7.21$

The rejection region requires $\alpha = .05$ in the upper tail of the F distribution with $v_1 = k = 2$ and $v_2 = n - (k+1) = 20 - (2+1) = 17$. From Table IX, Appendix A, $F_{.05} = 3.59$. The rejection region is $F > 3.59$.

Since the observed value of the test statistic falls in the rejection region ($F = 7.21 > 3.59$), H_0 is rejected. There is sufficient evidence to indicate the model is useful in predicting y at $\alpha = .05$.

The test statistic can also be calculated by $\dfrac{\text{MS(Model)}}{\text{MS(Error)}} = \dfrac{64,165}{8,883} = 7.22$.
From the printout, $F = 7.22$.

d. From the printout, the p-value is $p = .005$. The probability of observing a test statistic of 7.22 or anything higher if H_0 is true is .005. This is very unusual if H_0 is true. There is strong evidence to reject H_0 for $\alpha > .005$.

12.19 a. Some preliminary calculations are:

$$SSE = \sum(y_i - \hat{y}_i)^2 = 12.35, \ df = n - (k+1) = 20 - (2+1) = 17$$

$$SS(Total) = \sum(y - \bar{y})^2 = 24.44, \ df = n - 1 = 20 - 1 = 19$$

$$SS(Model) = SS(Total) - SSE = 24.44 - 12.35 = 12.09, \ df = k = 2$$

$$MS(Model) = \frac{SS(Model)}{k} = \frac{12.09}{2} = 6.045$$

$$MS(Error) = \frac{SSE}{n-(k+1)} = \frac{12.35}{17} = .72647$$

$$F = \frac{MS(Model)}{MS(Error)} = \frac{6.045}{.72647} = 8.321$$

$$R^2 = 1 - \frac{SSE}{SS(Total)} = 1 - \frac{12.35}{24.44} = .4947$$

The test statistic could also be calculated by:

$$F = \frac{R^2/k}{(1-R^2)/[n-(k+1)]} = \frac{.4947/2}{(1-.4947)/17} = 8.32$$

The analysis of variance table is:

Source	df	SS	MS	F
Model	2	12.09	6.045	8.321
Error	17	12.35	.72647	
Total	19	24.44		

b. H_0: $\beta_1 = \beta_2 = 0$
H_a: At least one $\beta_i \neq 0, \ i = 1, 2$

The test statistic is $F = \dfrac{MS(Model)}{MS(Error)} = \dfrac{6.045}{.72647} = 8.321$

The rejection region requires $\alpha = .05$ in the upper tail of the F distribution with df = $v_1 = k = 2$ and $v_2 = n - (k+1) = 17$. From Table IX, Appendix A, $F_{.05} = 3.59$. The rejection region is $F > 3.59$.

Since the observed value of the test statistic falls in the rejection region ($F = 8.321 > 3.59$), H_0 is rejected. There is sufficient evidence to indicate the model is useful in predicting y at $\alpha = .05$.

12.21 a. From the problem, the least squares prediction equation for y is

$$\hat{y} = ????? + .110x_1 + .065x_2 + .540x_3 - .009x_4 - .150x_5 - .027x_6.$$

b. To test for overall model adequacy, we test:

H_0: $\beta_1 = \beta_2 = \beta_3 = \beta_4 = \beta_5 = \beta_6 = 0$
H_a: At least one $\beta_i \neq 0$, for $i = 1, 2, 3, 4, 5,$ or 6

c. From the problem, the test statistic is $F = 32.47$ and the p-value is $p < .001$. Since the p-value is less than α ($p = .001 < .01$), H_0 is rejected. There is sufficient evidence that at least one of the independent variables (total population, population density, population concentration, population growth, farm land and agricultural change) is useful in predicting the urban/rural rating at $\alpha = .01$.

d. From the problem $R^2 = .44$. 44% of the sample variability of the urban/rural ratings around the mean is explained by the linear relationship between the urban/rural rating and the 6 independent variables (total population, population density, population concentration, population growth, farm land and agricultural change).

From the problem $R^2_{adj} = .43$. 43% of the sample variability of the urban/rural ratings around the mean is explained by the linear relationship between the urban/rural rating and the 6 iindependent variables (total population, population density, population concentration, population growth, farm land and agricultural change), adjusting for the sample size and the number of independent variables in the model.

e. To determine if population growth contributes to the prediction of urban/rural rating, the null hypothesis is:

H_0: $\beta_4 = 0$

f. To determine if population growth contributes to the prediction of urban/rural rating, we test:

H_0: $\beta_4 = 0$
H_a: $\beta_4 \neq 0$

No test statistic is given. However, the p-value associated with the test statistic is $p = .860$. Since the p-value is greater than α ($p = .860 > .01$), H_0 is not rejected. There is insufficient evidence of a linear relationship between urban/rural rating and population growth, holding all the other variables constant for any $\alpha < .860$.

12.23 a. $R^2 = .912$. 91.2% of the total variability of equivalent widths is explained by the model containing the four independent variables.

$R_a^2 = .894$. This statistic has a similar interpretation to that of R^2, but is adjusted for both the sample size n and the number of β parameters in the model.

The R_a^2 statistic is the preferred measure of model fit because it takes into account the sample size and the number of β parameters.

b. H_0: $\beta_1 = \beta_2 = \beta_3 = \beta_4 = 0$
H_a: At least one $\beta_i \neq 0$, $i = 1, 2, 3, 4$

The test statistic is $F = 51.720$ with p-value $< .000$. Since the p-value is so small, H_0 is rejected for any $\alpha \geq .000$. There is sufficient evidence to indicate that the model is adequate.

12.25 a. To determine if crime prevalence is positively related to density, we test:

H_0: $\beta_1 = 0$
H_a: $\beta_1 > 0$

The test statistic is $t = 3.88$.

The p-value is $p < .01/2 = .005$. Since the p-value is so small, there is strong evidence to indicate that the crime prevalence is positively related to density for $\alpha > .005$.

b. No. The tests are not independent of each other. If we conduct a series of t-tests to determine whether the independent variables are contributing to the predictive relationship, we would very likely make one or more errors in deciding which terms to retain in the model and which to exclude.

c. To test the utility of the model, we test:

H_0: $\beta_1 = \beta_2 = \beta_3 = \cdots = \beta_{18} = 0$
H_a: At least one $\beta_i \neq 0$, $i = 1, 2, 3, \ldots, 18$

The test statistic is:

$$F = \frac{R^2/k}{(1-R^2)/[n-(k+1)]} = \frac{.411/18}{(1-.411)/[313-(18+1)]} = 11.397$$

The rejection region requires $\alpha = .05$ in the upper tail of the F distribution with $v_1 = k = 18$ and $v_2 = n - (k+1) = 313 - (18+1) = 294$. From Table IX, Appendix A, $F_{.05} \approx 1.57$. The rejection region is $F > 1.57$.

Since the observed value of the test statistic falls in the rejection region ($F = 11.397 > 1.57$), H_0 is rejected. There is sufficient evidence that the model is useful in predicting crime prevalence at $\alpha = .05$.

12.27 a. $R^2 = .362$. 36.2% of the variability in the AC scores can be explained by the model containing the variables self-esteem score, optimism score, and group cohesion score.

b. To test the utility of the model, we test:

H_0: $\beta_1 = \beta_2 = \beta_3 = 0$
H_a: At least one $\beta_i \neq 0$, $i = 1, 2, 3$

The test statistic is:

$$F = \frac{R^2/k}{(1-R^2)[n-(k+1)]} = \frac{.362/3}{(1-.362)/[31-(3+1)]} = 5.11$$

The rejection region requires $\alpha = .05$ in the upper tail of the F distribution with $v_1 = k = 3$ and $v_2 = n - (k + 1) = 31 - (3 + 1) = 27$. From Table IX, Appendix A, $F_{.05} = 2.96$. The rejection region is $F > 2.96$.

Since the observed value of the test statistic falls in the rejection region ($F = 5.11 > 2.96$), H_0 is rejected. There is sufficient evidence that the model is useful in predicting AC score at $\alpha = .05$.

12.29   a.   The hypothesized model is:

$$E(y) = \beta_0 + \beta_1 x_1 + \beta_2 x_2 + \beta_3 x_3 + \beta_4 x_4 + \beta_5 x_5$$

$\beta_0 = y$-intercept. It has no interpretation in this model.

$\beta_1 = $ difference in the mean salaries between males and females, all other variables held constant.

$\beta_2 = $ difference in the mean salaries between whites and nonwhites, all other variables held constant.

$\beta_3 = $ change in the mean salary for each additional year of education, all other variables held constant.

$\beta_4 = $ change in the mean salary for each additional year of tenure with firm, all other variables held constant.

$\beta_5 = $ change in the mean salary for each additional hour worked per week, all other variables held constant.

   b.   The least squares equation is:

$$\hat{y} = 15.491 + 12.774x_1 + .713x_2 + 1.519x_3 + .32x_4 + .205x_5$$

$\hat{\beta}_0 = $ estimate of the y-intercept. It has no interpretation in this model.

$\hat{\beta}_1$:   We estimate the difference in the mean salaries between males and females to be $12.774, all other variables held constant.

$\hat{\beta}_2$:   We estimate the difference in the mean salaries between whites and nonwhites to be
$.713, all other variables held constant.

$\hat{\beta}_3$:   We estimate the change in the mean salary for each additional year of education to be $1.519, all other variables held constant.

$\hat{\beta}_4$:   We estimate the change in the mean salary for each additional year of tenure with firm to be $.320, all other variables held constant.

$\hat{\beta}_5$: We estimate the change in the mean salary for each additional hour worked per week to be \$.205, all other variables held constant.

c. $R^2 = .240$. 24% of the total variability of salaries is explained by the model containing gender, race, educational level, tenure with firm, and number of hours worked per week.

To determine if the model is useful for predicting annual salary, we test:

H_0: $\beta_1 = \beta_2 = \beta_3 = \beta_4 = \beta_5 = 0$
H_a: At least one $\beta_i \neq 0$

The test statistic is $F = \dfrac{R^2/k}{(1-R^2)[n-(k+1)]} = \dfrac{.24/5}{(1-.24)/[191-(5+1)]} = 11.68$

The rejection region requires $\alpha = .05$ in the upper tail of the F distribution with numerator df $= k = 5$ and denominator df $= n - (k+1) = 191 - (5+1) = 185$. From Table IX, Appendix A, $F_{.05} \approx 2.21$. The rejection region is $F > 2.21$.

Since the observed value of the test statistic falls in the rejection region ($F = 11.68 > 2.21$), H_0 is rejected. There is sufficient evidence to indicate the model containing gender, race, educational level, tenure with firm, and number of hours worked per week is useful for predicting annual salary for $\alpha = .05$.

d.   To determine if male managers are paid more than female managers, we test:

H_0: $\beta_1 = 0$
H_a: $\beta_1 > 0$

The p-value given for the test $p < .05/2 = .025$. Since the p-value is less than $\alpha = .05$, there is evidence to reject H_0. There is evidence to indicate male managers are paid more than female managers holding all other variables constant for $\alpha > .025$.

e.   The salary paid an individual depends on many factors other than gender. Thus, in order to adjust for other factors influencing salary, we include them in the model.

12.31   a.   From the printout, the 95% prediction interval is (1,759.7, 4,275.4). We are 95% confident that the actual annual salary for a 45-year old vendor who works 10 hours per day will be between \$1,759.70 and \$4,275.40.

b.   From the printout, the 95% confidence interval is (2,620.3, 3,414.9). We are 95% confident that the mean annual salary for 45-year old vendors who work 10 hours per day will be between \$2,620.30 and \$3,414.90.

c.   Yes. The prediction interval for the actual value will always be wider than the confidence interval for the mean value. The actual values are more variable than the mean values.

12.33   The 95% prediction interval is (90.69, 158.57). With 95% confidence, we conclude that the equivalent width for an individual quasar with a redshift of 3.07, line flux of -13.56, line luminosity of 45.30, and AB_{1450} of 19.59 will be between 90.69 and 158.57.

Multiple Regression   **295**

12.35   The first order model is:

$$E(y) = \beta_0 + \beta_1 x_1 + \beta_2 x_2 + \beta_3 x_5$$

We want to find a 95% prediction interval for the actual voltage when the volume fraction of the disperse phase is at the high level ($x_1 = 80$), the salinity is at the low level ($x_2 = 1$), and the amount of surfactant is at the low level ($x_5 = 2$).

Using MINITAB, the output is:

```
The regression equation is
y = 0.993 - 0.243 x1 + 0.142 x2 + 0.385 x5

Predictor           Coef        StDev          T           P
Constant          0.9326       0.2482        3.76       0.002
x1              -0.024272     0.004900       -4.95       0.000
x2               0.14206      0.07573        1.88       0.080
x5               0.38457      0.09801        3.92       0.001


S = 0.4796      R-Sq = 66.6%      R-Sq(adj) = 59.9%

Analysis of Variance

Source           DF          SS          MS          F          P
Regression        3        6.8701      2.2900       9.95       0.001
Residual         15        3.4509      0.2301
Error
Total            18       10.3210

Sourc      DF      Seq SS
e
x1          1       1.4016
x2          1       1.9263
x5          1       3.5422

Unusual Observations
Obs       x1        y        Fit    StDev Fit   Residual   St Resid
 3       40.0     3.200     2.068    0.239       1.132      2.72R

R denotes an observation with a large standardized residual

Predicted Values

   Fit     StDev Fit        95.0%   CI           95.0%    PI
 -0.098      0.232     ( -0.592,   0.396)     (  -1.233,  10308)
```

The 95% prediction interval is (−1.233, 1.038). We are 95% confident that the actual voltage is between −1.233 and 1.038 when the volume fraction of the disperse phase is at the high level ($x_1 = 80$), the salinity is at the low level ($x_2 = 1$), and the amount of surfactant is at the low level ($x_5 = 2$).

12.37   a. $E(y) = \beta_0 + \beta_1 x_1 + \beta_2 x_2 + \beta_3 x_1 x_2$

        b. $E(y) = \beta_0 + \beta_1 x_1 + \beta_2 x_2 + \beta_3 x_3 + \beta_4 x_1 x_2 + \beta_5 x_1 x_3 + \beta_6 x_2 x_3$

12.39   a.   The response surface is a twisted surface in three-dimensional space.

        b.   For $x_1 = 0$, $E(y) = 3 + 0 + 2x_2 - 0x_2 = 3 + 2x_2$
             For $x_1 = 1$, $E(y) = 3 + 1 + 2x_2 - 1x_2 = 4 + x_2$
             For $x_1 = 2$, $E(y) = 3 + 2 + 2x_2 - 2x_2 = 5$

The plot of the lines is:

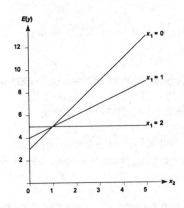

c. The lines are not parallel because interaction between x_1 and x_2 is present. Interaction between x_1 and x_2 means that the effect of x_2 on y depends on what level x_1 takes on.

d. For $x_1 = 0$, as x_2 increases from 0 to 5, $E(y)$ increases from 3 to 13.
For $x_1 = 1$, as x_2 increases from 0 to 5, $E(y)$ increases from 4 to 9.
For $x_1 = 2$, as x_2 increases from 0 to 5, $E(y) = 5$.

e. For $x_1 = 2$ and $x_2 = 4$, $E(y) = 5$
For $x_1 = 0$ and $x_2 = 5$, $E(y) = 13$

Thus, $E(y)$ changes from 5 to 13.

12.41 a. The prediction equation is:

$$\hat{y} = -2.55 + 3.82x_1 + 2.63x_2 - 1.29x_1x_2$$

b. The response surface is a twisted plane, since the equation contains an interaction term.

c. For $x_2 = 1$, $= -2.55 + 3.82x_1 + 2.63(1) - 1.29x_1(1)$
$= .08 + 2.53x_1$
For $x_2 = 3$, $= -2.55 + 3.82x_1 + 2.63(3) - 1.29x_1(3)$
$= 5.34 - .05x_1$
For $x_2 = 5$, $= -2.55 + 3.82x_1 + 2.63(5) - 1.29x_1(5)$
$= 10.6 - 2.63x_1$

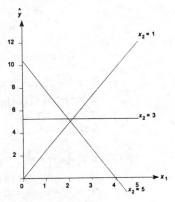

d. If x_1 and x_2 interact, the effect of x_1 on y is different at different levels of x_2. When $x_2 = 1$, as x_1 increases, $\hat{y}$ also increases. When $x_2 = 5$, as x_1 increases, $\hat{y}$ decreases.

e. The hypotheses are:

H_0: $\beta_3 = 0$
H_a: $\beta_3 \neq 0$

f. The test statistic is $t = \dfrac{\hat{\beta}_3}{s_{\hat{\beta}_3}} = \dfrac{-1.285}{.159} = -8.06$

The rejection region requires $\alpha/2 = .01/2 = .005$ in each tail of the t distribution with df $= n - (k + 1) = 15 - (3 + 1) = 11$. From Table VI, Appendix A, $t_{.005} = 3.106$. The rejection region is $t < -3.106$ or $t > 3.106$.

Since the observed value of the test statistic falls in the rejection region ($t = -8.06 < -3.106$), H_0 is rejected. There is sufficient evidence to indicate that x_1 and x_2 interact at $\alpha = .01$.

12.43   a.   The phrase "x_1 and x_2 interact" means that the effect of attitude toward drinking (x_1) on frequency of drinking alcoholic beverages (y) depends on the level of social support (x_2).

b.   To determine if attitude and social support interact, we test:

H_0: $\beta_3 = 0$
H_a: $\beta_3 \neq 0$

c.   Since the p-value is so small ($p < .001$), H_0 is rejected for any value of $\alpha > .001$. There is sufficient evidence to indicate that attitude toward drinking and social support interact to affect frequency of drinking.

12.45   a.   A model including the interaction term is:

$E(y) = \beta_0 + \beta_1 x_1 + \beta_2 x_2 + \beta_3 x_1 x_2$

b.   To determine if the effect of treatment on spelling score depends on disease intensity, we test:

H_0: $\beta_3 = 0$
H_a: $\beta_3 \neq 0$

The test statistic is $t = 1.6$.

The p-value is $p = .02$. Since the p-value is less than α ($p = .02 < .05$), H_0 is rejected. There is sufficient evidence to indicate that the effect of treatment on spelling score depends on disease intensity at $\alpha = .05$.

c.   Since the two variables interact, the main effects may be covered up by the interaction effect. Thus, tests on main effects should not be made. Also, the interpretation of the coefficients of the main effects should be interpreted with caution. Since the independent variables interact, the effect of one independent variable on the dependent variable depends on the level of the second independent variable.

12.47   a.   By including the interaction terms, it implies that the relationship between voltage and volume fraction of the disperse phase depend on the levels of salinity and surfactant concentration.

A possible sketch of the relationship is:

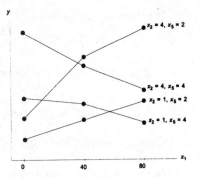

b. Using SAS, the printout is:

Model: MODEL1
Dependent Variable: Y

Analysis of Variance

Source	DF	Sum of Squares	Mean Square	F Value	Prob>F
Model	5	7.01028	1.40206	5.505	0.0061
Error	13	3.31073	0.25467		
C Total	18	10.32101			

Root MSE	0.50465	R-square	0.6792		
Dep Mean	0.97684	Adj R-sq	0.5558		
C.V.	51.66138				

Parameter Estimates

Variable	DF	Parameter Estimate	Standard Error	T for H0: Parameter=0	Prob > |T|
INTERCEP	1	0.905732	0.28546326	3.173	0.0073
X1	1	-0.022753	0.00831751	-2.736	0.0170
X2	1	0.304719	0.23660006	1.288	0.2202
X5	1	0.274741	0.22704807	1.210	0.2478
X1X2	1	-0.002804	0.00378998	-0.740	0.4725
X1X5	1	0.001579	0.00394692	0.400	0.6956

Obs	X1	X2	X5	Dep Var Y	Predict Value	Std Err Predict	Lower95% Predict	Upper95% Predict	Residual
1	40	1	2	0.6400	0.8640	0.185	-0.2969	2.0248	-0.2240
2	80	1	4	0.8000	0.7701	0.309	-0.5082	2.0484	0.0299
3	40	4	4	3.2000	2.1174	0.264	0.8869	3.3480	1.0826
4	80	4	2	0.4800	0.2092	0.305	-1.0645	1.4828	0.2708
5	40	1	4	1.7200	1.5398	0.309	0.2617	2.8178	0.1802
6	80	1	2	0.3200	-0.0320	0.283	-1.2820	1.2181	0.3520
7	40	4	2	0.6400	1.4416	0.292	0.1824	2.7009	-0.8016
8	80	4	4	0.6800	1.0113	0.298	-0.2553	2.2779	-0.3313
9	40	1	2	0.1200	0.8640	0.185	-0.2969	2.0248	-0.7440
10	80	1	4	0.8800	0.7701	0.309	-0.5082	2.0484	0.1099
11	40	4	4	2.3200	2.1174	0.264	0.8869	3.3480	0.2026
12	80	4	2	0.4000	0.2092	0.305	-1.0645	1.4828	0.1908
13	40	1	4	1.0400	1.5398	0.309	0.2617	2.8178	-0.4998
14	80	1	2	0.1200	-0.0320	0.283	-1.2820	1.2181	0.1520
15	40	4	2	1.2800	1.4416	0.292	0.1824	2.7009	-0.1616
16	80	4	4	0.7200	1.0113	0.298	-0.2553	2.2779	-0.2913
17	0	0	0	1.0800	0.9057	0.285	-0.3468	2.1583	0.1743
18	0	0	0	1.0800	0.9057	0.285	-0.3468	2.1583	0.1743
19	0	0	0	1.0400	0.9057	0.285	-0.3468	2.1583	0.1343

Sum of Residuals		0
Sum of Squared Residuals		3.3107
Predicted Resid SS (Press)		6.5833

The fitted regression line is:

$$\hat{y} = .906 - .023x_1 + .350x_2 + .275x_5 - .003x_1x_2 + .002x_1x5$$

To determine if the model is useful, we test:

$H_0: \beta_1 = \beta_2 = \beta_3 = \beta_4 = \beta_5 = 0$
$H_a:$ At least one $\beta_i \neq 0$, for $i = 1, 2, ..., 5$

The test statistic is $F = 5.505$.

The rejection region requires $\alpha = .05$ in the upper tail of the F-distribution with $v_1 = k = 5$ and $v_2 = n - (k + 1) = 19 - (5 + 1) = 13$. From Table VIII, Appendix B, $F_{.05} = 3.03$. The rejection region is $F > 3.03$.

Since the observed value of the test statistic falls in the rejection region ($F = 5.505 > 3.03$), H_0 is rejected. There is sufficient evidence to indicate the model is useful for predicting voltage at $\alpha = .05$.

$R^2 = .6792$. Thus, 67.92% of the sample variation of voltage is explained by the model containing the three independent variables and two interaction terms.

The estimate of the standard deviation is $s = .505$.

Comparing this model to that fit in Exercise 12.13, the model in Exercise 12.13 appears to fit the data better. The model in Exercise 12.13 has a higher R^2 (.7710 vs .6792) and a smaller estimate of the standard deviation (.464 vs .505).

c. $\hat{\beta}_0 = .906$. This is simply the estimate of the y-intercept.

$\hat{\beta}_1 = -.023$. For each unit increase in disperse phase volume, we estimate that the mean voltage will decrease by .023 units, holding salinity and surfactant concentration at 0.

$\hat{\beta}_2 = .305$. For each unit increase in salinity, we estimate that the mean voltage will increase by .305 units, holding disperse phase volume and surfactant concentration at 0.

$\hat{\beta}_3 = .275$. For each unit increase in surfactant concentration, we estimate that the mean voltage will increase by .275 units, holding disperse phase volume and salinity at 0.

$\hat{\beta}_4 = -.003$. This estimates the difference in the slope of the relationship between voltage and disperse phase volume for each unit increase in salinity, holding surfactant concentration constant.

$\hat{\beta}_5 = .002$. This estimates the difference in the slope of the relationship between voltage and disperse phase volume for each unit increase in surfactant concentration, holding salinity constant.

12.49   a. H_0: $\beta_2 = 0$
             H_a: $\beta_2 \neq 0$

The test statistic is $t = \dfrac{\hat{\beta}_2 - 0}{s_{\hat{\beta}_2}} = \dfrac{.47 - 0}{.15} = 3.133$

The rejection region requires $\alpha/2 = .05/2 = .025$ in each tail of the t distribution with df $= n - (k + 1) = 25 - (2 + 1) = 22$. From Table VI, Appendix A, $t_{.025} = 2.074$. The rejection region is $t < -2.074$ or $t > 2.074$.

Since the observed value of the test statistic falls in the rejection region ($t = 3.133 > 2.074$), H_0 is rejected. There is sufficient evidence to indicate the true relationship is given by the quadratic model at $\alpha = .05$.

   b. H_0: $\beta_2 = 0$
        H_a: $\beta_2 > 0$

The test statistic is the same as in part **a**, $t = 3.133$.

The rejection region requires $\alpha = .05$ in the upper tail of the t distribution with df $= 22$. From Table VI, Appendix A, $t_{.05} = 1.717$. The rejection region is $t > 1.717$.

Since the observed value of the test statistic falls in the rejection region ($t = 3.133 > 1.717$), H_0 is rejected. There is sufficient evidence to indicate the quadratic curve opens upward at $\alpha = .05$.

12.51   a.

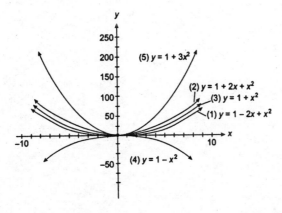

   b.   It moves the graph to the right ($-2x$) or to the left ($+2x$) compared to the graph of $y = 1 + x^2$.

   c.   It controls whether the graph opens up ($+x^2$) or down ($-x^2$). It also controls how steep the curvature is, i.e., the larger the absolute value of the coefficient of x^2, the narrower the curve is.

12.53 a.

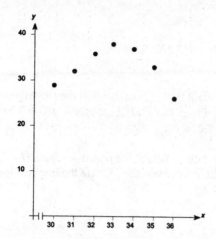

b.   If information was available only for x = 30, 31, 32, and 33, we would suggest a first-order model where $\beta_1 > 0$. If information was available only for x = 33, 34, 35, and 36, we would again suggest a first-order model where $\beta_1 < 0$. If all the information was available, we would suggest a second-order model.

12.55 a. $E(y) = \beta_0 + \beta_1 x_1 + \beta_2 x_2 + \beta_3 x_1 x_2 + \beta_4 x_1^2 + \beta_5 x_2^2$

b. $\beta_4 x_1^2$ and $\beta_5 x_2^2$

12.57 a.   A scattergram of the data is:

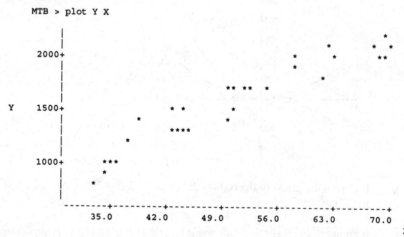

b.   From the plot, it looks like there is a linear relationship between IgG and maximal oxygen uptake. There could also be a slight downward curve to the data, requiring a quadratic model. The quadratic model is:

$$E(y) = \beta_0 + \beta_1 x + \beta_2 x^2$$

c. To determine if the model provides information for the prediction of IgG, we test:

H_0: $\beta_1 = \beta_2 = 0$
H_a: At least one $\beta_i \neq 0$, $i = 1, 2$

The test statistic is $F = 203.16$.

The rejection region requires $\alpha = .05$ in the upper tail of the F distribution with $v_1 = k = 2$ and $v_2 = n - (k + 1) = 30 - (2 + 1) = 27$. From Table IX, Appendix A, $F_{.05} \approx 3.35$. The rejection region is $F > 3.35$.

Since the observed value of the test statistic falls in the rejection region ($F = 203.16 > 3.35$), H_0 is rejected. There is sufficient evidence to indicate the model provides information for the prediction of IgG at $\alpha = .05$.

d. The observed value of the test statistic in part c is p-value = 0.0000. Since the p-value is less than α ($p = .0000 < .05$), H_0 is rejected. There is sufficient evidence to indicate that the model provides information for the prediction of IgG at $\alpha = .05$.

e. To determine if the second-order term contributes information for the prediction of IgG, we test:

H_0: $\beta_2 = 0$
H_a: $\beta_2 \neq 0$

The test statistic is $t = -3.53$.

The rejection region requires $\alpha/2 = .05/2$ in each tail of the t distribution with df $= n - (k + 1) = 30 - (2 + 1) = 27$. From Table VI, Appendix A, $t_{.025} = 2.052$. The rejection region is $t > -2.052$ or $t > 2.052$.

Since the observed value of the test statistic falls in the rejection region ($t = -3.53 -2.052$), H_0 is rejected. There is sufficient evidence to indicate that the second-order terms provides information for the prediction of IgG at $\alpha = .05$.

f. The observed value of the test statistic in part c is p-value = 0.0022. Since the p-value is less than α ($p = .0022 < .05$), H_0 is rejected. There is sufficient evidence to indicate that the second-order term provides information for the prediction of IgG at $\alpha = .05$.

12.59 a. The quadratic model would be:

$$E(y) = \beta_0 + \beta_1 x + \beta_2 x^2$$

b. From the plot, the β_2 would be positive because the points appear to form an upward curve.

c. This value of r^2 applies to the linear model, not the quadratic model. The linear model is:

$$E(y) = \beta_0 + \beta_1 x$$

12.61   a.   The shape of the graph is not a straight line.

        b.   The shape of the graph is not a parabola.

        c.   The graph includes a reversal of curvature, so the cubic model would be the most appropriate.

12.63   The model is $E(y) = \beta_0 + \beta_1 x_1 + \beta_2 x_2$

        where $x_1 = \begin{cases} 1 & \text{if the variable is at level 2} \\ 0 & \text{otherwise} \end{cases}$ $x_2 = \begin{cases} 1 & \text{if the variable is at level 3} \\ 0 & \text{otherwise} \end{cases}$

        β_0 = mean value of y when qualitative variable is at level 1.
        β_1 = difference in mean value of y between level 2 and level 1 of qualitative variable.
        β_2 = difference in mean value of y between level 3 and level 1 of qualitative variable.

12.65   a.   The least squares prediction equation is:

             $$\hat{y} = 80 + 16.8x_1 + 40.4x_2$$

        b. β_1 is the difference in the mean value of the dependent variable between level 2 and level 1 of the independent variable.

             β_2 is the difference in the mean value of the dependent variable between level 3 and level 1 of the independent variable.

        c.   The hypothesis H_0: $\beta_1 = \beta_2 = 0$ is the same as H_0: $\mu_1 = \mu_2 = \mu_3$.

             The hypothesis H_a: At least one of the parameters β_1 and β_2 differs from 0 is the same as H_a: At least one mean (μ_1, μ_2, or μ_3) is different.

        d.   The test statistic is $F = \dfrac{\text{MSR}}{\text{MSE}} = \dfrac{2059.5}{83.3} = 24.72$

             The rejection region requires $\alpha = .05$ in the upper tail of the test statistic with numerator df $= k = 2$ and denominator df $= n - (k + 1) = 15 - (2 + 1) = 12$. From Table IX, Appendix A, $F_{.05} = 3.89$. The rejection region is $F > 3.89$.

             Since the observed value of the test statistic falls in the rejection region ($F = 24.72 > 3.89$), H_0 is rejected. There is sufficient evidence to indicate at least one of the means is different at $\alpha = .05$.

12.67   a.   The model is $E(y) = \beta_0 + \beta_1 x_1 + \beta_2 x_2$

             where $x_1 = \begin{cases} 1 & \text{if political status is democratic} \\ 0 & \text{otherwise} \end{cases}$

             $x_2 = \begin{cases} 1 & \text{is political status is dictatorship} \\ 0 & \text{otherwise} \end{cases}$

b. β_0 = mean value of level of assissination risk (y) for political status communist.

β_1 = difference in mean value of level of assissination risk (y) between political status democratic and political status communist.

β_2 = difference in mean value of level of assissination risk (y) between political status dictatorship and political status communist.

12.69 a.   The model is $E(y) = \beta_0 + \beta_1 x_1 + \beta_2 x_2 + \beta_3 x_3$

where $x_1 = \begin{cases} 1 & \text{if location 2} \\ 0 & \text{otherwise} \end{cases}$ $x_2 = \begin{cases} 1 & \text{if location 3} \\ 0 & \text{otherwise} \end{cases}$ $x_3 = \begin{cases} 1 & \text{if location 4} \\ 0 & \text{otherwise} \end{cases}$

b.   The mean growth rate of the fish at location 1 is β_0

c.   The mean growth rate of the fish at location 4 is $\beta_0 + \beta_3$

d.   The difference in mean growth rates between locations 2 and 3 is
$(\beta_0 + \beta_1) - (\beta_0 + \beta_2) = \beta_1 - \beta_2$

12.71 a.   Let $\begin{cases} 1 & \text{if pond is enriched} \\ 0 & \text{if otherwise} \end{cases}$

The model is $E(y) = \beta_0 + \beta_1 x_1$

b. β_0 = mean mosquito larvae in the natural pond
β_1 = difference in the mean mosquito larvae between the enriched pond and the natural pond

c.   To determine if the mean larval density for the enriched pond exceeds the mean for the natural pond, we test:

H_0: $\beta_1 = 0$
H_a: $\beta_1 > 0$

d.   The p-value for the global F test is $p = .004$. Thus, the p-value for this test is $p = .004/2 = .002$. Since the p-value is so small, there is evidence to reject H_0. There is sufficient evidence to indicate that the mean larval densities differ for the enriched pond and the natural pond at $\alpha > .002$.

12.73 a.   Let $x_1 = \begin{cases} 1 & \text{if adult female with offspring} \\ 0 & \text{if otherwise} \end{cases}$

$x_2 = \begin{cases} 1 & \text{if independent sub - adult female} \\ 0 & \text{if otherwise} \end{cases}$

$x_3 = \begin{cases} 1 & \text{if adult male} \\ 0 & \text{if otherwise} \end{cases}$

$x_4 = \begin{cases} 1 & \text{if independent sub - adult male} \\ 0 & \text{if otherwise} \end{cases}$

The model would be: $E(y) = \beta_0 + \beta_1 x_1 + \beta_2 x_2 + \beta_3 x_3 + \beta_4 x_4$

b. $\hat{\beta}_0 = \hat{\mu}_1 = 38$

   $\hat{\beta}_1 = \hat{\mu}_2 - \hat{\mu}_1 = 16 - 38 = -22$

   $\hat{\beta}_2 = \hat{\mu}_3 - \hat{\mu}_1 = 89 - 38 = 51$

   $\hat{\beta}_3 = \hat{\mu}_4 - \hat{\mu}_1 = 43 - 38 = 5$

   $\hat{\beta}_4 = \hat{\mu}_5 - \hat{\mu}_1 = 58 - 38 = 20$

c. To determine if the mean percent use of HTZ differs among the grizzly bear classes, we test:

   H_0: $\beta_1 = \beta_2 = \beta_3 = \beta_4 = 0$
   H_a: At least one $\beta_i \neq 0$, $i = 1, 2, 3, 4$

d. The p-value is .15. Since the p-value is not small, there is no evidence to indicate the mean percent use of HTZ differs among the grizzly bear classes for $\alpha < .15$.

12.75 a. The first-order model is $E(y) = \beta_0 + \beta_1 x_1$

b. The new model is $E(y) = \beta_0 + \beta_1 x_1 + \beta_2 x_2 + \beta_3 x_3$

   where $x_2 = \begin{cases} 1 & \text{if level 2} \\ 0 & \text{otherwise} \end{cases}$ $\qquad x_3 = \begin{cases} 1 & \text{if level 3} \\ 0 & \text{otherwise} \end{cases}$

c. To allow for interactions, the model is:

   $E(y) = \beta_0 + \beta_1 x_1 + \beta_2 x_2 + \beta_3 x_3 + \beta_4 x_1 x_2 + \beta_5 x_1 x_3$

d. The response lines will be parallel if $\beta_4 = \beta_5 = 0$

e. There will be one response line if $\beta_2 = \beta_3 = \beta_4 = \beta_5 = 0$

12.77 a. When $x_2 = x_3 = 0$, $E(y) = \beta_0 + \beta_1 x_1$
   When $x_2 = 1$ and $x_3 = 0$, $E(y) = \beta_0 + \beta_1 x_1 + \beta_2$
   When $x_2 = 0$ and $x_3 = 1$, $E(y) = \beta_0 + \beta_1 x_1 + \beta_3$

b. For level 1, $x_2 = x_3 = 0$. $\hat{y} = 44.8 + 2.2x_1 + 9.4(0) + 15.6(0) = 44.8 + 2.2x_1$.

   For level 2, $x_2 = 1$ and $x_3 = 0$. $\hat{y} = 44.8 + 2.2x_1 + 9.4(1) + 15.6(0) = 44.8 + 2.2x_1 + 9.4$
   $= 54.2 + 2.2x_1$.

   For level 3, $x_2 = 0$ and $x_3 = 1$. $\hat{y} = 44.8 + 2.2x_1 + 9.4(0) + 15.6(1) = 44.8 + 2.2x_1 + 15.6$
   $= 60.4 + 2.2x_1$.

12.79   The model is $E(y) = \beta_0 + \beta_1 x_1 + \beta_2 x_1^2 + \beta_3 x_2 + \beta_4 x_3 + \beta_5 x_4$
where x_1 is the quantitative variable and

$$x_2 = \begin{cases} 1 & \text{if level 2 of qualitative variable} \\ 0 & \text{otherwise} \end{cases}$$

$$x_3 = \begin{cases} 1 & \text{if level 3 of qualitative variable} \\ 0 & \text{otherwise} \end{cases}$$

$$x_4 = \begin{cases} 1 & \text{if level 4 of qualitative variable} \\ 0 & \text{otherwise} \end{cases}$$

12.81   a.   For obese smokers, $x_2 = 0$. The equation of the hypothesized line relating mean REE to time after smoking for obese smokers is:

$$E(y) = \beta_0 + \beta_1 x_1 + \beta_2(0) + \beta_3 x_1(0) = \beta_0 + \beta_1 x_1$$

The slope of the line is β_1.

     b.   For normal smokers, $x_2 = 1$. The equation of the hypothesized line relating mean REE to time after smoking for normal smokers is:

$$E(y) = \beta_0 + \beta_1 x_1 + \beta_2(1) + \beta_3 x_1(1) = (\beta_0 + \beta_2) + (\beta_1 + \beta_3)x_1$$

The slope of the line is $\beta_1 + \beta_3$.

     c.   The reported p-value is .044. Since the p-value is so small, there is evidence to indicate that interaction between time and weight is present for $\alpha > .044$.

12.83   a.   The first-order model is $E(y) = \beta_0 + \beta_1 x_1 + \beta_2 x_2 + \beta_3 x_1 x_2$.

     b.   The graphs will vary. However, for each value of slope face, the slopes of the straight-line relationship between mean lead level and elevation will vary. The lines will not be parallel. The line for slope face = west will have a slope equal to β_2. The line for slope face = east will have a slope equal to $\beta_2 + \beta_3$.

     c.   For slope face = East, the slope of the line is $\beta_2 + \beta_3$. Thus, for each foot increase in elevation, the mean lead level will increase by $\beta_2 + \beta_3$.

d. Using SAS, the results of the analysis are:

The REG Procedure

Model: MODEL1
Dependent Variable: LEAD

Analysis of Variance

Source	DF	Sum of Squares	Mean Square	F Value	Pr > F
Model	3	20.23864	6.74621	0.26	0.8567
Error	66	1738.08227	26.33458		
Corrected Total	69	1758.32091			

Root MSE	5.13172	R-Square	0.0115	
Dependent Mean	6.83994	Adj R-Sq	-0.0334	
Coeff Var	75.02582			

Parameter Estimates

Variable	DF	Parameter Estimate	Standard Error	t Value	Pr > |t|
Intercept	1	2.38487	5.39314	0.44	0.6598
ELEVAT	1	0.00181	0.00214	0.84	0.4014
SLOPE	1	3.20146	7.66988	0.42	0.6777
ES	1	-0.00133	0.00303	-0.44	0.6629

To determine if the overall model is useful for predicting lead level, we test:

H_0: $\beta_1 = \beta_2 = \beta_3 = 0$
H_a: At least one $\beta_i \neq 0$, $i = 1, 2, 3$

From the printout, the test statistic is $F = .26$ and the p-value is $p = .8567$. Since the p-value is greater than α ($p = .8567 > .10$), H_0 is not rejected. There is insufficient evidence to indicate that the overall model is useful for predicting lead level at $\alpha = .10$.

12.85 a. $E(y) = \beta_0 + \beta_1 x_1 + \beta_2 x_2$

where y = blood cell count

$$x_1 = \begin{cases} 1 & \text{if a woman exercises regularly} \\ 0 & \text{otherwise} \end{cases}$$ x_2 = amount of iron supplement

b. $E(y) = \beta_0 + \beta_1 x_1 + \beta_2 x_2 + \beta_3 x_1 x_2$

where the variables are the same as in part a.

12.87 The models in parts a and b are nested:

The complete model is $E(y) = \beta_0 + \beta_1 x_1 + \beta_2 x_2$
The reduced model is $E(y) = \beta_0 + \beta_1 x_1$

The models in parts a and d are nested.

The complete model is $E(y) = \beta_0 + \beta_1 x_1 + \beta_2 x_2 + \beta_3 x_1 x_2$
The reduced model is $E(y) = \beta_0 + \beta_1 x_1 + \beta_2 x_2$

The models in parts **a** and **e** are nested.

The complete model is $E(y) = \beta_0 + \beta_1 x_1 + \beta_2 x_2 + \beta_3 x_1 x_2 + \beta_4 x_1^2 + \beta_5 x_2^2$
The reduced model is $E(y) = \beta_0 + \beta_1 x_1 + \beta_2 x_2$

The models in parts **b** and **c** are nested.

The complete model is $E(y) = \beta_0 + \beta_1 x_1 + \beta_2 x_1^2$
The reduced model is $E(y) = \beta_0 + \beta_1 x_1$

The models in parts **b** and **d** are nested.

The complete model is $E(y) = \beta_0 + \beta_1 x_1 + \beta_2 x_2 + \beta_3 x_1 x_2$
The reduced model is $E(y) = \beta_0 + \beta_1 x_1$

The models in parts **b** and **e** are nested.

The complete model is $E(y) = \beta_0 + \beta_1 x_1 + \beta_2 x_2 + \beta_3 x_1 x_2 + \beta_4 x_1^2 + \beta_5 x_2^2$
The reduced model is $E(y) = \beta_0 + \beta_1 x_1$

The models in parts **c** and **e** are nested.

The complete model is $E(y) = \beta_0 + \beta_1 x_1 + \beta_2 x_2 + \beta_3 x_1 x_2 + \beta_4 x_1^2 + \beta_5 x_2^2$
The reduced model is $E(y) = \beta_0 + \beta_1 x_1 + \beta_2 x_1^2$

The models in parts **d** and **e** are nested.

The complete model is $E(y) = \beta_0 + \beta_1 x_1 + \beta_2 x_2 + \beta_3 x_1 x_2 + \beta_4 x_1^2 + \beta_5 x_2^2$
The reduced model is $E(y) = \beta_0 + \beta_1 x_1 + \beta_2 x_2 + \beta_3 x_1 x_2$

12.89   a.   The least squares prediction equation for the complete model is:

$$\hat{y} = 14.575 - .611 x_1 + .439 x_2 - .080 x_3 - .064 x_4$$

The least squares prediction equation for the reduced model is:

$$\hat{y} = 13.968 - .642 x_1 + .396 x_2$$

b. $SSE_R = 160.44$ and $SSE_C = 152.66$

The sum of the squared deviations from the mean for the complete model is 152.66 while the sum of the squared deviations from the mean for the reduced model is 160.44.

c.   There are five β parameters in the complete model and three in the reduced model.

d.   The hypotheses are:

H_0: $\beta_3 = \beta_4 = 0$
H_a: At least one $\beta_i \neq 0$, $i = 3, 4$

e.	The test statistic is $F = \dfrac{(SSE_R - SSE_C)/(k-g)}{SSE_C/[n-(k+1)]}$

$$= \frac{(160.44 - 152.66)/(4-2)}{152.66/[20-(4+1)]} = \frac{3.89}{10.1773} = .38$$

The rejection region requires $\alpha = .05$ in the upper tail of the F distribution with numerator df $= k - g = 4 - 2 = 2$ and denominator df $= n - (k+1) = 20 - (4+1) = 15$. From Table IX, Appendix A, $F_{.05} = 3.68$. The rejection region is $F > 3.68$.

Since the observed value of the test statistic does not fall in the rejection region ($F = .38$ $\not> 3.68$), H_0 is not rejected. There is insufficient evidence to indicate the complete model is better than the reduced model at $\alpha = .05$.

f.	The p-value $= P(F \geq .38)$. With numerator df $= 2$ and denominator df $= 15$, $P(F \geq .38) > .10$ from Table VIII, Appendix A.

12.91	a.	$E(y) = \beta_0 + \beta_1 x_1 + \beta_2 x_2 + \beta_3 x_3 + \beta_4 x_4 + \beta_5 x_5 + \beta_6 x_6 + \beta_7 x_7 + \beta_8 x_8 + \beta_9 x_9$
			$+ \beta_{10} x_{10} + \beta_{11} x_{11}$

	b.	$E(y) = \beta_0 + \beta_1 x_1 + \beta_2 x_2 + \beta_3 x_3 + \beta_4 x_4 + \beta_5 x_5 + \beta_6 x_6 + \beta_7 x_7$
			$+ \beta_8 x_8 + \beta_9 x_9 + \beta_{10} x_{10} + \beta_{11} x_{11}$
			$+ \beta_{12} x_1 x_9 + \beta_{13} x_1 x_{10} + \beta_{14} x_1 x_{11}$
			$+ \beta_{15} x_2 x_9 + \beta_{16} x_2 x_{10} + \beta_{17} x_2 x_{11}$
			$+ \beta_{18} x_3 x_9 + \beta_{19} x_3 x_{10} + \beta_{20} x_3 x_{11}$
			$+ \beta_{21} x_4 x_9 + \beta_{22} x_4 x_{10} + \beta_{23} x_4 x_{11}$
			$+ \beta_{24} x_5 x_9 + \beta_{25} x_5 x_{10} + \beta_{26} x_5 x_{11}$
			$+ \beta_{27} x_6 x_9 + \beta_{28} x_6 x_{10} + \beta_{29} x_6 x_{11}$
			$+ \beta_{30} x_7 x_9 + \beta_{31} x_7 x_{10} + \beta_{32} x_7 x_{11}$
			$+ \beta_{33} x_8 x_9 + \beta_{34} x_8 x_{10} + \beta_{35} x_8 x_{11}$

	c.	To test for interaction, we would use the hypothesis:

		H_0: $\beta_{12} = \beta_{13} = \beta_{14} = ... = \beta_{35} = 0$
		H_a: At least one $\beta_i \neq 0$, $i = 12, 13, 14, ..., 35$

		We would compare the complete model in part **b** to the reduce model in part **a**.

12.93	a.	The complete second order model is:

		$E(y) = \beta_0 + \beta_1 x_1 + \beta_2 x_1^2 + \beta_3 x_2 + \beta_4 x_1 x_2 + \beta_5 x_1^2 x_2$

		where $x_1 =$ age
		$\begin{cases} 1 & \text{if current} \\ 0 & \text{otherwise} \end{cases}$

	b.	To determine if the quadratic terms are important, we test:

		H_0: $\beta_2 = \beta_5 = 0$

c. To determine if the interaction terms are important, we test:

H_0: $\beta_4 = \beta_5 = 0$

d. Test for part b:

The test statistic is:

$$F = \frac{(SSE_R - SSE_C)/(k-g)}{SSE_C/[n-(k+1)]} = \frac{3,689,526 - 3,618,994/(5-3)}{82,249.862} = .429$$

Since no α is given, we will use $\alpha = .05$. The rejection region requires $\alpha = .05$ in the upper tail of the F distribution with $v_1 = 2$ numerator degrees of freedom and $v_2 = 44$ denominator degrees of freedom. From Table IX, Appendix A, $F_{.05} \approx 3.23$. The rejection region is $F > 3.23$.

Since the observed value of the test statistic does not fall in the rejection region ($F = .429 \ngtr 3.23$), H_0 is not rejected. There is insufficient evidence to indicate the quadratic terms are important for predicting market value at $\alpha = .05$.

Test for part c:

The test statistic is:

$$F = \frac{(SSE_R - SSE_C)/(k-g)}{SSE_C/[n-(k+1)]} = \frac{3,723,332 - 3,618,994/(5-3)}{82,249.862} = .634$$

The rejection region is the same as in previous test. Reject H_0 if $F > 3.23$.

Since the observed value of the test statistic does not fall in the rejection region ($F = .634 \ngtr 3.23$), H_0 is not rejected. There is insufficient evidence to indicate the interactive terms are important for predicting market value at $\alpha = .05$.

12.95   a. The hypothesized alternative model is

$E(y) = \beta_0 + \beta_1 x_1 + \beta_2 x_2 + \beta_3 x_3 + \beta_4 x_4 + \beta_5 x_5 + \beta_6 x_6 + \beta_7 x_7 + \beta_8 x_8 + \beta_9 x_9 + \beta_{10} x_{10}$

b. The null hypothesis would be H_0: $\beta_3 = \beta_4 = \beta_5 = \beta_6 = \beta_7 = \beta_8 = \beta_9 = \beta_{10} = 0$

c. The test was statistically significant. Thus, H_0 was rejected. There is sufficient evidence to indicate that at least one of the "control" variables contributes to the prediction of SAT-Math scores.

d. $R^2_{adj} = .79$. 79% of the sample variability of the SAT-Math scores around their means is explained by the proposed model relating SAT-Math scores to the 10 independent variables, adjusting for the sample size and the number of β parameters in the model.

e. For confidence coefficient .95, $\alpha = .05$ and $\alpha/2 = .05/2 = .025$. From Table VI, Appendix A, with df $= n - (k + 1) = 3{,}492 - (10 + 1) = 3{,}481$, $t_{.025} = 1.96$. The 95% confidence interval is:

$$\hat{\beta}_2 \pm t_{.025} s_{\hat{\beta}_2} \Rightarrow 14 \pm 1.96(3) \Rightarrow 14 \pm 5.88 \Rightarrow (8.12,\ 19.88)$$

We are 95% confident that the difference in the mean SAT-Math scores between students who were coached and those who were not is between 8.12 and 19.88 points, holding all the other variables constant.

f. Yes. From Exercise 12.66, the confidence interval for β_2 was (13.12, 24.88). In part e, the confidence interval for β_2 was (8.12, 19.88). Even though coaching is significant in both models, the change in the mean SAT-Math scores is not as great if the control variables are added to the model.

g. The complete model would be:

$$E(y) = \beta_0 + \beta_1 x_1 + \beta_2 x_2 + \beta_3 x_3 + \beta_4 x_4 + \beta_5 x_5 + \beta_6 x_6 + \beta_7 x_7 + \beta_8 x_8 + \beta_9 x_9 + \beta_{10} x_{10} + \beta_{11} x_1 x_2$$
$$+ \beta_{12} x_3 x_2 + \beta_{13} x_4 x_2 + \beta_{14} x_5 x_2 + \beta_{15} x_6 x_2 + \beta_{16} x_7 x_2 + \beta_{17} x_8 x_2 + \beta_{18} x_9 x_2 + \beta_{19} x_{10} x_2$$

h. The null hypothesis would be $H_0: \beta_{11} = \beta_{12} = \beta_{13} = \beta_{14} = \beta_{15} = \beta_{16} = \beta_{17} = \beta_{18} = \beta_{19} = 0$. To perform this test, you would fit the complete model specified in part **g**. You would also fit the reduced model specified in part **a**. Then, you would perform the test comparing the complete and reduced models.

12.97 a. To determine whether the rate of increase of emotional distress with experience is different for the two groups, we test:

$H_0: \beta_4 = \beta_5 = 0$
H_a: At least one $\beta_i \neq 0$, $i = 4, 5$

b. To determine whether there are differences in mean emotional distress levels that are attributable to exposure group, we test:

$H_0: \beta_3 = \beta_4 = \beta_5 = 0$
H_a: At least one $\beta_i \neq 0$, $i = 3, 4, 5$

c. To determine whether there are differences in mean emotional distress levels that are attributable to exposure group, we test:

$H_0: \beta_3 = \beta_4 = \beta_5 = 0$
H_a: At least one $\beta_i \neq 0$, $i = 3, 4, 5$

The test statistic is $F = \dfrac{(\text{SSE}_R - \text{SSE}_C)/(k - g)}{\text{SSE}_C/[n - (k+1)]} = \dfrac{(795.23 - 783.9)/(5 - 2)}{783.9/[200 - (5+1)]} = .93$

The rejection region requires $\alpha = .05$ in the upper tail of the F distribution with $v_1 = k - g = 5 - 2 = 3$ and $v_2 = n - (k + 1) = 200 - (5 + 1) = 194$. From Table IX, Appendix A, $F_{.05} \approx 2.60$. The rejection region is $F > 2.60$.

Since the observed value of the test statistic does not fall in the rejection region ($F = .93 \not> 2.60$), H_0 is not rejected. There is insufficient evidence to indicate that there are differences in mean emotional distress levels that are attributable to exposure group at $\alpha = .05$.

12.99   a.   In Step 1, all one-variable models are fit to the data. These models are of the form:

$E(y) = \beta_0 + \beta_1 x_i$

Since there are 7 independent variables, 7 models are fit. (Note: There are actually only 6 independent variables. One of the qualitative variables has three levels and thus two dummy variables. Some statistical packages will allow one to bunch these two variables together so that they are either both in or both out. In this answer, we are assuming that each x_i stands by itself.

b.   In Step 2, all two-varirable models are fit to the data, where the variable selected in Step 1 is one of the variables. These models are of the form:

$E(y) = \beta_0 + \beta_1 x_1 + \beta_2 x_i$

Since there are 6 independent variables remaining, 6 models are fit.

c.   In Step 3, all three-variable models are fit to the data, where the variables selected in Step 2 are two of the variables. These models are of the form:

$E(y) = \beta_0 + \beta_1 x_1 + \beta_2 x_2 + \beta_3 x_i$

Since there are 5 independent variables remaining, 5 models are fit.

d.   The procedure stops adding independent variables when none of the remaining variables, when added to the model, have a p-value less than some predetermined value. This predetermined value is usually $\alpha = .05$.

e.   Two major drawbacks to using the final stepwise model as the "best" model are:

(1)   An extremely large number of single β parameter t-tests have been conducted. Thus, the probability is very high that one or more errors have been made in including or excluding variables.

(2)   Often the variables selected to be included in a stepwise regression do not include the high-order terms. Consequently, we may have initially omitted several important terms from the model.

12.101 a. All of the independent variables are continuous except District. Four dummy variables
were created for District before using the stepwise regression. Also, note that several
observations had missing values. Thus, only 217 observations were used in the
analysis. Since Ratio was computed from the ratio of Price and DOT estimate, it was
not used as a predictor variable. Using SAS and $\alpha = .10$ for keeping variables in the
model, the Stepwise regression results are:

```
                    The REG Procedure
                      Model: MODEL1
                Dependent Variable: PRICE

             Stepwise Selection: Step 1

   Variable DOTEST Entered: R-Square = 0.9763 and C(p) = 14.2985


                    Analysis of Variance

                              Sum of        Mean
Source                 DF     Squares       Square     F Value    Pr > F

Model                   1    687136244    687136244    8856.79    <.0001
Error                 215     16680334        77583
Corrected Total       216    703816578


             Parameter    Standard
Variable     Estimate      Error     Type II SS  F Value   Pr > F

Intercept    33.24768    22.19263       174129     2.24    0.1356
DOTEST        0.90637     0.00963    687136244  8856.79    <.0001

           Bounds on condition number: 1, 1
-----------------------------------------------------------------------

             Stepwise Selection: Step 2


   Variable STATUS Entered: R-Square = 0.9774 and C(p) = 5.9631


                    Analysis of Variance

                              Sum of        Mean
Source                 DF     Squares       Square     F Value    Pr > F

Model                   2    687894714    343947357    4622.87    <.0001
Error                 214     15921865        74401
Corrected Total       216    703816578

                    The REG Procedure
                      Model: MODEL1
                Dependent Variable: PRICE

             Stepwise Selection: Step 2

             Parameter    Standard
Variable     Estimate      Error     Type II SS  F Value   Pr > F

Intercept   -12.92808    26.10499        18247     0.25    0.6209
DOTEST        0.91233     0.00961    669924194  9004.21    <.0001
STATUS      132.16624    41.39440       758469    10.19    0.0016

       Bounds on condition number: 1.0392, 4.1569
-----------------------------------------------------------------------
```

Stepwise Selection: Step 3

Variable DAYS Entered: R-Square = 0.9779 and C(p) = 3.4115

Analysis of Variance

Source	DF	Sum of Squares	Mean Square	F Value	Pr > F
Model	3	688228730	229409577	3134.76	<.0001
Error	213	15587848	73182		
Corrected Total	216	703816578			

Variable	Parameter Estimate	Standard Error	Type II SS	F Value	Pr > F
Intercept	-59.79987	33.93608	227239	3.11	0.0795
DOTEST	0.88616	0.01552	238455655	3258.37	<.0001
STATUS	139.40952	41.19370	838166	11.45	0.0008
DAYS	0.35761	0.16739	334016	4.56	0.0338

Bounds on condition number: 2.7653, 19.698

All variables left in the model are significant at the 0.1000 level.

The REG Procedure
Model: MODEL1
Dependent Variable: PRICE

Stepwise Selection: Step 3

No other variable met the 0.1500 significance level for entry into the model.

Summary of Stepwise Selection

Step	Variable Entered	Variable Removed	Number Vars In	Partial R-Square	Model R-Square	C(p)	F Value	Pr > F
1	DOTEST		1	0.9763	0.9763	14.2985	8856.79	<.0001
2	STATUS		2	0.0011	0.9774	5.9631	10.19	0.0016
3	DAYS		3	0.0005	0.9779	3.4115	4.56	0.0338

From the results, only three independent variables are selected to predict price using the stepwise regression and $\alpha = .05$. These variables are DOT estimate, Status, and Days.

b. $\hat{\beta}_0 = -59.79987$. This value has no meaning because an observation with all the independent variables equal to 0 is not in the observed range.

$\hat{\beta}_1 = .88616$. For each dollar increase in the DOT estimate, the mean price is estimated to increase by .88616 dollars, holding all other variables constant.

$\hat{\beta}_2 = 139.40952$. The mean price is estimated to be 139.40952 dollars higher for fixed bids than for competitive bids, holding all other variables constant.

$\hat{\beta}_3 = .35761$. For each additional day to complete work, the mean price is estimated to increase by .35761 dollars, holding all other variables constant.

c.   Because of the large number of t tests performed in the stepwise regression, the probability is very high that one or more errors have been made in including and excluding variables. Also, we have not considered any higher order terms.

12.103 Yes. x_2 and x_4 are highly correlated (.93), as well as x_4 and x_5 (.86). When highly correlated independent variables are present in a regression model, the results can be confusing. The researcher may want to include only one of the variables.

12.105 When independent variables that are highly correlated with each other are included in a regression model, the results may be confusing. Highly correlated independent variables contribute overlapping information in the prediction of the dependent variable. The overall global test can indicate that the model is useful in predicting the dependent variable, while the individual t-tests on the independent variables can indicate that none of the independent variables are significant. This happens because the individual t-tests tests for the significance of an independent variable after the other independent variables are taken into account. Usually, only one of the independent variables that are highly correlated with each other are included in the regression model.

12.107 a.   Using MINITAB, the analysis results are:

```
The regression equation is
MAX-VO2 = - 4.77 - 0.0352 Age + 0.0516 Height - 0.0234 Weight + 0.0345 Depth

Predictor        Coef        StDev           T           P
Constant      -4.7747       0.8628       -5.53       0.003
Age          -0.03521      0.01539       -2.29       0.071
Height       0.051637     0.006215        8.31       0.000
Weight       -0.02342      0.01343       -1.74       0.142
Depth         0.03449      0.08524        0.40       0.702

S = 0.03721     R-Sq = 96.7%      R-Sq(adj) = 94.1%

Analysis of Variance

Source              DF           SS           MS          F          P
Regression           4     0.206037     0.051509      37.20      0.001
Residual Error       5     0.006923     0.001385
Total                9     0.212960

Source        DF      Seq SS
Age            1     0.003950
Height         1     0.196252
Weight         1     0.005609
Depth          1     0.000227
```

The least squares prediction equation is:

$$\hat{y} = -4.77 - .0352x_1 + .0516x_2 - .0234x_3 + .0345x_4.$$

b.   One possible explanation could be that weight is correlated with some of the other independent variables: age, height, or at chest depth. If independent variables are highly correlated, the sign of the estimated parameters may be different than expected.

c. Again, the small t value associated with chest depth could be due to the fact that chest depth is correlated with one or more of the other independent variables: age, height, or weight.

To determine if there is a positive correlation between chest depth and maximum oxygen uptake, we test:

H_0: $\beta_4 = 0$
H_a: $\beta_4 > 0$

From the printout, the test statistic is $t = .40$ and the p-value is $p = .702/2 = .351$. Since the p-value is not small, H_0 is not rejected. There is insufficient evidence to indicate a positive correlation between chest depth and maximum oxygen uptake for $\alpha < .351$.

We would expect there to be a positive correlation between these two variables. This test is not significant because chest depth could be highly correlated with some of the other independent variables in the model.

d. Using MINITAB, the correlation coefficients between all pairs of independent variables are as follows:

```
         Age     Height   Weight
Height   0.327
         0.356

Weight   0.231    0.790
         0.521    0.007

Depth    0.166    0.791    0.881
         0.647    0.006    0.001

Cell Contents: Correlation
               P-Value
```

The correlations between height and weight ($r = .790$), height and depth ($r = .791$), and weight and depth ($r = .881$) are very high. Because the independent variables are highly correlated to each other, including the correlated variables in the same model can lead to these results.

12.109 a. In Exercise 12.101, the three independent variables selected using the stepwise procedure are DOT estimate, Status, and Days. Since there were many missing values in the data, only 217 observations were used in the stepwise regression analysis. To stay consistent, we will use only these 217 observations. Using SAS, the correlations are:

```
The CORR Procedure

        4  Variables:    PRICE    DOTEST    STATUS    DAYS


                        Simple Statistics

Variable       N        Mean     Std Dev        Sum     Minimum     Maximum

PRICE        217        1127        1805     244486    23.00000       10480
DOTEST       217        1206        1968     261783    22.00000       10744
STATUS       217     0.29493     0.45707   64.00000           0     1.00000
DAYS         217   213.38710   182.85968      46305    30.00000   900.00000
```

```
                    Pearson Correlation Coefficients, N = 217
                            Prob > |r| under H0: Rho=0

                        PRICE         DOTEST        STATUS          DAYS

        PRICE         1.00000        0.98808      -0.15979       0.79940
                                     <.0001        0.0185        <.0001

        DOTEST        0.98808        1.00000      -0.19431       0.79744
                      <.0001                       0.0041        <.0001

        STATUS       -0.15979       -0.19431       1.00000      -0.20366
                      0.0185         0.0041                      0.0026

        DAYS          0.79940        0.79744      -0.20366       1.00000
                      <.0001         <.0001        0.0026
```

From the pairwise correlations, is appears that the correlation between Days and DOT estimate ($r = .79744$) is fairly large. There might be a problem with multicollinearity with both of these variables in the model. However, when fitting the independent variables one at a time, the parameter estimates for DOT estimate and Days do not change much. Therefore, there probably is not a problem with multicollinearity. Thus, one can use all three variables in the model.

b.   Using SAS, the fitted full interaction model results are:

```
The REG Procedure
                            Model: MODEL1
                       Dependent Variable: PRICE

                          Analysis of Variance

                                 Sum of          Mean
    Source            DF         Squares         Square     F Value    Pr > F

    Model              6       689573044      114928841     1694.46    <.0001
    Error            210        14243534          67826
    Corrected Total  216       703816578

              Root MSE             260.43493    R-Square     0.9798
              Dependent Mean      1126.66359    Adj R-Sq     0.9792
              Coeff Var             23.11559

                          Parameter Estimates

                         Parameter      Standard
    Variable      DF      Estimate        Error      t Value    Pr > |t|

    Intercept      1      -9.24852       40.30694      -0.23      0.8187
    DOTEST         1       0.86618        0.03554      24.37      <.0001
    STATUS         1       5.32761       63.97853       0.08      0.9337
    DAYS           1       0.14195        0.18918       0.75      0.4539
    DOT_STAT       1       0.20858        0.07604       2.74      0.0066
    DOT_DAY        1       0.00004264     0.00005855    0.73      0.4673
    STAT_DAY       1      -0.04285        0.48301      -0.09      0.9294
```

c.   Using SAS, the some of the residual plots are:

Plot of stres*yhat.   Symbol used is '*'.

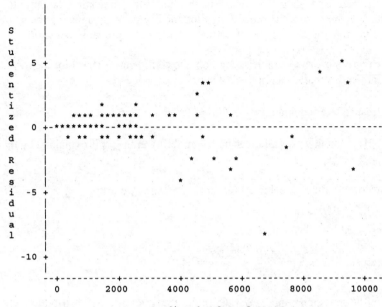

NOTE: 160 obs hidden.

Plot of zsc*resid.   Symbol used is '*'.

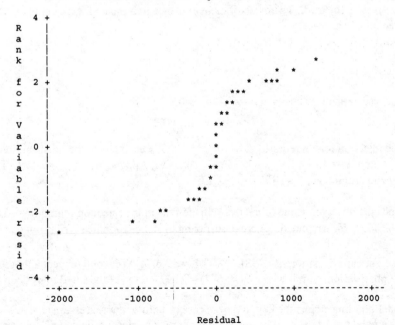

NOTE: 185 obs hidden.

From the plot of the standardized residual versus yhat, we see that as the value of yhat increases, the spread of the residuals also increases. Thus, there appears to be a violation of the assumption of constant variance. In addition, there appear to be several standardized residuals greater than 3 in absolute value, indicating the presence of outliers. From the normal probability plot, the line does not appear to be straight, indicating that the assumption of normal error terms is probably not valid.

12.111 In multiple regression, as in simple regression, the confidence interval for the mean value of y is narrower than the prediction interval of a particular value of y.

12.113 a. The least squares equation is $\hat{y} = 90.1 - 1.836x_1 + .285x_2$

b. $R^2 = .916$. About 91.6% of the sample variability in the y's is explained by the model $E(y) = \beta_0 + \beta_1 x_1 + \beta_2 x_2$

c. To determine if the model is useful for predicting y, we test:

H_0: $\beta_1 = \beta_2 = 0$
H_a: At least 1 $\beta_i \neq 0$, $i = 1, 2$

The test statistic is $F = \dfrac{MSR}{MSE} = \dfrac{7400}{114} = 64.91$

The rejection region requires $\alpha = .05$ in the upper tail of the F distribution with $v_1 = k = 2$ and $v_2 = n - (k + 1) = 15 - (2 + 1) = 12$. From Table IX, Appendix A, $F_{.05} = 3.89$. The rejection region is $F > 3.89$.

Since the observed value of the test statistic falls in the rejection region ($F = 64.91 > 3.89$), H_0 is rejected. There is sufficient evidence to indicate the model is useful for predicting y at $\alpha = .05$.

d. H_0: $\beta_1 = 0$
H_a: $\beta_1 \neq 0$

The test statistic is $t = \dfrac{\hat{\beta}_1}{s_{\hat{\beta}_1}} = \dfrac{-1.836}{.367} = -5.01$

The rejection region requires $\alpha/2 = .05/2 = .025$ in each tail of the t distribution with df $= n - (k + 1) = 15 - (2 + 1) = 12$. From Table VI, Appendix A, $t_{.025} = 2.179$. The rejection region is $t < -2.179$ or $t > 2.179$.

Since the observed value of the test statistic falls in the rejection region ($t = -5.01 < -2.179$), H_0 is rejected. There is sufficient evidence to indicate β_1 is not 0 at $\alpha = .05$.

e. The standard deviation is $\sqrt{MSE} = \sqrt{114} = 10.677$. We would expect about 95% of the observations to fall within $2(10.677) = 21.354$ units of the fitted regression line.

12.115 The model-building step is the key to the success or failure of a regression analysis. If the model is a good model, we will have a good predictive model for the dependent variable y. If the model is not a good model, the predictive ability will not be of much use.

12.117 $E(y) = \beta_0 + \beta_1 x_1 + \beta_2 x_2 + \beta_3 x_3$

where $x_1 = \begin{cases} 1, & \text{if level 2} \\ 0, & \text{otherwise} \end{cases}$ $x_2 = \begin{cases} 1, & \text{if level 3} \\ 0, & \text{otherwise} \end{cases}$ $x_3 = \begin{cases} 1, & \text{if level 4} \\ 0, & \text{otherwise} \end{cases}$

12.119 The stepwise regression method is used to try to find the best model to describe a process. It is a screening procedure that tries to select a small subset of independent variables from a large set of independent variables that will adequately predict the dependent variable. This method is useful in that it can eliminate some unimportant independent variables from consideration.

12.121 Even though SSE = 0, we cannot estimate σ^2 because there are no degrees of freedom corresponding to error. With three data points, there are only two degrees of freedom available. The degrees of freedom corresponding to the model is $k = 2$ and the degrees of freedom corresponding to error is $n - (k + 1) = 3 - (2 + 1) = 0$. Without an estimate for σ^2, no inferences can be made.

12.123 a. To determine if the AD score is positively related to assertiveness level, once age and length of disability are accounted for, we test:

H_0: $\beta_1 = 0$
H_a: $\beta_1 > 0$

The test statistic is $t = 5.96$.

The p-value is $p = .0001/2 = .00005$. Since the p-value is less than α ($p = .00005 < .05$), H_0 is rejected. There is sufficient evidence to indicate that the AD score is positively related to assertiveness level, once age and length of disability are accounted for with $\alpha = .05$.

b. To determine if age is related to assertiveness level, once AD score and length of disability are accounted for, we test:

H_0: $\beta_2 = 0$
H_a: $\beta_2 \neq 0$

The test statistic is $t = 0.01$.

The p-value is $p = .9620$. Since the p-value is less than α ($p = .9620 \nleq .05$), H_0 is not rejected. There is insufficient evidence to indicate that Age score is related to assertiveness level, once AD score and length of disability are accounted for with $\alpha = .05$.

c.   To determine if length of disability is positively related to assertiveness level, once AD score and age are accounted for, we test:

H_0: $\beta_3 = 0$
H_a: $\beta_3 > 0$

The test statistic is $t = 1.91$.

The p-value is $p = .0576/2 = .0288$. Since the p-value is less than α ($p = .0288 < .05$), H_0 is rejected. There is sufficient evidence to indicate that length of disability is positively related to assertiveness level, once the AD score and age are accounted for with $\alpha = .05$.

12.125 a.   The first order model for $E(y)$ as a function of the first five independent variables is:

$$E(y) = \beta_0 + \beta_1 x_1 + \beta_2 x_2 + \beta_3 x_3 + \beta_4 x_4 + \beta_5 x_5$$

b.   To test the utility of the model, we test:

H_0: $\beta_1 = \beta_2 = \beta_3 = \beta_4 = \beta_5 = 0$
H_a: At least one $\beta_i \neq 0$, $i = 1, 2, 3, 4, 5$

The test statistic is $F = 34.47$.

The p-value is $p < .001$. Since the p-value is so small, there is sufficient evidence to indicate the model is useful for predicting GSI at $\alpha > .001$.

$R^2 = .469$.   46.9% of the variability in the GSI scores is explained by the model including the first five independent variables.

c.   The first order model for $E(y)$ as a function of the first seven independent variables is:

$$E(y) = \beta_0 + \beta_1 x_1 + \beta_2 x_2 + \beta_3 x_3 + \beta_4 x_4 + \beta_5 x_5 + \beta_6 x_6 + \beta_7 x_7$$

d. $R^2 = .603$ 60.3% of the variability in the GSI scores is explained by the model including the first seven independent variables.

e.   Since the p-values associated with the variables DES and PDEQ-SR are both less than .001, there is evidence that both variables contribute to the prediction of GSI, adjusted for all the other variables already in the model for $\alpha > .001$.

12.127 a. x_1 = number of years a patient has smoked = quantitative
x_2 = sex of the patient = qualitative

where $x_2 = \begin{cases} 1 \text{ if male} \\ 0 \text{ otherwise} \end{cases}$

b. $E(y) = \beta_0 + \beta_1 x_1 + \beta_2 x_2$

c.   The estimated difference is given by the coefficient of x_2, $\hat{\beta}_2 = .09$.

d. The estimated mean lung damage score for a female patient ($x_2 = 0$) who has smoked for 15 years is obtained by substituting appropriate values of the independent variables into the estimated regression equation.

$$\hat{y} = -.85 + .65(15) + .09(0) = 8.90$$

e. The required model would be a full second-order model.

$$E(y) = \beta_0 + \beta_1 x_1 + \beta_2 x_2 + \beta_3 x_1^2 + \beta_4 x_1 x_2 + \beta_5 x_1^2 x_2$$

12.129 a. $\hat{\beta}_0 = 39.05 =$ the estimate of the y-intercept

$\hat{\beta}_1 = -5.41$. We estimate that the mean operating margin will decrease by 5.41% for each additional increase of 1 unit of x_1, the state population divided by the total number of inns in the state (with all other variables held constant).

$\hat{\beta}_2 = 5.86$. We estimate that the mean operating margin will increase by 5.86% for each additional increase of 1 unit of x_2, the room rate (with all other variables held constant).

$\hat{\beta}_3 = -3.09$. We estimate that the mean operating margin will decrease by 3.09% for each additional increase of 1 unit of x_3, the square root of the median income of the area (with all other variables held constant).

$\hat{\beta}_4 = 1.75$. We estimate that the mean operating margin will increase by 1.75% for each additional increase of 1 unit of x_4, the number of college students within four miles of the inn (with all other variables held constant).

b. $R^2 = .51$. 51% of the variability in the operating margins can be explained by the model containing these four independent variables.

c. To determine if the model is adequate, we test:

H_0: $\beta_1 = \beta_2 = \beta_3 = \beta_4 = 0$
H_a: At least one $\beta_i \neq 0$, $i = 1, 2, 3, 4$

The test statistic is

$$F = \frac{R^2/k}{(1-R^2)/[n-(k+1)]} = \frac{.51/4}{(1-.51)/[57-(4+1)]} = 13.53$$

The rejection region requires $\alpha = .05$ in the upper tail of the F distribution with $v_1 = k = 4$ and $v_2 = n - (k+1) = 57 - (4+1) = 52$. From Table IX, Appendix A, $F_{.05} \approx 2.55$. The rejection region is $F > 2.55$.

Since the observed value of the test statistic falls in the rejection region ($F = 13.53 > 2.55$), H_0 is rejected. There is sufficient evidence that the model is useful in predicting operating margins at $\alpha = .05$.

12.131 a. Let x_1 = ASO scale, x_2 = Burns scale, and x_3 = Rotter scale. The first order model relating depression (Zung scale) to self-acceptance (ASO scale), perfectionism (Burns scale), and reinforcement (Rotter scale) is:

$$E(y) = \beta_0 + \beta_1 x_1 + \beta_2 x_2 + \beta_3 x_3$$

b. $R^2 = .70$.   70% of the variability in the depression scores is explained by the model including self-acceptance, perfectionism, and reinforcement.

c. To test if the model is useful for predicting depression, we test:

H_0: $\beta_1 = \beta_2 = \beta_3 = 0$
H_a: At least one $\beta_i \neq 0, i = 1, 2, 3$

The test statistic is $F = \dfrac{R^2/k}{(1-R^2)/[n-(k+1)]} = \dfrac{.70/3}{(1-.70)/[76-(3+1)]} = 56$

The rejection region requires $\alpha = .05$ in the upper tail of the F distribution with $v_1 = k = 3$ and $v_2 = n - (k+1) = 76 - (3+1) = 72$. From Table IX, Appendix A, $F_{.05} \approx 2.76$. The rejection region is $F > 2.76$.

Since the observed value of the test statistic falls in the rejection region ($F = 56 > 2.76$), H_0 is rejected. There is sufficient evidence that the model is useful in predicting depression at $\alpha = .05$.

d. The p-value for the perfectionism variable is .87. Since the p-value is so large, H_0 is not rejected. There is insufficient evidence to indicate perfectionism is useful for predicting depression, adjusting for self-acceptance and reinforcement already being in the model at $\alpha < .87$.

12.133 a. Since in the proposed model the funniness of the joke should increase and then decrease, the value of β_2 is expected to be negative.

b. To determine if the quadratic model relating pain to funniness rating is useful, we test:

H_0: $\beta_1 = \beta_2 = 0$
H_a: At least one $\beta_i \neq 0, i = 1, 2$

The test statistic is $F = 1.60$.

The rejection region requires $\alpha = .05$ in the upper tail of the F distribution with $v_1 = k = 2$ and $v_2 = n - (k+1) = 32 - (2+1) = 29$. From Table IX, Appendix A, $F_{.05} = 3.33$. The rejection region is $F > 3.33$.

Since the observed value of the test statistic does not fall in the rejection region ($F = 1.60 \not> 3.33$), H_0 is not rejected. There is insufficient evidence to indicate that the quadratic model relating pain to funniness rating is useful at $\alpha = .05$.

c. To determine if the quadratic model relating aggression/hostility to funniness rating is useful, we test:

H_0: $\beta_1 = \beta_2 = 0$
H_a: At least one $\beta_i \neq 0$, $i = 1, 2$

The test statistic is $F = 1.61$.

The rejection region requires $\alpha = .05$ in the upper tail of the F distribution with $v_1 = k = 2$ and $v_2 = n - (k + 1) = 32 - (2 + 1) = 29$. From Table IX, Appendix A, $F_{.05} = 3.33$. The rejection region is $F > 3.33$.

Since the observed value of the test statistic does not fall in the rejection region ($F = 1.61 - 3.33$), H_0 is not rejected. There is insufficient evidence to indicate that the quadratic model relating aggression/hostility to funniness rating is useful at $\alpha = .05$.

12.135 a.

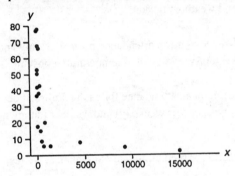

There is a curvilinear trend.

b. From MINITAB, the output is:

The regression equation is y = 42.2 - 0.0114x + 0.000001 xsq

Predictor	Coef	StDev		P
Constant	42.247	5.712	7.4	0.000
x	-0.011404	0.005053	-2.2	0.037
xsq	0.00000061	0.00000037	1.6	0.115

S = 21.81   R-Sq = 34.9%   R-Sq(adj) = 27.2%

Analysis of Variance

Source	DF	SS	MS	F	P
Regression	2	4325.4	2162.7	4.55	0.026
Residual Error	17	8085.5	475.6		
Total	19	12410.9			

Source	D	Seq SS
x		3013.3
xsq		1312.1

Unusual Observations

Obs	x1	y	Fit	StDev Fit	Residual	St Resid
16	9150	4.60	-11.21	16.24	15.81	1.09 x
17	15022	2.20	8.09	21.40	-5.89	-1.41 x

X denotes an observation whose X value gives it large influence.

The fitted model is $\hat{y} = 42.2 - .0114x + .00000061x^2$

c. To determine if a curvilinear relationship exists, we test:

H_0: $\beta_2 = 0$
H_a: $\beta_2 \neq 0$

From MINITAB, the test statistic is $t = 1.66$ with p-value $= .115$. Since the p-value is greater than $\alpha = .05$, do not reject H_0. There is insufficient evidence to indicate that a curvilinear relationship exists between dissolved phosphorus percentage and soil loss at $\alpha = .05$.

12.137 a. $\hat{\beta}_0 = -105$ has no meaning because $x_3 = 0$ is not in the observable range. $\hat{\beta}_0$ is simply the y-intercept.

$\hat{\beta}_1 = 25$. The estimated difference in mean attendance between weekends and weekdays is 25, temperature and weather constant.

$\hat{\beta}_2 = 100$. The estimated difference in mean attendance between sunny and overcast days is 100, type of day (weekend or weekday) and temperature constant.

$\hat{\beta}_3 = 10$. The estimated change in mean attendance for each additional degree of temperature is 10, type of day (weekend or weekday) and weather (sunny or overcast) held constant.

b. To determine if the model is useful for predicting daily attendance, we test:

H_0: $\beta_1 = \beta_2 = \beta_3 = 0$
H_a: At least one $\beta_i \neq 0$, $i = 1, 2, 3$

The test statistic is $F = \dfrac{R^2/k}{(1-R^2)/[n-(k+1)]} = \dfrac{.65/3}{(1-.65)/[30-(3+1)]} = 16.10$

The rejection region requires $\alpha = .05$ in the upper tail of the F distribution with numerator df $= k = 3$ and denominator df $= n - (k+1) = 30 - (3+1) = 26$. From Table IX, Appendix A, $F_{.05} \approx 2.98$. The rejection region is $F > 2.98$.

Since the observed value of the test statistic falls in the rejection region ($F = 16.10 > 2.98$), H_0 is rejected. There is sufficient evidence to indicate the model is useful for predicting daily attendance at $\alpha = .05$.

c. To determine if mean attendance increases on weekends, we test:

H_0: $\beta_1 = 0$
H_a: $\beta_1 > 0$

The test statistic is $t = \dfrac{\hat{\beta}_1 - 0}{s_{\hat{\beta}_1}} = \dfrac{25 - 0}{10} = 2.5$

The rejection region requires $\alpha = .10$ in the upper tail of the t distribution with df $= n - (k + 1) = 30 - (3 + 1) = 26$. From Table VI, Appendix A, $t_{.10} = 1.316$. The rejection region is $t > 1.316$.

Since the observed value of the test statistic falls in the rejection region ($t = 2.5 > 1.316$), H_0 is rejected. There is sufficient evidence to indicate the mean attendance increases on weekends at $\alpha = .10$.

d.   Sunny $\Rightarrow x_2 = 1$, Weekday $\Rightarrow x_1 = 0$, Temperature 95° $\Rightarrow x_3 = 95$

$$\hat{y} = -105 + 25(0) + 100(1) + 10(95) = 945$$

e.   We are 95% confident that the actual attendance for sunny weekdays with a temperature of 95° is between 645 and 1245.

12.139 a.   To determine whether the impact of college track on verbal achievement score depends on sector, we would test to see if the interaction term is significant:

$$H_0: \ \beta_3 = 0$$
$$H_a: \ \beta_3 \neq 0$$

b.   For public school students not on a college program, $x_1 = 0$ and $x_2 = 0$. Then $E(y) = \beta_0 + \beta_1(0) + \beta_2(0) + \beta_3(0)(0) = \beta_0$.

c.   For public school students on a college program, $x_1 = 1$ and $x_2 = 0$. Then $E(y) = \beta_0 + \beta_1(1) + \beta_2(0) + \beta_3(1)(0) = \beta_0 + \beta_1$.

d.   Subtracting part **a** from part **b**, we get $\beta_0 + \beta_1 - \beta_0 = \beta_1$. Here β_1 is the difference between the means of college-track and regular-track students in the public schools.

e.   For Catholic school students not on a college program, $x_1 = 0$ and $x_2 = 1$. Then $E(y) = \beta_0 + \beta_1(0) + \beta_2(1) + \beta_3(0)(1) = \beta_0 + \beta_2$.

f.   For Catholic school students on a college program, $x_1 = 1$ and $x_2 = 1$. Then $E(y) = \beta_0 + \beta_1(1) + \beta_2(1) + \beta_3(1)(1) = \beta_0 + \beta_1 + \beta_2 + \beta_3$.

g.   Subtracting part **d** from part **e**, we get $\beta_0 + \beta_1 + \beta_2 + \beta_3 - (\beta_0 + \beta_2) = \beta_1 + \beta_3$. Here $\beta_1 + \beta_3$ is the difference between the means of college-track and regular-track students in the Catholic schools.

Categorical Data Analysis

13.1   a.   With df = 10, $\chi^2_{.05} = 18.3070$

     b.   With df = 50, $\chi^2_{.990} = 29.7067$

     c.   With df = 16, $\chi^2_{.10} = 23.5418$

     d.   With df = 50, $\chi^2_{.005} = 79.4900$

13.3   a.   The rejection region requires $\alpha = .05$ in the upper tail of the χ^2 distribution with df = $k - 1 = 3 - 1 = 2$. From Table VII, Appendix A, $\chi^2_{.05} = 5.99147$. The rejection region is $\chi^2 > 5.99147$.

     b.   The rejection region requires $\alpha = .10$ in the upper tail of the χ^2 distribution with df = $k - 1 = 5 - 1 = 4$. From Table VII, Appendix A, $\chi^2_{.10} = 7.77944$. The rejection region is $\chi^2 > 7.77944$.

     c.   The rejection region requires $\alpha = .01$ in the upper tail of the χ^2 distribution with df = $k - 1 = 4 - 1 = 3$. From Table VII, Appendix A, $\chi^2_{.10} = 11.3449$. The rejection region is $\chi^2 > 11.3449$.

13.5   The sample size n will be large enough so that for every cell the expected cell count $E(n_i)$ will be equal to 5 or more.

13.7   Some preliminary calculations are:

     If the probabilities are the same, $p_{1,0} = p_{2,0} = p_{3,0} = p_{4,0} = .25$

$$E(n_1) = np_{1,0} = 205(.25) = 51.25$$
$$E(n_2) = E(n_3) = E(n_4) = 205(.25) = 51.25$$

     a.   To determine if the multinomial probabilities differ, we test:

          H_0: $p_1 = p_2 = p_3 = p_4 = .25$
          H_a: At least one of the probabilities differs from .25

          The test statistic is $\chi^2 = \sum \dfrac{[n_i - E(n_i)]^2}{E(n_i)}$

$$= \frac{(43 - 51.25)^2}{51.25} + \frac{(56 - 51.25)^2}{51.25} + \frac{(59 - 51.25)^2}{51.25} + \frac{(47 - 51.25)^2}{51.25} = 3.293$$

The rejection region requires $\alpha = .05$ in the upper tail of the χ^2 distribution with df = k 1 = 4 − 1 = 3. From Table VII, Appendix A, $\chi^2_{.05} = 7.81473$. The rejection region is χ^2 > 7.81473.

Since the observed value of the test statistic does not fall in the rejection region ($\chi^2 = 3.293 − 7.81473$), H_0 is not rejected. There is insufficient evidence to indicate the multinomial probabilities differ at $\alpha = .05$.

b.   The Type I error is concluding the multinomial probabilities differ when, in fact, they do not.

The Type II error is concluding the multinomial probabilities are equal, when, in fact, they are not.

13.9   Some preliminary calculations are:

$E(n_1) = np_{1,0} = 400(.2) = 80$
$E(n_2) = np_{2,0} = 400(.4) = 160$
$E(n_3) = np_{3,0} = 400(.1) = 40$
$E(n_4) = np_{4,0} = 400(.3) = 120$

To determine if probabilities differ from the hypothesized values, we test:

H_0: $p_1 = .2, p_2 = .4, p_3 = .1$, and $p_4 = .3$
H_a: At least one of the probabilities differs from its hypothesized value

The test statistic is $\chi^2 = \sum \dfrac{[n_i - E(n_i)]^2}{E(n_i)}$

$$= \frac{(70-80)^2}{80} + \frac{(196-160)^2}{160} + \frac{(46-40)^2}{40} + \frac{(88-120)^2}{120} = 18.78$$

The rejection region requires $\alpha = .05$ in the upper tail of the χ^2 distribution with df = $k - 1 =$ 4 − 1 = 3. From Table VII, Appendix A, $\chi^2_{.05} = 7.81473$. The rejection region is $\chi^2 >$ 7.81473.

Since the observed value of the test statistic falls in the rejection region ($\chi^2 = 18.78 >$ 7.81473), H_0 is rejected. There is sufficient evidence to indicate the probabilities differ from their hypothesized values at $\alpha = .05$.

13.11 a.   The qualitative variable of interest in this problem is the type of pottery found. There are 4 levels of the variable: burnished, monochrome, painted, and other.

b.   If all 4 types of pottery occur with equal probabilities, the values of p_1, p_2, p_3, and p_4 are all .25.

c.   To determine if one type of pottery is more likely to occur at the site than any other, we test:

H_0: $p_1 = p_2 = p_3 = p_4 = .25$
H_a: At least one $p_i \neq .25$ for $i = 1, 2, 3, 4$

d. From the printout, the test statistic is $\chi^2 = 436.59$ and the p-value is $p = .0000$. Since the p-value is so small, H_0 is rejected. There is sufficient evidence to indicate at least one type of pottery is more likely to occur at the site than another for any $\alpha > .0000$.

e. From the printout, the p-value is $p = .0000$. Thus, $P(\chi^2 \geq 436.59) = .0000$ when H_0 is true. Since this p-value is less than α ($p = .0000 < .10$), H_0 is rejected. There is sufficient evidence to indicate at least one type of pottery is more likely to occur at the site than another for $\alpha = .10$.

13.13 To determine if the true percentages of ADEs in the five "cause" categories are different, we test:

H_0: $p_1 = p_2 = p_3 = p_4 = p_5 = .2$
H_a: At least one p_i differs from .2, $i = 1, 2, 3, 4, 5$

The test statistic is $\chi^2 = 16$ (from the printout).

The p-value of the test is $p = .003$.

Since the p-value is less than α ($p = .003 < .10$), H_0 is rejected. There is sufficient evidence to indicate that at least one percentage of ADEs in the five "cause" categories is different at $\alpha = .10$.

13.15 a. Some preliminary calculations are:

$E(n_1) = np_{1,0} = 85(.26) = 22.1$
$E(n_2) = np_{2,0} = 85(.30) = 25.5$
$E(n_3) = np_{3,0} = 85(.11) = 9.35$
$E(n_4) = np_{4,0} = 85(.14) = 11.9$
$E(n_5) = np_{5,0} = 85(.19) = 16.15$

To determine of probabilities differ from the hypothesized values, we test:

H_0: $p_1 = .26, p_2 = .30, p_3 = .11, p_4 = .14, p_5 = .19$
H_a: At least one of the probabilities differs from its hypothesized value.

The test statistic is $\chi^2 = \sum \dfrac{\left[n_i - E(n_i)\right]^2}{E(n_i)}$

$$= \frac{(32-22.1)^2}{22.1} + \frac{(26-25.5)^2}{25.5} + \frac{(15-9.35)^2}{9.35} + \frac{(6-11.9)^2}{11.9} + \frac{(6-16.15)^2}{16.15}$$

$$= 17.16$$

The rejection region requires $\alpha = .05$ in the upper tail of the χ^2 distribution with $df = k - 1 = 5 - 1 = 4$. From Table VII, Appendix A, $\chi^2_{.05} = 11.0705$. The rejection region is $\chi^2 > 11.0705$.

Since the observed value of the test statistic falls in the rejection region ($\chi^2 = 17.16 > 11.0705$), reject H_0. There is sufficient evidence to indicate the probabilities differ from their hypothesized values at $\alpha = .05$.

13.17   a. $df = (r-1)(c-1) = (5-1)(5-1) = 16$. From Table VII, Appendix A, $\chi^2_{.05} = 26.2962$. The rejection region is $\chi^2 > 26.2962$.

   b. $df = (r-1)(c-1) = (3-1)(6-1) = 10$. From Table VII, Appendix A, $\chi^2_{.10} = 15.9871$. The rejection region is $\chi^2 > 15.9871$.

   c. $df = (r-1)(c-1) = (2-1)(3-1) = 2$. From Table VII, Appendix A, $\chi^2_{.01} = 9.21034$. The rejection region is $\chi^2 > 9.21034$.

13.19   a.   To convert the frequencies to percentages, divide the numbers in each column by the column total and multiply by 100. Also, divide the row totals by the overall total and multiply by 100. The column totals are 25, 64, and 78, while the row totals are 96 and 71. The overall sample size is 165. The table of percentages are:

	Column			
	1	**2**	**3**	
Row 1	$\frac{9}{25} \cdot 100 = 36\%$	$\frac{34}{64} \cdot 100 = 53.1\%$	$\frac{53}{78} \cdot 100 = 67.9\%$	$\frac{96}{167} \cdot 100 = 57.5\%$
2	$\frac{16}{25} \cdot 100 = 64\%$	$\frac{30}{64} \cdot 100 = 46.9\%$	$\frac{25}{78} \cdot 100 = 32.1\%$	$\frac{71}{167} \cdot 100 = 42.5\%$

   b.

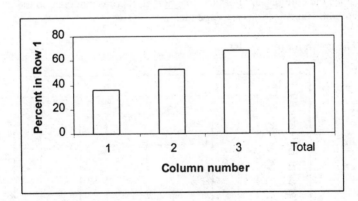

   c.   If the rows and columns are independent, the row percentages in each column would be close to the row total percentages. This pattern is not evident in the plot, implying the rows and columns are not independent.

13.21 To convert the frequencies to percentages, divide the numbers in each column by the column total and multiply by 100. Also, divide the row totals by the overall total and multiply by 100.

		B B_1	B_2	B_3	Totals
Row	A_1	$\frac{40}{134}\cdot 100 = 29.9\%$	$\frac{72}{163}\cdot 100 = 44.2\%$	$\frac{42}{142}\cdot 100 = 29.6\%$	$\frac{154}{439}\cdot 100 = 35.1\%$
	A_2	$\frac{63}{134}\cdot 100 = 47.0\%$	$\frac{53}{163}\cdot 100 = 32.5\%$	$\frac{70}{142}\cdot 100 = 49.3\%$	$\frac{186}{439}\cdot 100 = 42.4\%$
	A_3	$\frac{31}{134}\cdot 100 = 23.1\%$	$\frac{38}{163}\cdot 100 = 23.3\%$	$\frac{30}{142}\cdot 100 = 21.1\%$	$\frac{99}{439}\cdot 100 = 22.6\%$

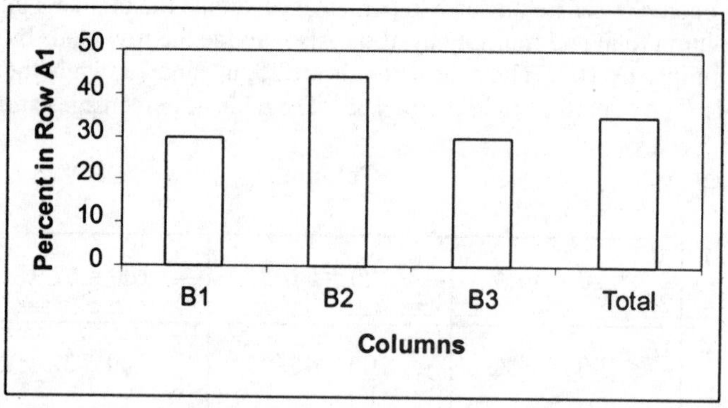

Since the columns are not the same height, there is evidence to indicate that the row and column classifications are dependent.

b. The graph for row A2 is:

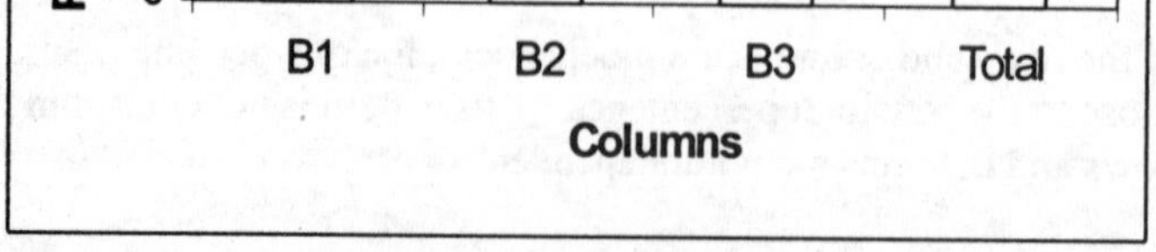

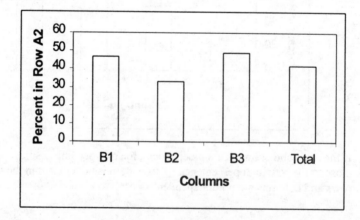

Since the columns are not the same height, there is evidence to indicate that the row and column classifications are dependent.

c. The graph for row A3 is:

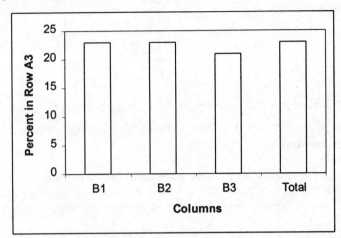

Since the columns are almost the same height, there is evidence to indicate that the row and column classifications are independent.

13.23 a. The 2×2 contingency table is:

LLD

Genetic Trait	Yes	No	Total
Yes	21	15	36
No	150	306	456
Total	171	321	492

b. Some preliminary calculations are:

$$\hat{E}(n_{11}) = \frac{r_1 c_1}{n} = \frac{36(171)}{492} = 12.51 \qquad \hat{E}(n_{12}) = \frac{r_1 c_2}{n} = \frac{36(321)}{492} = 23.49$$

$$\hat{E}(n_{21}) = \frac{r_2 c_1}{n} = \frac{456(171)}{492} = 158.49 \qquad \hat{E}(n_{22}) = \frac{r_2 c_2}{n} = \frac{456(321)}{492} = 297.51$$

To determine if the genetic trait occurs at a higher rate in LDD patients than in the controls, we test:

H_0: Genetic trait and LDD group are independent
H_a: Genetic trait and LDD group are dependent

The test statistic is $\chi^2 = \sum\sum \dfrac{[n_{ij} - \hat{E}(n_{ij})]^2}{\hat{E}(n_{ij})} = \dfrac{(21 - 12.51)^2}{12.51} + \dfrac{(15 - 23.49)}{23.49}$

$$+ \dfrac{(150 - 158.49)^2}{158.49} + \dfrac{(306 - 297.51)^2}{297.51} = 9.52$$

The rejection region requires $\alpha = .01$ in the upper tail of the χ^2 distribution with df = $(r-1)(c-1) = (2-1)(2-1) = 1$. From Table VIII, Appendix A, $\chi^2_{.01} = 6.63490$. The rejection region is $\chi^2 > 6.63491$.

Since the observed value of the test statistic falls in the rejection region ($\chi^2 = 9.52 > 6.63491$), H_0 is rejected. There is sufficient evidence to indicate that the genetic trait occurs at a higher rate in LDD patients than in the controls at $\alpha = .01$.

c.   To construct a bar graph, we will first compute the proportion of LDD/Control patients that have the genetic trait. Of the LLD patients, $21 / 171 = .123$ have the trait. For the Controls, $15 / 321 = .043$. The bar graph is:

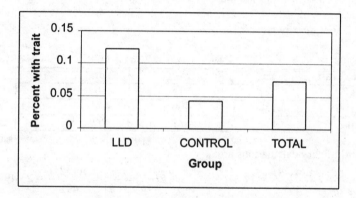

Because the bars are not close to being the same height, it indicates that the genetic trait appears at a higher rate among LLD patients than among the controls. This supports the conclusion of the test in part **b**.

13.25   Some preliminary calculations are:

$\hat{E}(n_{11}) = \dfrac{r_1 c_1}{n} = \dfrac{62(106)}{275} = 23.90$ $\hat{E}(n_{12}) = \dfrac{r_1 c_2}{n} = \dfrac{62(112)}{275} = 25.25$

$\hat{E}(n_{13}) = \dfrac{r_1 c_3}{n} = \dfrac{62(57)}{275} = 12.85$ $\hat{E}(n_{21}) = \dfrac{r_2 c_1}{n} = \dfrac{59(106)}{275} = 22.74$

$\hat{E}(n_{22}) = \dfrac{r_2 c_2}{n} = \dfrac{59(112)}{275} = 24.03$ $\hat{E}(n_{23}) = \dfrac{r_2 c_3}{n} = \dfrac{59(57)}{275} = 12.23$

$$\hat{E}(n_{31}) = \frac{r_3 c_1}{n} = \frac{59(106)}{275} = 22.74 \qquad \hat{E}(n_{32}) = \frac{r_3 c_2}{n} = \frac{59(112)}{275} = 24.03$$

$$\hat{E}(n_{33}) = \frac{r_3 c_3}{n} = \frac{59(57)}{275} = 12.23 \qquad \hat{E}(n_{41}) = \frac{r_4 c_1}{n} = \frac{95(106)}{275} = 36.62$$

$$\hat{E}(n_{42}) = \frac{r_4 c_2}{n} = \frac{95(112)}{275} = 38.69 \qquad \hat{E}(n_{43}) = \frac{r_4 c_3}{n} = \frac{95(57)}{275} = 19.69$$

To determine if political strategy of ethnic groups depends on world region, we test:

H_0: Political strategy and world region are independent
H_a: Political strategy and world region are dependent

The test statistic is $\chi^2 = \sum \sum \dfrac{\left[n_{ij} - \hat{E}(n_{ij}) \right]^2}{\hat{E}(n_{ij})} = \dfrac{(24-23.90)^2}{23.90} + \dfrac{(31-25.25)^2}{25.25}$

$+ \dfrac{(7-12.85)^2}{12.85} + \dfrac{(32-22.74)^2}{22.74} + \dfrac{(23-24.03)^2}{24.03} + \dfrac{(4-12.23)^2}{12.23} + \dfrac{(11-22.74)^2}{22.74}$

$+ \dfrac{(22-24.03)^2}{24.03} + \dfrac{(26-12.23)^2}{12.23} + \dfrac{(39-36.62)^2}{36.62} + \dfrac{(36-38.69)^2}{38.69} + \dfrac{(20-19.69)^2}{19.69}$

$= 35.409$

The rejection region requires $\alpha = .10$ in the upper tail of the χ^2 distribution with df $= (r-1)(c-1) = (4-1)(3-1) = 6$. From Table VIII, Appendix A, $\chi^2_{.10} = 10.6446$. The rejection region is $\chi^2 > 10.6446$.

Since the observed value of the test statistic falls in the rejection region ($\chi^2 = 35.409 > 10.6446$), H_0 is rejected. There is sufficient evidence to indicate that the political strategy of the ethnic groups depends on world region at $\alpha = .10$.

To graph the data, we will first compute the percent of observations in each category of Political Strategy for each World Region. To do this, we divide each cell frequency by the row total and then multiply by 100%. Using SAS, a graph of the data is:

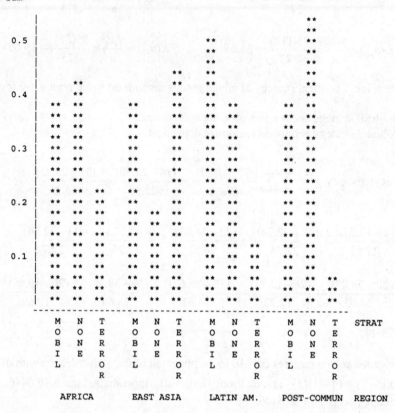

The graph supports the results of the test above. The percents in each level of Political Strategy differs for the different values of World Region.

13.27   Some preliminary calculations are:

$$\hat{E}(n_{11}) = \frac{r_1 c_1}{n} = \frac{12(10)}{24} = 5 \qquad \hat{E}(n_{12}) = \frac{r_1 c_2}{n} = \frac{12(14)}{24} = 7$$

$$\hat{E}(n_{21}) = \frac{r_2 c_1}{n} = \frac{12(10)}{24} = 5 \qquad \hat{E}(n_{22}) = \frac{r_2 c_2}{n} = \frac{12(14)}{24} = 7$$

To determine if a relationship exists between food choice and whether or not chickadees fed on gypsy moth eggs, we test:

H_0: Food choice and whether or not chickadees fed on gypsy moth eggs are independent
H_a: Food choice and whether or not chickadees fed on gypsy moth eggs are dependent

The test statistic is $\chi^2 = \sum\sum \dfrac{\left[n_{ij} - \hat{E}(n_{ij})\right]^2}{\hat{E}(n_{ij})}$

$$= \frac{(2-5)^2}{5} + \frac{(10-7)^2}{7} + \frac{(8-5)^2}{5} + \frac{(4-7)^2}{7} = 6.171$$

The rejection region requires $\alpha = .10$ in the upper tail of the χ^2 distribution with df $=$ $(r-1)(c-1) = (2-1)(2-1) = 1$. From Table VII, Appendix A, $\chi^2_{.10} = 2.70554$. The rejection region is $\chi^2 > 2.70554$.

Since the observed value of the test statistic falls in the rejection region ($\chi^2 = 6.171 >$ 2.70554), H_0 is rejected. There is sufficient evidence to indicate that a relationship exists between food choice and whether or not chickadees fed on gypsy moth eggs at $\alpha = .10$.

13.29   a.   From the printout, the expected cell counts range from 0.4154 to 18.092. There are 7 out of 16 cells or $7/16 = .4375$ of the cells with expected cell counts less than 5. In order for the test to be valid, all of the cells should have expected cell counts of 5 or more. Thus, we should not proceed with the analysis.

   b.   Combining the High and Very High categories, the new table is:

		CAHS LEVEL		
		Low	Medium	High/Very High
SHSS: C LEVEL	**Low**	32	14	2
	Medium	11	14	6
	High/Very High	6	16	29

   c.   The expected cell counts are:

$$\hat{E}(n_{11}) = \frac{r_1 c_1}{n} = \frac{48(49)}{130} = 18.09 \qquad \hat{E}(n_{21}) = \frac{r_2 c_1}{n} = \frac{31(49)}{130} = 11.68$$

$$\hat{E}(n_{12}) = \frac{r_1 c_2}{n} = \frac{48(44)}{130} = 16.25 \qquad \hat{E}(n_{22}) = \frac{r_2 c_2}{n} = \frac{31(44)}{130} = 10.49$$

$$\hat{E}(n_{13}) = \frac{r_1 c_3}{n} = \frac{48(37)}{130} = 13.66 \qquad \hat{E}(n_{23}) = \frac{r_2 c_3}{n} = \frac{31(37)}{130} = 8.82$$

$$\hat{E}(n_{31}) = \frac{r_3 c_1}{n} = \frac{51(49)}{130} = 19.22 \qquad \hat{E}(n_{32}) = \frac{r_3 c_2}{n} = \frac{51(44)}{130} = 17.26$$

$$\hat{E}(n_{33}) = \frac{r_3 c_3}{n} = \frac{51(37)}{130} = 14.52$$

Since all of the expected cell counts are now 5 or greater, the assumption is met.

   d.   To determine whether CAHS Levels and SHSS:C Levels are dependent, we test:

H_0: CAHS Levels and SHSS:C Levels are independent
H_a: CAHS Levels and SHSS:C Levels are dependent

The test statistic is $\chi^2 = \sum\sum \dfrac{\left[n_{ij} - \hat{E}(n_{ij})\right]^2}{\hat{E}(n_{ij})}$

$$= \frac{(32-18.09)^2}{18.09} + \frac{(14-16.25)^2}{16.25} + \frac{(2-13.66)^2}{13.66} + \frac{(11-11.68)^2}{11.68}$$

$$+ \frac{(14-10.49)^2}{10.49} + \frac{(6-8.82)^2}{8.82} + \frac{(6-19.22)^2}{19.22} + \frac{(16-17.26)^2}{17.26}$$

$$+ \frac{(29-14.52)^2}{14.52} = 46.70$$

The rejection region requires $\alpha = .05$ in the upper tail of the χ^2 distribution with df $= (r-1)(c-1) = (3-1)(3-1) = 4$. From Table VII, Appendix A, $\chi^2_{.05} = 9.48773$. The rejection region is $\chi^2 > 9.48773$.

Since the observed value of the test statistic falls in the rejection region ($\chi^2 = 46.70 > 9.48773$), H_0 is rejected. There is sufficient evidence to indicate that CAHS Levels and SHSS:C Levels are dependent at $\alpha = .05$.

13.31   a.   Some preliminary calculations are:

$$\hat{E}(n_{11}) = \frac{r_1 c_1}{n} = \frac{31(24)}{38} = 19.58 \qquad\qquad \hat{E}(n_{12}) = \frac{r_1 c_2}{n} = \frac{31(14)}{38} = 11.42$$

$$\hat{E}(n_{21}) = \frac{r_2 c_1}{n} = \frac{7(24)}{38} = 4.42 \qquad\qquad \hat{E}(n_{22}) = \frac{r_2 c_2}{n} = \frac{7(14)}{38} = 2.58$$

To determine whether the vaccine is effective in treating the MN strain of HIV, we test:

H_0: Vaccine and MN strain are independent
H_a: Vaccine and MN strain are dependent

The test statistic is $\chi^2 = \sum\sum \dfrac{\left[n_{ij} - \hat{E}(n_{ij})\right]^2}{\hat{E}(n_{ij})} = \dfrac{(22-19.58)^2}{19.58} + \dfrac{(9-11.42)^2}{11.42}$

$$+ \frac{(2-4.42)^2}{4.42} + \frac{(5-2.58)^2}{2.58} = 4.407$$

The rejection region requires $\alpha = .05$ in the upper tail of the χ^2 distribution with df $= (r-1)(c-1) = (2-1)(2-1) = 1$. From Table VIII, Appendix A, $\chi^2_{.05} = 3.84146$. The rejection region is $\chi^2 > 3.84146$.

Since the observed value of the test statistic falls in the rejection region ($\chi^2 = 4.407 > 3.84146$), H_0 is rejected. There is sufficient evidence to indicate that the vaccine is effective in treating the MN strain of HIV at $\alpha = .05$.

b. The necessary assumptions are:

1. The n observed counts are a random sample from the population of interest.

2. The sample size, n, will be large enough so that, for every cell, the expected count, $E(n_{ij})$, will be equal to 5 or more.

For this example, the second assumption is violated. Two of the expected cell counts are less than 5. What we have computed for the test statistic may not have a χ^2 distribution.

c. For this contingency table:

$$p = \frac{\binom{7}{2}\binom{31}{22}}{\binom{38}{24}} = \frac{\frac{7!}{2!5!}\frac{31!}{22!9!}}{\frac{38!}{24!14!}} = .0438$$

d. If vaccine and MN strain are independent, then the proportion of positive results should be relatively the same for both patient groups. In the two tables presented, the proportion of positive results for the vaccinated group is smaller than the proportion for the original table.

For the first table,

$$p = \frac{\binom{7}{1}\binom{31}{23}}{\binom{38}{24}} = \frac{\frac{7!}{1!6!}\frac{31!}{23!8!}}{\frac{38!}{24!14!}} = .0057$$

For the second table:

$$p = \frac{\binom{7}{0}\binom{31}{24}}{\binom{38}{24}} = \frac{\frac{7!}{0!7!}\frac{31!}{24!7!}}{\frac{38!}{24!14!}} = .0003$$

e. The p-value for Fisher's exact test is $p = .0438 + .0057 + .0003 = .0498$. Since the p-value is small, there is evidence to reject H_0. There is sufficient evidence to indicate that the vaccine is effective in treating the MN strain of HIV at $\alpha > .0498$.

13.33 a. Some preliminary calculations are:

If all the categories are equally likely,

$$p_{1,0} = p_{2,0} = p_{3,0} = p_{4,0} = p_{5,0} = .2$$

$$E(n_1) = E(n_2) = E(n_3) = E(n_4) = E(n_5) = np_{1,0} = 150(.2) = 30$$

To determine if the categories are not equally likely, we test:

H_0: $p_1 = p_2 = p_3 = p_4 = p_5 = .2$
H_a: At least one probability is different from .2

The test statistic is $\chi^2 = \sum \dfrac{\left[n_i - E(n_i)\right]^2}{E(n_i)}$

$$= \frac{(28-30)^2}{30} + \frac{(35-30)^2}{30} + \frac{(33-30)^2}{30} + \frac{(25-30)^2}{30} + \frac{(29-30)^2}{30} = 2.133$$

The rejection region requires $\alpha = .10$ in the upper tail of the χ^2 distribution with df = $k - 1 = 5 - 1 = 4$. From Table VII, Appendix A, $\chi^2_{.10} = 7.77944$. The rejection region is $\chi^2 > 7.77944$.

Since the observed value of the test statistic does not fall in the rejection region ($\chi^2 = 2.133 \ngtr 7.77944$), H_0 is not rejected. There is insufficient evidence to indicate the categories are not equally likely at $\alpha = .10$.

b. $\hat{p}_2 = \dfrac{35}{150} = .233$

For confidence coefficient .90, $\alpha = .10$ and $\alpha/2 = .05$. From Table IV, Appendix A, $z_{.05} = 1.645$. The confidence interval is:

$$\hat{p}_2 \pm z_{.05} \sqrt{\frac{\hat{p}_2 \hat{q}_2}{n_2}} \Rightarrow .233 \pm 1.645 \sqrt{\frac{.233(.767)}{150}} \Rightarrow .233 \pm .057 \Rightarrow (.176, .290)$$

13.35 a. No. In a multinomial experiment, each trial results in one of $k = 8$ possible outcomes. In this experiment, each trial cannot result in one of all of the eight possible outcomes. The people are assigned to a particular diet before the experiment. Once the people are assigned to a particular diet, there are only two possible outcomes, cancer tumors or no cancer tumors.

b. The expected cell counts are:

$$\hat{E}(n_{11}) = \frac{r_1 c_1}{n} = \frac{80(30)}{120} = 20 \qquad \hat{E}(n_{21}) = \frac{r_2 c_1}{n} = \frac{40(30)}{120} = 10$$

$$\hat{E}(n_{12}) = \frac{r_1 c_2}{n} = \frac{80(30)}{120} = 20 \qquad \hat{E}(n_{22}) = \frac{r_2 c_2}{n} = \frac{40(30)}{120} = 10$$

$$\hat{E}(n_{13}) = \frac{r_1 c_3}{n} = \frac{80(30)}{120} = 20 \qquad \hat{E}(n_{23}) = \frac{r_2 c_3}{n} = \frac{40(30)}{120} = 10$$

$$\hat{E}(n_{14}) = \frac{r_1 c_4}{n} = \frac{80(30)}{120} = 20 \qquad \hat{E}(n_{24}) = \frac{r_2 c_4}{n} = \frac{40(30)}{120} = 10$$

c. The test statistic is $\chi^2 = \sum\sum \dfrac{\left[n_{ij} - \hat{E}(n_{ij})\right]^2}{\hat{E}(n_{ij})}$

$$= \frac{(27-20)^2}{20} + \frac{(20-20)^2}{20} + \frac{(19-20)}{20} + \frac{(14-20)^2}{20} + \frac{(3-10)^2}{10} + \frac{(10-10)^2}{10}$$

$$+ \frac{(11-10)^2}{10} + \frac{(16-10)^2}{10} = 12.9$$

d. To determine if diet and presence/absence of cancer are independent, we test:

H_0: Diet and presence/absence of cancer are independent
H_a: Diet and presence/absence of cancer are dependent

The test statistic is $\chi^2 = 12.9$ (from part c).

The rejection region requires $\alpha = .05$ in the upper tail of the χ^2 distribution with df = $(r-1)(c-1) = (2-1)(4-1) = 3$. From Table VII, Appendix A, $\chi^2_{.05} = 7.81473$. The rejection region is $\chi^2 > 7.81473$.

Since the observed value of the test statistic falls in the rejection region ($\chi^2 = 12.9 > 7.81473$), H_0 is rejected. There is sufficient evidence to indicate that diet and presence/absence of cancer are dependent for $\alpha = .05$.

e. $\hat{p}_1 = 27/30 = .9$; $\hat{p}_2 = 20/30 = .667$

For confidence coefficient .95, $\alpha = .05$ and $\alpha/2 = .05/2 = .025$. From Table IV, Appendix A, $z_{.025} = 1.96$. The confidence interval is:

$$(\hat{p}_1 - \hat{p}_2) \pm z_{.025}\sqrt{\frac{\hat{p}_1\hat{q}_1}{n_1} + \frac{\hat{p}_2\hat{q}_2}{n_1}} \Rightarrow (.900 - .667) \pm 1.96\sqrt{\frac{.900(.100)}{30} + \frac{.667(.333)}{30}}$$

$$\Rightarrow .233 \pm .200 \Rightarrow (.033, .433)$$

We are 95% confident that the difference in the percentage of rats on a high fat/no fiber diet with cancer and the percentage of rats on a high fat/fiber diet with cancer is between 3.3% and 43.3%.

13.37   a. Some preliminary calculations:

$E(n_1) = np_{1,0} = 28(.25) = 7$
$E(n_2) = np_{2,0} = 28(.25) = 7$
$E(n_3) = np_{3,0} = 28(.25) = 7$
$E(n_4) = np_{4,0} = 28(.25) = 7$

To determine if the true proportions of intellectually disabled elderly patients in each of the hearing loss categories differ, we test:

H_0: $p_1 = p_2 = p_3 = p_4 = .25$
H_a: At least one p_i differs from .25, $i = 1, 2, 3, 4$

The test statistic is $\chi^2 = \sum \dfrac{[n_i - E(n_i)]^2}{E(n_i)} = \dfrac{(7-7)^2}{7} + \dfrac{(7-7)^2}{7} + \dfrac{(9-7)^2}{7} + \dfrac{(5-7)^2}{7}$

$$= 1.143$$

The rejection region requires $\alpha = .05$ in the upper tail of the χ^2 distribution with df $= k - 1 = 4 - 1 = 3$. From Table VII, Appendix A, $\chi^2_{.05} = 7.81473$. The rejection region is $\chi^2 > 7.81473$.

Since the observed value of the test statistic does not fall in the rejection region ($\chi^2 = 1.143 \not> 7.81473$), H_0 is not rejected. There is insufficient evidence to indicate that at least one of the true proportions of intellectually disabled elderly patients in each of the hearing loss categories differs at $\alpha = .05$.

b.　　$\hat{p} = \dfrac{x}{n} = \dfrac{5}{28} = .179$

For confidence coefficient .90; $\alpha = .10$ and $\alpha/2 = .10/2 = .05$. From Table IV, Appendix A, $z_{.05} = 1.645$. The confidence interval is

$\hat{p} \pm z_{.05}\sqrt{\dfrac{\hat{p}\hat{q}}{n}}$

$\Rightarrow .179 \pm 1.645\sqrt{\dfrac{(.179)(1-.179)}{28}}$

$\Rightarrow .179 \pm .119$

$\Rightarrow (.060, .298)$

We are 90% confident that the true proportion of disabled elderly patients with severe hearing loss is between .060 and .298.

13.39　To determine if there is a dependence between a son's choice of occupation and his father's occupation, we test:

H_0: Son's choice of occupation and his father's occupation are independent
H_a: Son's choice of occupation and his father's occupation are dependent.

The test statistic is $\chi^2 = \sum\sum \dfrac{[n_{ij} - \hat{E}(n_{ih})]^2}{\hat{E}(n_{ij})} = 180.8739$ (from the printout).

The p-value is $p < .0001$. Since the p-value is less than α ($p < .0001 < .05$), H_0 is rejected. There is sufficient evidence to indicate a dependence between a son's choice of occupation and his father's occupation at $\alpha = .05$.

13.41　Some preliminary calculations are:

If the strategies are equally preferred, $p_{1,0} = p_{2,0} = p_{3,0} = p_{4,0} = p_{5,0} = .2$

$E(n_1) = E(n_2) = E(n_3) = E(n_4) = E(n_5) = np = 100(.2) = 20$

To determine if there is a preference for one or more of the strategies, we test:

H_0: $p_1 = p_2 = p_3 = p_4 = p_5 = .2$
H_a: At least one probability differs from .2

The test statistic is $\chi^2 = \sum \dfrac{[n_i - E(n_i)]^2}{E(n_i)}$

$$= \frac{(17-20)^2}{20} + \frac{(27-20)^2}{20} + \frac{(22-20)^2}{20} + \frac{(15-20)^2}{20} + \frac{(19-20)^2}{20} = 4.4$$

The rejection region requires $\alpha = .05$ in the upper tail of the χ^2 distribution with df $= k - 1 =$ $5 - 1 = 4$. From Table VII, Appendix A, $\chi^2_{.05} = 9.48773$. The rejection region is $\chi^2 >$ 9.48773.

Since the observed value of the test statistic does not fall in the rejection region ($\chi^2 = 4.4 \not> $ 9.48773), H_0 is not rejected. There is insufficient evidence to indicate one of the strategies is preferred over any of the others at $\alpha = .05$.

13.43 To determine if a relationship exists between eye and hair color, we test:

H_0: Hair and eye color are independent
H_a: Hair and eye color are dependent

From the printout, the test statistic is $\chi^2 = 9.350$ and the p-value is $p = .002$. Since the p-value is less than α ($p = .002 < .10$), H_0 is rejected. There is sufficient evidence to indicate that hair color and eye color are related at $\alpha = .05$.

13.45 a. Calculate $\hat{E}(n_{ij}) = \dfrac{r_i c_j}{n}$ for $i = 1, 2, 3, 4, 5$, and $j = 1, 2, 3, 4, 5$.

The expected values for each cell in the table is summarized below.

Observed and Estimated Expected (in Parentheses) Counts

		Wife's Lifestyle					
		Pleasing	Outdoing	Avoiding	Control	Achieving	Totals
	Pleasing	9 (5.88)	6 (6.70)	6 (6.21)	3 (8.17)	9 (6.04)	33
	Outdoing	7 (8.20)	11 (9.34)	12 (8.65)	11 (11.39)	5 (8.43)	46
	Avoiding	8 (6.77)	8 (7.71)	6 (7.15)	11 (9.41)	5 (6.96)	38
Husband's Lifestyle	Control	8 (9.09)	10 (10.35)	7 (9.59)	15 (12.62)	11 (9.34)	51
	Achieving	4 (6.06)	6 (6.90)	7 (6.40)	10 (8.42)	7 (6.23)	34
	Totals	36	41	38	50	37	202

The test is: H_0: The wife's lifestyle types and the husband's lifestyle types are independent.

H_a: The wife's lifestyle types and the husband's lifestyle types are dependent.

The test statistic is $\chi^2 = \sum\sum \dfrac{\left[n_{ij} - \hat{E}(n_{ij})\right]^2}{\hat{E}(n_{ij})}$

$$= \dfrac{(9-5.88)^2}{5.88} + \dfrac{(6-6.70)^2}{6.70} + \cdots + \dfrac{(10-8.42)^2}{8.42} + \dfrac{(7-6.23)^2}{6.23}$$

$$= 13.715$$

Since no α is given we will use $\alpha = .05$. The rejection region requires $\alpha = .05$ in the upper tail of the χ^2 distribution with df $= (r-1)(c-1) = 4(4) = 16$. From Table VII, Appendix A, $\chi^2_{.05} = 26.2962$. The rejection region is $\chi^2 > 26.2962$.

Since the observed value of the test statistic does not fall in the rejection region ($\chi^2 = 13.715 \leq 26.2962$), H_0 is not rejected. There is insufficient evidence to indicate that the wife's lifestyle types and the husband's lifestyle types are dependent at $\alpha = .05$.

b. There are two tests:

Wife

To determine if there are differences in the proportions of wives who fall into the five personality profiles, we test:

H_0: $p_1 = p_2 = p_3 = p_4 = p_5 = .20$
H_a: At least one of the proportions differs from .20.

where
p_1 = proportion of all wives that are pleasing
p_2 = proportion of all wives that are outstanding
p_3 = proportion of all wives that are avoiding
p_4 = proportion of all wives that are controlling
p_5 = proportion of all wives that are achieving

$E(n_1) = E(n_2) = E(n_3) = E(n_4) = E(n_5) = 202(.2) = 40.4$

The test statistic is $\chi^2 = \sum \dfrac{[n_i - E(n_i)]^2}{E(n_i)}$

$$= \dfrac{(36-40.4)^2}{40.4} + \dfrac{(41-40.4)^2}{40.4} + \dfrac{(38-40.4)^2}{40.4} + \dfrac{(50-40.4)^2}{40.4} + \dfrac{(37-40.4)^2}{40.4}$$

$$= 3.198$$

Since no α is given, we will use $\alpha = .05$. The rejection region is $\chi^2 > 9.48773$ for df $= k - 1 = 5 - 1 = 4$. Since the observed value of the test statistic does not fall in the rejection region ($\chi^2 = 3.198 \not> 9.48773$), do not reject H_0. There is insufficient evidence to indicate that a difference exists in the proportions of wives who fall in the five personality profiles at $\alpha = .05$.

Husband

To determine if there are differences in the proportions of wives who fall into the five personality profiles, we test:

H_0: $p_1 = p_2 = p_3 = p_4 = p_5 = .20$
H_a: At least one of the proportions differs from .20.

where
p_1 = proportion of all husbands that are pleasing
p_2 = proportion of all husbands that are outdoing
p_3 = proportion of all husbands that are avoiding
p_4 = proportion of all husbands that are controlling
p_5 = proportion of all husbands that are achieving

$$E(n_1) = E(n_2) = E(n_3) = E(n_4) = E(n_5) = 202(.2) = 40.4$$

The test statistic is $\chi^2 = \sum \dfrac{[n_i - E(n_i)]^2}{E(n_i)}$

$$= \frac{(33 - 40.4)^2}{40.4} + \frac{(46 - 40.4)^2}{40.4} + \frac{(38 - 40.4)^2}{40.4} + \frac{(51 - 40.4)^2}{40.4} + \frac{(34 - 40.4)^2}{40.4}$$
$$= 6.069$$

Using $\alpha = .05$, in the rejection region is $\chi^2 > 9.48773$ for df $= k - 1 = 5 - 1 = 4$. Since the observed value of the test statistic does not fall in the rejection region ($\chi^2 = 6.069 \ngtr 9.48773$), do not reject H_0. There is insufficient evidence to indicate that a difference exists in the proportions of husbands who fall in the five personality profiles at $\alpha = .05$.

13.47   a.   From the data, the contingency table is:

	Surgery	Radiation	Total
Cancer Controlled	21	15	36
Cancer not Controlled	2	3	5
Total	23	18	41

b. $\hat{E}(n_{11}) = \dfrac{r_1 c_1}{n} = \dfrac{36(23)}{41} = 20.20$ $\hat{E}(n_{21}) = \dfrac{r_1 c_2}{n} = \dfrac{36(18)}{41} = 15.80$

  $\hat{E}(n_{21}) = \dfrac{r_2 c_1}{n} = \dfrac{5(23)}{41} = 2.80$ $\hat{E}(n_{22}) = \dfrac{r_2 c_2}{n} = \dfrac{5(18)}{41} = 2.20$

c. In order for the chi-square test to be valid, the expected cell counts must be at least 5. In this case, two of the four expected cell counts are less than 5.

d. The p-value is $p = .187$. Since the p-value is so large, there is no evidence to indicate that the rate of controlling larynx cancer is dependent on type of treatment for any $\alpha \leq .10$.

e. The test statistic is $\chi^2 = \sum\sum \dfrac{\left[n_{ij} - \hat{E}(n_{ih})\right]^2}{\hat{E}(n_{ij})}$

$$= \dfrac{(21-20.2)^2}{20.2} + \dfrac{(15-15.8)^2}{15.8} + \dfrac{(2-2.8)^2}{2.8} + \dfrac{(3-2.2)^2}{2.2} = .592$$

From Table VII, Appendix A, with df $= (r-1)(c-1) = (2-1)(2-1) = 1$, $P(\chi^2 > .592) > .10$. This agrees with the p-value that is associated with the Fisher's exact test. The conclusions for the two tests would be identical. However, there are many values greater than .10. We cannot be sure how close these two p-values are unless we find the exact probability which we cannot find with the table provided in the text.

13.49 a. $\chi^2 = \sum \dfrac{[n_i - E(n_i)]^2}{E(n_i)}$

$$= \dfrac{(26-23)^2}{23} + \dfrac{(146-136)^2}{136} + \dfrac{(361-341)^2}{341} + \dfrac{(143-136)^2}{136} + \dfrac{(13-23)^2}{23}$$
$$= 9.647$$

b. From Table VII, Appendix A, with df $= 5$, $\chi^2_{.05} = 11.0705$

c. No. $\chi^2 = 9.647 \ngtr 11.0705$. Do not reject H_0. There is insufficient evidence to indicate the salary distribution is nonnormal for $\alpha = .05$.

d. The p-value $= P(\chi^2 \geq 9.647)$.

Using Table VII, Appendix A, with df $= 5$,

$.05 < P(\chi^2 \geq 9.647) < .10$.

Nonparametric Statistics

14.1   The sign test is preferred to the *t*-test when the population from which the sample is selected is not normal.

14.3   a. $P(x \geq 6) = 1 \leq P(x - 5) = 1 - .937 = .063$

      b. $P(x \geq 5) = 1 \leq P(x - 4) = 1 - .500 = .500$

      c. $P(x \geq 8) = 1 \leq P(x - 7) = 1 - .996 = .004$

      d. $P(x \geq 10) = 1 \leq P(x - 9) = 1 - .849 = .151$

$$\mu = np = 15(.5) = 7.5 \text{ and } \sigma = \sqrt{npq} = \sqrt{15(.5)(.5)} = 1.9365$$

$$P(x \geq 10) \approx P\left(z \geq \frac{(10 - .5) - 7.5}{1.9365}\right) = P(z \geq 1.03) = .5 - .3485 = .1515$$

      e. $P(x \geq 15) = 1 \leq P(x \leq 14) = 1 - .788 = .212$

$$\mu = np = 25(.5) = 12.5 \text{ and } \sigma = \sqrt{npq} = \sqrt{25(.5)(.5)} = 2.5$$

$$P(x \geq 15) \approx P\left(z \geq \frac{(15 - .5) - 12.5}{2.5}\right) = P(z \geq .80) = .5 - .2881 = .2119$$

14.5   To determine if the median is greater than 80, we test:

$H_0: \eta = 80$
$H_a: \eta > 80$

The test statistic is S = number of measurements greater than 80 = 16.

The *p*-value = $P(x \geq 16)$ where *x* is a binomial random variable with $n = 25$ and $p = .5$. From Table II,

$$p\text{-value} = P(x \geq 16) = 1 - P(x \leq 15) = 1 - .885 = .115$$

Since the *p*-value = $.115 > \alpha = .10$, H_0 is not rejected. There is insufficient evidence to indicate the median is greater than 80 at $\alpha = .10$.

We must assume the sample was randomly selected from a continuous probability distribution.

Note: Since $n \geq 10$, we could use the large-sample approximation.

14.7   a.   To determine if the median daily ammonia concentration for all afternoon drive-time days exceeds 1.5 ppm, we test:

H_0: $\eta = 1.5$
H_a: $\eta > 1.5$

       b.   The test statistic is S = number of measurements greater than 1.5 = 3.

       c.   The p-value = $P(x \geq 3)$ where x is a binomial random variable with $n = 8$ and $p = .5$. From Table II, Appendix A,

$$P(x \geq 3) = 1 - P(x \leq 2) = 1 - .145 = .855.$$

       d.   Sine the p-value is greater than α ($p = .855 > .05$), H_0 is not rejected. There is insufficient evidence to indicate that the median daily ammonia concentration for all afternoon drive-time days exceeds 1.5 ppm at $\alpha = .05$.

14.9   a.   To determine whether the median biting rate is higher in bright, sunny weather, we test:

H_0: $\eta = 5$
H_a: $\eta > 5$

       b.   The test statistic is $z = \dfrac{(S - .5) - .5n}{.5\sqrt{n}} = \dfrac{(95 - .5) - .5(122)}{.5\sqrt{122}} = 6.07$
(where S = number of observations greater than 5)

The p-value is $p = P(z \geq 6.07)$. From Table IV, Appendix A, $p = P(z \geq 6.07) \approx 0.0000$.

       c.   Since the observed p-value is less than α ($p = 0.0000 < .01$), H_0 is rejected. There is sufficient evidence to indicate that the median biting rate in bright, sunny weather is greater than 5 at $\alpha = .01$.

14.11   a.   To determine whether the median is less than \$2.25, we test:

H_0: $\eta = 2.25$
H_a: $\eta < 2.25$

The test statistic is $z = \dfrac{(S - .5) - .5n}{.5\sqrt{n}}$

$$= \dfrac{(50 - .5) - .5(80)}{.5\sqrt{80}}$$

$$= 2.12$$

(where S = number of observations less than 2.25)

The p-value is $p = P(z \geq 2.12)$. From Table IV, Appendix A, $p = P(z \geq 2.12) = .5 - .4830 = .0170$. Therefore, reject H_0 for $\alpha > .017$ and conclude that the median amount spent for hamburgers at lunch at McDonald's is less than \$2.25.

b. No; The sample was taken from only two restaurants in Boston.

c. The distribution of the prices is continuous. Nothing is assumed about the shape of the probability distribution.

14.13 a. The hypotheses are:

H_0: Two sampled populations have identical distributions
H_a: The probability distribution for population A is shifted to the left of that for B

b. First, we rank all the data:

A		B	
Observation	Rank	Observation	Rank
37	8	65	13
40	9	35	6.5
33	3.5	47	11
29	2	52	12
42	10		
33	3.5		
35	6.5		
28	1		
34	5		
	$T_1 = 48.5$		$T_2 = 42.5$

The test statistic is $T_2 = 42.5$ because $n_2 < n_1$.

The rejection region is $T_2 \geq 39$ from Table XII, Appendix A, with $n_1 = 9$, $n_2 = 4$ and $\alpha = .05$.

Since the observed value of the test statistic falls in the rejection region ($T_2 = 42.5 \geq 39$), H_0 is rejected. There is sufficient evidence to indicate the distribution for population B is shifted to the right of the distribution for population A at $\alpha = .05$.

14.15

Sample from Population 1 (A)	Rank	Sample from Population 2 (B)	Rank
15	13	5	2.5
10	8.5	12	10.5
12	10.5	9	6.5
16	14	9	6.5
13	12	8	4.5
8	4.5	4	1
	$T_1 = 62.5$	5	2.5
		10	8.5
			$T_2 = 42.5$

a. H_0: The two sampled populations have identical probability distributions
   H_a: The probability distribution for population A is shifted to the left or to the right of that for B

The test statistic is $T_1 = 62.5$ since sample A has the smallest number of measurements.

The null hypothesis will be rejected if $T_1 \leq T_L$ or $T_1 \geq T_U$ where T_L and T_U correspond to $\alpha = .05$ (two-tailed), $n_1 = 6$ and $n_2 = 8$. From Table XII, Appendix A, $T_L = 29$ and $T_U = 61$.

Reject H_0 if $T_1 \leq 29$ or $T_1 \geq 61$.

Since $T_1 = 62.5 \geq 61$, we reject H_0 and conclude there is sufficient evidence to indicate population A is shifted to the left or right of population B at $\alpha = .05$.

b. H_0: The two sampled populations have identical probability distributions
   H_a: The probability distribution for population A is shifted to the right of population B

The test statistic remains $T_1 = 62.5$.

The null hypothesis will be rejected if $T_1 \geq T_U$ where T_U corresponds to $\alpha = .05$ (one-tailed), $n_1 = 6$ and $n_2 = 8$. From Table XII, Appendix A, $T_U = 58$.

Reject H_0 if $T_1 \geq 58$.

Since $T_1 = 62.5 \geq 58$, we reject H_0 and conclude there is sufficient evidence to indicate population A is shifted to the right of population B at $\alpha = .05$.

14.17 a. To determine if the distribution of FNE scores for bulimic students is shifted above the corresponding distribution for female students with normal eating habits, we test:

H_0: The distribution of FNE scores for bulimic students is the same as the corresponding distribution for female students with normal eating habits

H_a: The distribution of FNE scores for bulimic students is shifted above the corresponding distribution for female students with normal eating habits

b. The data ranked are:

Bulimic Students	Rank	Normal Students	Rank
21	21.5	13	8.5
13	8.5	6	1
10	4.5	16	13.5
20	19.5	13	8.5
25	25	8	3
19	17	19	17
16	13.5	23	23
21	21.5	18	15
24	24	11	6
13	8.5	19	17
14	11	7	2
		10	4.5
$T_1 = 174.5$		15	12
		20	19.5
		$T_2 = 150.5$	

c. The sum of the ranks of the 11 FNE scores for bulimic students is $T_1 = 174.5$.

d. The sum of the ranks of the 14 FNE scores for normal students is $T_2 = 150.5$.

$$\text{The test statistic is } z = \frac{T_1 - \dfrac{n_1(n_1 + n_2 + 1)}{2}}{\sqrt{\dfrac{n_1 n_2(n_1 + n_2 + 1)}{12}}} = \frac{174.5 - \dfrac{11(11 + 14 + 1)}{2}}{\sqrt{\dfrac{11(14)(11 + 14 + 1)}{12}}} = 1.72$$

The rejection region requires $\alpha = .10$ in the upper tail of the z distribution. From Table IV, Appendix A, $z_{.10} = 1.28$. The rejection region is $z > 1.28$.

f. Since the observed value of the test statistic falls in the rejection region ($z = 1.72 > 1.28$), H_0 is rejected. There is sufficient evidence to indicate that the distribution of FNE scores for bulimic students is shifted above the corresponding distribution for female students with normal eating habits at $\alpha = .10$.

14.19 a.

Deaf Children	Rank	Hearing Children	Rank
2.75	18	1.15	1
3.14	19	1.65	6
3.23	20	1.43	4
2.30	15	1.83	8.5
2.64	17	1.75	7
1.95	10	1.23	2
2.17	13	2.03	12
2.45	16	1.64	5
1.83	8.5	1.96	11
2.23	14	1.37	3
	$T_1 = 150.5$		$T_2 = 59.5$

H_0: The probability distributions of eye movement rates are identical for deaf and hearing children

H_a: The probability distribution of eye movement rates for deaf children lies to the right of that for hearing children

The test statistic can be either T_1 or T_2 since the sample sizes are equal. Use $T_1 = 150.5$.

The null hypothesis will be rejected if $T_1 \geq T_U$ where $\alpha = .05$ (one-tailed), $n_1 = 10$ and $n_2 = 10$. From Table XII, Appendix A, $T_U = 127$.

Reject H_0 if $T_1 \geq 127$.

Since $T_1 = 150.5 \geq 127$, we reject H_0. There is sufficient evidence to indicate that deaf children have greater visual acuity than hearing children at $\alpha = .05$.

b. H_0: The probability distributions of eye movement rates are identical for deaf and hearing children

H_a: The probability distribution of eye movement rates for deaf children lies to the right of that for hearing children

The test statistic is $z = \dfrac{T_1 - \dfrac{n_1(n_1 + n_2 + 1)}{2}}{\sqrt{\dfrac{n_1 n_2 (n_1 + n_2 + 1)}{12}}} = \dfrac{150.5 - \dfrac{10(10 + 10 + 1)}{2}}{\sqrt{\dfrac{10(10)(10 + 10 + 1)}{12}}} = 3.44$

The rejection region requires $\alpha = .05$ in the upper tail of the z distribution. From Table IV, Appendix A, $z_{.05} = 1.645$. The rejection region is $z > 1.645$.

Since the observed value of the test statistic falls in the rejection region $(3.44 > 1.645)$, H_0 is rejected. There is sufficient evidence to support the psychologist's claim that deaf children have greater visual acuity than hearing children at $\alpha = .05$.

The results agree with those found in part a.

14.21 a.

Private Sector	Rank	Private Sector	Rank
2.58%	10	5.40%	15
5.05	13	2.55	9
.05	1	9.00	16
2.10	5	10.55	17
4.30	12	1.02	2
2.25	6	5.11	14
2.50	8	12.42	18
1.94	4	1.67	3
2.33	7	3.33	11
	$T_1 = 66$		$T_2 = 105$

To determine if the two populations have identical distributions, we test:

H_0: The two sampled sectors have identical probability distributions
H_a: The probability distribution for the public sector is shifted to the right of the private sector.

The test statistic is $T_2 = 105$.

The rejection region is $T_2 \geq 105$ from Table XII, Appendix A, with $n_1 = 9$, $n_2 = 9$ and $\alpha = .05$. Since the observed value of the test statistic falls in the rejection region ($T_2 = 105 \geq 105$), H_0 is rejected. We can conclude that the probability distribution for the public sector is shifted to the right of the private sector at $\alpha = .05$.

b. The p-value is approximately equal to .05 because the test statistic is equal to the critical value T_U.

c. 1. The two samples are random and independent.
   2. The two probability distributions from which the samples are drawn are continuous.

14.23 a. Since the data can take on a very limited number of values, the data are probably not from a normal distribution.

b. Using SAS, the output for the Wilcoxon test is:

```
                    The NPAR1WAY Procedure

            Wilcoxon Scores (Rank Sums) for Variable GSHWS
                    Classified by Variable COND

                     Sum of      Expected       Std Dev        Mean
    COND      N      Scores      Under H0       Under H0       Score
    ---------------------------------------------------------------------
    ATIPS    98      8039.0      11123.0       476.284142     82.030612
    TIPS    128     17612.0      14528.0       476.284142    137.593750

               Average scores were used for ties.


                    Wilcoxon Two-Sample Test

               Statistic              8039.0000

               Normal Approximation
               Z                        -6.4741
               One-Sided Pr <  Z         <.0001
               Two-Sided Pr > |Z|        <.0001

               t Approximation
               One-Sided Pr <  Z         <.0001
               Two-Sided Pr > |Z|        <.0001

          Z includes a continuity correction of 0.5.


                     Kruskal-Wallis Test

               Chi-Square              41.9273
               DF                            1
               Pr > Chi-Square          <.0001
```

To determine if there was a difference in the level of involvement in science homework assignments between TIPS and ATIPS students, we test:

H_0: The distributions of the involvement scores for science homework for the two groups have the same location

H_a: The distributions of the involvement scores for science homework for the TIPS students is shifted to the right or left of that for the ATIPS students

From the printout, the test statistic is $z = -6.4741$ and the p-value is $p < .0001$. Since the p-value is less than α ($p < .0001 < .05$), H_0 is rejected. There is sufficient evidence to indicate that there is a difference in the level of involvement in science homework assignments between TIPS and ATIPS students at $\alpha = .05$.

c. Using SAS, the output for the Wilcoxon test is:

```
                    The NPAR1WAY Procedure

         Wilcoxon Scores (Rank Sums) for Variable MTHHWS
                   Classified by Variable COND

                     Sum of      Expected      Std Dev        Mean
COND        N        Scores      Under H0      Under H0       Score
------------------------------------------------------------------------
ATIPS      98       10936.0      11123.0       472.485707    111.591837
TIPS      128       14715.0      14528.0       472.485707    114.960938

             Average scores were used for ties.


                  Wilcoxon Two-Sample Test

         Statistic                10936.0000

         Normal Approximation
         Z                           -0.3947
         One-Sided Pr <  Z            0.3465
         Two-Sided Pr > |Z|           0.6930

         t Approximation
         One-Sided Pr <  Z            0.3467
         Two-Sided Pr > |Z|           0.6934

     Z includes a continuity correction of 0.5.


                     Kruskal-Wallis Test

         Chi-Square                   0.1566
         DF                                1
         Pr > Chi-Square              0.6923
```

To determine if there was a difference in the level of involvement in mathematics homework assignments between TIPS and ATIPS students, we test:

H_o: The distributions of the involvement scores for mathematics homework for the two groups have the same location

H_a: The distributions of the involvement scores for mathematics homework for the TIPS students is shifted to the right or left of that for the ATIPS students

From the printout, the test statistic is $z = -.3947$ and the p-value is $p = .6930$. Since the p-value is not less than α ($p = .6930 > .05$), H_0 is not rejected. There is insufficient evidence to indicate that there is a difference in the level of involvement in mathematics homework assignments between TIPS and ATIPS students at $\alpha = .05$.

d. Using SAS, the output for the Wilcoxon test is:

```
                     The NPAR1WAY Procedure

          Wilcoxon Scores (Rank Sums) for Variable LAHWS
                    Classified by Variable COND

                          Sum of     Expected     Std Dev        Mean
          COND      N     Scores     Under H0     Under H0       Score
          --------------------------------------------------------------
          ATIPS    98    10496.0     11123.0    464.844149   107.102041
          TIPS    128    15155.0     14528.0    464.844149   118.398438

                  Average scores were used for ties.


                       Wilcoxon Two-Sample Test

              Statistic              10496.0000

              Normal Approximation
              Z                         -1.3478
              One-Sided Pr <  Z          0.0889
              Two-Sided Pr > |Z|         0.1777

              t Approximation
              One-Sided Pr <  Z          0.0895
              Two-Sided Pr > |Z|         0.1791

          Z includes a continuity correction of 0.5.


                        Kruskal-Wallis Test

              Chi-Square                 1.8194
              DF                              1
              Pr > Chi-Square            0.1774
```

To determine if there was a difference in the level of involvement in language arts homework assignments between TIPS and ATIPS students, we test:

H_0: The distributions of the involvement scores for language arts homework for the two groups have the same location

H_a: The distributions of the involvement scores for language arts homework for the TIPS students is shifted to the right or left of that for the ATIPS students

From the printout, the test statistic is $z = -1.3478$ and the p-value is $p = .1777$. Since the p-value is not less than α ($p = .1777 > .05$), H_0 is not rejected. There is insufficient evidence to indicate that there is a difference in the level of involvement in language arts homework assignments between TIPS and ATIPS students at $\alpha = .05$.

14.25 a. The hypotheses are:

H_0: The two sampled populations have identical probability distributions

H_a: The probability distributions for population A is shifted to the right of that for population B

b. Some preliminary calculations are:

| Treatment | | Difference | Rank of Absolute |
A	B	A – B	Difference
54	45	9	5
60	45	15	10
98	87	11	7
43	31	12	9
82	71	11	7
77	75	2	2.5
74	63	11	7
29	30	–1	1
63	59	4	4
80	82	–2	2.5
			$T_- = 3.5$

The test statistic is $T_- = 3.5$

The rejection region is $T_- \le 8$, from Table XIII, Appendix A, with $n = 10$ and $\alpha = .025$.

Since the observed value of the test statistic falls in the rejection region ($T_- = 3.5 \le 8$), H_0 is rejected. There is sufficient evidence to indicate the responses for A tend to be larger than those for B at $\alpha = .025$.

14.27 We assume that the probability distribution of differences is continuous so that the absolute differences will have unique ranks. Although tied (absolute) differences can be assigned average ranks, the number of ties should be small relative to the number of observations to assure validity.

14.29 a. H_0: The two sampled populations have identical probability distributions
   H_a: The probability distribution for population A is located to the right of that for population B

b. The test statistic is:

$$z = \frac{T_+ - \frac{n(n+1)}{4}}{\sqrt{\frac{n(n+1)(2n+1)}{24}}} = \frac{354 - \frac{30(30+1)}{4}}{\sqrt{\frac{30(30+1)(60+1)}{24}}} = \frac{121.5}{48.6184} = 2.499$$

The rejection region requires $\alpha = .05$ in the upper tail of the z distribution. From Table IV, Appendix A, $z = 1.645$. The rejection region is $z > 1.645$.

Since the observed value of the test statistic falls in the rejection region ($z = 2.499 > 1.645$), H_0 is rejected. There is sufficient evidence to indicate population A is located to the right of that for population B at $\alpha = .05$.

c. The p-value $= P(z \ge 2.499) = .5 - .4938 = .0062$ (using Table IV, Appendix A).

d. The necessary assumptions are:

1. The sample of differences is randomly selected from the population of differences.
2. The probability distribution from which the sample of paired differences is drawn is continuous.

14.31 a. The population of differences is probably not normal.

b. To compare the male students' attitudes toward their father and their attitudes toward their mother, we test:

H_0: The distributions of male students' attitudes toward their fathers has the same location as the distribution of male students' attitudes toward their mothers

H_a: The distributions of male students' attitudes toward their fathers is shifted to the right or left of the distribution of male students' attitudes toward their mothers

c.

Student	Attitude toward Father	Attitude toward Mother	Difference	Rank of Absolute Difference
1	2	3	-1	5.5
2	5	5	0	(eliminated)
3	4	3	1	5.5
4	4	5	-1	5.5
5	3	4	-1	5.5
6	5	4	1	5.5
7	4	5	-1	5.5
8	2	4	-2	11
9	4	5	-1	5.5
10	5	4	1	5.5
11	4	5	-1	5.5
12	5	4	1	5.5
13	3	3	0	(eliminated)

Positive rank sum $T_+ = 22$

d. The sum of the ranks of the positive differences is $T_+ = 22$. The sum of the ranks of the negative differences is $T_- = 44$.

e. The test statistic is $T =$ smaller of T_+ and T_- which is $T_+ = 22$.

Reject H_0 if $T_+ \leq T_0$ where T_0 is based on $\alpha = .05$ and $n = 11$ (two-tailed).

Reject H_0 if $T_+ \leq 11$ (From Table XIII, Appendix A.)

f. Since the test statistic does not fall in the rejection region ($T_+ = 22 > 11$), H_0 is not rejected. There is insufficient evidence to indicate a difference in the male students' attitudes toward their father and their attitudes toward their mother at $\alpha = .05$.

14.33   Some preliminary calculations are:

Set	Firstborn	Secondborn	Difference	Rank of Absolute Difference
1	86	88	−2	3
2	71	77	−6	7
3	77	76	1	1.5
4	68	64	4	4
5	91	96	−5	5.5
6	72	72	0	(eliminated)
7	77	65	12	10
8	91	90	1	1.5
9	70	65	5	5.5
10	71	80	−9	9
11	88	81	7	8
12	87	72	15	11

Negative rank sum $T_- = 24.5$
Positive rank sum $T_+ = 41.5$

To determine if the firstborn of a pair of twins is more aggressive than the other, we test:

H_0:   The probability distributions of the two populations are identical
H_a: The probability distribution of test scores for the firstborn is shifted to the right of the probability distribution of test scores for the secondborn

From the printout, the test statistic is $z = -.756$ and the p-value is $p = .449/2 = .2245$. Since the p-value is not less than α ($p = .2245 \not< .05$), H_0 is not rejected. There is insufficient evidence to conclude that the probability distribution of the test scores for the firstborn is shifted to the right of the probability distribution for the secondborn at $\alpha = .05$.

14.35   Some preliminary calculations are:

Mouse	X	Y	Difference	Rank of Absolute Difference
1	4.7	5.1	−.4	4
2	3.3	4.6	−1.3	7
3	8.5	8.7	−.2	2
4	3.9	3.6	.3	3
5	7.0	6.1	.9	6
6	4.7	4.1	.6	5
7	5.2	5.1	.1	1

Negative rank sum $T_- = 13$

To determine if one compound tends to produce higher blood sugar uptake readings than the other, we test:

H_0: The probability distributions of blood sugar readings are identical for the two compounds

H_a: The probability distribution of blood sugar readings for compound X is shifted to the right or left of the probability distribution for compound Y

The test statistic is $T = T_- = 13$

Reject H_0 if $T \leq T_0$ where T_0 is based on $\alpha = .10$ and $n = 7$ (two-tailed):

Reject H_0 if $T \leq 4$ (from Table XIII, Appendix A).

Since the observed value of the test statistic does not fall in the rejection region ($T = 13 \nleq 4$), do not reject H_0 at $\alpha = .10$. There is insufficient evidence to conclude that the probability distributions of the blood sugar readings are different for the two compounds at $\alpha = .10$.

14.37   a. H_0: The three probability distributions are identical
H_a: At least two of the three probability distributions differ in location

b.   The test statistic is $H = \dfrac{12}{n(n+1)} \sum \dfrac{R_j^2}{n_j} - 3(n+1)$

$$= \dfrac{12}{45(45+1)} \left[\dfrac{(235)^2}{15} + \dfrac{(439)^2}{15} + \dfrac{(361)^2}{15} \right] - 3(45+1) = 146.190 - 138 = 8.190$$

The rejection region requires $\alpha = .05$ in the upper tail of the χ^2 distribution with df $= p - 1 = 3 - 1 = 2$. From Table VII, Appendix A, $\chi_{.05}^2 = 5.99147$. The rejection region is $H > 5.99147$.

Since the observed value of the test statistic falls in the rejection region ($H = 8.19 > 5.99147$), reject H_0. There is sufficient evidence to indicate that at least two of the three probability distributions differ in location at $\alpha = .05$.

c.   The p-value $= P(\chi^2 \geq 8.19)$. At 2 degrees of freedom, $H = 8.19$ falls between $\chi_{.025}^2 = 7.37776$ and $\chi_{.010}^2 = 9.21034$; therefore, $.010 < p$-value $< .025$.

d.   The means are:

$$\bar{R}_A = \dfrac{R_A}{n_A} = \dfrac{235}{15} = 15.667 \qquad \bar{R}_B = \dfrac{R_B}{n_B} = \dfrac{439}{15} = 29.267$$

$$\bar{R}_C = \dfrac{R_C}{n_C} = \dfrac{361}{15} = 24.067 \qquad \bar{R} = \dfrac{1}{2}(n+1) = \dfrac{1}{2}(45+1) = 23$$

$$H = \dfrac{12}{n(n+1)} \sum n_j (\bar{R}_j - \bar{R})^2$$

$$= \dfrac{12}{45(45+1)} [15(15.667 - 23)^2 + 15(29.267 - 23)^2 + 15(24.067 - 23)^2]$$

$$= \dfrac{45}{45(46)} [1412.8] = 8.190$$

14.39 The distribution of H is approximately a χ^2 if the null hypothesis is true and if the sample sizes, n_j, for each of the p distributions is more than 5.

14.41 a. Some preliminary calculations are:

Programmed	Rank	Standard	Rank	Open	Rank
.9	5.5	1.0	7.5	1.7	15
1.5	13	.8	4	.5	1.5
.7	3	.9	5.5	1.6	14
1.1	9	1.2	10	1.4	11.5
.5	1.5	1.4	11.5	1.0	7.5
	$R_1 = 32$		$R_2 = 38.5$		$R_3 = 49.5$

b. The rank sums for the 3 methods are:

$R_1 = 32$, $R_2 = 38.5$, and $R_3 = 49.5$.

c. The test statistic is $H = \dfrac{12}{n(n+1)} \sum \dfrac{R_j^2}{n_j} - 3(n+1)$

$$= \dfrac{12}{15(15+1)}\left[\dfrac{(32)^2}{5} + \dfrac{(38.5)^2}{5} + \dfrac{(49.5)^2}{5}\right] - 3(15+1) = 49.565 - 48 = 1.565$$

d. To determine if the probability distributions of increases in reading level differ among the three methods, we test:

H_0: The probability distributions of the reading levels are the same for the three classes
H_a: At least two of the three reading levels differ in location

The rejection region requires $\alpha = .05$ in the upper tail of the χ^2 distribution with df $= p - 1 = 3 - 1 = 2$. From Table VII, Appendix A, $\chi^2_{.05} = 5.99147$. The rejection region is $H > 5.99147$.

Since the observed value of the test statistic does not fall in the rejection region, ($H = 1.565 \not> 5.99147$), do not reject H_0. There is insufficient evidence to indicate a difference in the reading levels among the three methods at $\alpha = .05$.

14.43 a. To determine if the Group B subjects exercise for longer durations than the Group A subjects, we test:

H_0: The two sampled populations have identical probability distributions
H_a: The group B probability distribution is shifted to the right of that of the probability distribution of Group A

b. The test statistic is $H = 5.1429$.

The p-value is $p = .0233$. Since the p-value is so small, H_0 would be rejected for any $\alpha > .0233$. There is sufficient evidence to indicate that the Group B subjects exercise for longer durations than the Group A subjects for $\alpha > .0233$.

c. For the Kruskal-Wallis *H*-test to be valid, the following assumptions must be met:

1. The two samples are random and independent
2. There are five or more measurements in each sample.
3. The two probability distributions from which the samples are drawn are continuous.

For this problem, the second assumption is not met. One of the samples contained only four observations.

Since there are only two populations, the Wilcoxon Rank Sum test can be used. For this test, the only necessary assumptions are:

1. The two samples are random and independent.
2. The two probability distributions from which the samples are drawn are continuous.

14.45 a. Using MINITAB, the histograms of the three data sets are:

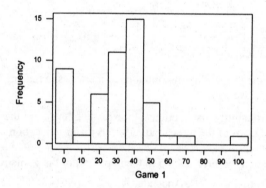

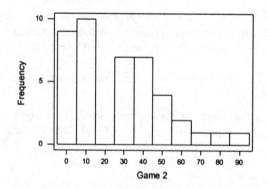

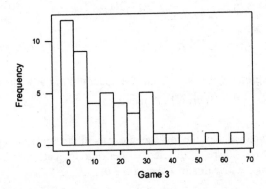

Game 3

None of these three graphs look approximately normal. Thus, the assumption of normality necessary for ANOVA is likely violated.

b.   Using MINITAB, the results of the Kruskal-Wallis analysis are:

```
Kruskal-Wallis Test on Percent

Game          N      Median    Ave Rank         Z
1            50      32.000       83.1        2.88
2            42      26.000       73.0        0.59
3            47       9.000       53.3       -3.49
Overall     139                   70.0

H = 13.61   DF = 2   P = 0.001
H = 13.66   DF = 2   P = 0.001 (adjusted for ties)
```

To determine if the percentages of names recalled are different for the three retrieval methods, we test:

H_0: The distributions of the three retrieval methods are identical
H_a: At least two of the three retrieval methods have probability distributions that differ in location

From the printout, the test statistic is $H = 13.66$ and the p-value is $p = .001$. Since the p-value is less than α ($p = .001 < .05$), H_0 is rejected. There is sufficient evidence to indicate that the percentages of names recalled are different for the three retrieval methods at $\alpha = .05$.

14.47   a.   The number of blocks, b, is 6.

         b. H_0: The probability distributions for the four treatments are identical
               H_a: At least two of the probability distributions differ in location

c. The test statistic is $F_r = \dfrac{12}{bp(p+1)} \sum R_j^2 - 3b(p+1)$

$$= \dfrac{12}{6(4)(4+1)} [11^2 + 21^2 + 21^2 + 7^2] - 3(6)(4+1) = 105.2 - 90 = 15.2$$

The rejection region requires $\alpha = .10$ in the upper tail of the χ^2 distribution with df $= p - 1 = 4 - 1 = 3$. From Table VII, Appendix A, $\chi^2_{.10} = 6.25139$. The rejection region is $F_r > 6.25139$.

Since the observed value of the test statistic falls in the rejection region ($F_r = 15.2 > 6.25139$), reject H_0. There is sufficient evidence to indicate a difference in the location of at least two of the four treatments at $\alpha = .10$.

d. The p-value $= P(F_r \ge 15.2) = P(\chi^2 \ge 15.2)$. With 3 degrees of freedom, $F_r = 15.2$ falls above $\chi^2_{.005}$; therefore, p-value $< .005$.

e. The means are:

$$\bar{R}_A = \dfrac{R_A}{b} = \dfrac{11}{6} = 1.833 \qquad\qquad \bar{R}_B = \dfrac{R_B}{b} = \dfrac{21}{6} = 3.5$$

$$\bar{R}_C = \dfrac{R_C}{b} = \dfrac{21}{6} = 3.5 \qquad\qquad \bar{R}_D = \dfrac{R_D}{b} = \dfrac{7}{6} = 1.167$$

$$\bar{R} = \dfrac{1}{2}(p+1) = \dfrac{1}{2}(4+1) = 2.5$$

$$F_r = \dfrac{12}{p(p+1)} \sum b(\bar{R}_j - \bar{R})^2$$

$$= \dfrac{12}{(4)(4+1)} [6(1.833 - 2.5)^2 + 6(3.5 - 2.5)^2 + 6(3.5 - 2.5)^2 + 6(1.167 - 2.5)^2]$$

$$= .6[25.3307] = 15.2$$

14.49 $R_1 = 16 \qquad R_2 = 7 \qquad R_3 = 23 \qquad R_4 = 14$

To determine if at least two of the treatment probability distributions differ in location, we test:

H_0: The probability distributions of the four treatments are identical
H_a: At least two of the probability distributions differ in location

The test statistic is $F_r = \dfrac{12}{bp(p+1)} \sum R_j^2 - 3b(p+1)$

$$= \dfrac{12}{6(4)(4+1)} [16^2 + 7^2 + 23^2 + 14^2] - 3(6)(4+1) = 103 - 90 = 13$$

The rejection region requires $\alpha = .05$ in the upper tail of the χ^2 distribution with df $= p - 1 - 4 - 1 = 3$. From Table VII, Appendix A, $\chi^2_{.05} = 7.84173$. The rejection region is $F_r > 7.81473$.

Since the observed value of the test statistic falls in the rejection region ($F_r = 13 > 7.81473$), reject H_0. There is sufficient evidence to indicate a difference in the location for at least two of the probability distributions at $\alpha = .05$.

14.51   a.   From the printout, the rank sums are: $R_1 = 27.0$, $R_2 = 32.5$, $R_3 = 29.0$, and $R_4 = 31.5$.

b. $F_r = \dfrac{12}{bp(p+1)} \sum R_j^2 - 3b(p+1) = \dfrac{12}{12(4)(4+1)}(27^2 + 32.5^2 + 29^2 + 31.5^2) - 3(12)(4+1) = .925$

c.   From the printout, $F_r = S = .93$ and the p-value is $p = .819$.

d.   To determine if the atlas theme ranking distributions of the four groups differ, we test:

H_0: The probability distributions of the atlas theme rankings are the same for the four groups
H_a: The probability distributions of the atlas theme rankings differ in location

The test statistic is $F_r = .925$ and the p-value is $p = .819$. Since the p-value is so large, there is no evidence to reject H_0 for any reasonable value of α. There is insufficient evidence to indicate that the atlas theme ranking distributions of the four groups differ.

14.53   Some preliminary calculations are:

Boxer	M1	Rank	R1	Rank	M5	Rank	R5	Rank
1	1243	1	1244	2	1291	4	1262	3
2	1147	2	1053	1	1169	3	1177	4
3	1247	1	1375	4	1309	2	1321	3
4	1274	2	1235	1	1290	4	1285	3
5	1177	2	1139	1	1233	3	1238	4
6	1336	2	1313	1	1366	4	1362	3
7	1238	1	1279	4	1275	3	1261	2
8	1261	2	1152	1	1289	4	1266	3
		$R_1=13$		$R_2=15$		$R_3=27$		$R_4=25$

To compare the punching power means of the four interventions, we test:

H_0: The four probability distributions of punching power are the same
H_a: At least two of the probability distributions of punching power differ in location

The test statistic is

$$F_r = \dfrac{12}{bp(p+1)} \sum R_j^2 - 3b(p+1) = \dfrac{12}{8(4)(4+1)}(13^2 + 15^2 + 27^2 + 25^2) - 3(8)(4+1) = 11.1$$

Since no α was given, we will use $\alpha = .05$. The rejection region requires $\alpha = .05$ in the upper tail of the χ^2 distribution with df $= p - 1 = 4 - 1 = 3$. From Table VII, Appendix A, $\chi^2_{.05} = 7.81473$. The rejection region is $F_r > 7.81473$.

Since the observed value of the test statistic falls in the rejection region ($F_r = 11.1 > 7.81473$), H_0 is rejected. There is sufficient evidence to indicate that the punching power means of the four interventions differ at $\alpha = .05$.

This is the same result that we obtained in Exercise 10.48.

14.55 Some preliminary calculations are:

Location	Temephos	Rank	Malsathion	Rank	Fenitrothion	Rank	Fenthion	Rank	Chlorpyrifos	Rank
Anguilla	4.6	5	1.2	1	1.5	2.5	1.8	4	1.5	2.5
Antigua	9.2	5	2.9	3	2.0	1.5	7.0	4	2.0	1.5
Dominica	7.8	5	1.4	1	2.4	2	4.2	4	4.1	3
Guyana	1.7	2	1.9	4	2.2	5	1.5	1	1.8	3
Jamaica	3.4	3	3.7	4	2.0	2	1.5	1	7.1	5
St. Lucia	6.7	4	2.7	1.5	2.7	1.5	4.8	3	8.7	5
Suriname	1.4	1	1.9	3	2.0	4	2.1	5	1.7	2
	$R_1 = 13$		$R_2 = 15$		$R_3 = 18.5$		$R_4 = 22$		$R_5 = 22$	

To determine if the resistance ratio distributions of the 5 insecticides differ, we test:

H_0: The distributions of the 5 insecticide ratios are the same
H_a: At least two of the distributions of insecticide ratios differ

The test statistic is $F_r = \dfrac{12}{bp(p+1)}\sum R_j^2 - 3b(p+1)$

$$= \dfrac{12}{7(5)(5+1)}(25^2 + 17.5^2 + 18.5^2 + 22^2 + 22^2) - 3(7)(5+1) = 2.086$$

Since no α was given, we will use $\alpha = .05$. The rejection region requires $\alpha = .05$ in the upper tail of the χ^2 distribution with df $= p - 1 = 5 - 1 = 4$. From Table VII, Appendix A, $\chi_{.05}^2 = 9.48773$. The rejection region is $F_r > 9.48773$.

Since the observed value of the test statistic does not fall in the rejection region ($F_r = 2.086 \ngtr 9.48773$), H_0 is not rejected. There is insufficient evidence to indicate that the resistance ratio distributions of the 5 insecticides differ at $\alpha = .05$.

14.57 a. For $n = 22$, $P(r_s > .508) = .01$

b. For $n = 28$, $P(r_s > .448) = .01$

c. For $n = 10$, $P(r_s \le .648) = 1 - .025 = .975$

d. For $n = 8$, $P(r_s < -.738 \text{ or } r_s > .738) = 2(.025) = .05$

14.59 Since there are no ties, we will use the shortcut formula.

a. Some preliminary calculations are:

x Rank (u_i)	y Rank (v_i)	$d_i = u_i - v_i$	d_i^2
3	2	1	1
5	4	1	1
2	5	−3	9
1	1	0	0
4	3	1	1
		Total =	12

$$r_s = 1 - \dfrac{6\sum d_i^2}{n(n^2 - 1)} = 1 - \dfrac{6(12)}{5(5^2 - 1)} = 1 - .6 = .4$$

b.

x Rank (u_i)	y Rank (v_i)	$d_i = u_i - v_i$	d_i^2
2	3	-1	1
3	4	-1	1
4	2	2	4
5	1	4	16
1	5	-4	16
			Total = 38

$$r_s = 1 - \frac{6 \sum d_i^2}{n(n^2 - 1)} = 1 - \frac{6(38)}{5(5^2 - 1)} = 1 - 1.9 = -.9$$

c.

x Rank (u_i)	y Rank (v_i)	$d_i = u_i - v_i$	d_i^2
1	2	-1	1
4	1	3	9
2	3	-1	1
3	4	-1	1
			Total = 12

$$r_s = 1 - \frac{6 \sum d_i^2}{n(n^2 - 1)} = 1 - \frac{6(12)}{4(4^2 - 1)} = 1 - 1.2 = -.2$$

d.

x Rank (u_i)	y Rank (v_i)	$d_i = u_i - v_i$	d_i^2
2	1	1	1
5	3	2	4
4	5	-1	1
3	2	1	1
1	4	-3	9
			Total = 16

$$r_s = 1 - \frac{6 \sum d_i^2}{n(n^2 - 1)} = 1 - \frac{6(16)}{5(5^2 - 1)} = 1 - .8 = .2$$

14.61   a.   Some preliminary calculations are:

Blood Lactate Level	Rank u	Perceived Recovery	Rank v	u^2	v^2	uv
3.8	3	7	1.5	9	2.25	4.5
4.2	5.5	7	1.5	30.25	2.25	8.25
4.8	7	11	3	49	9	21
4.1	4	12	5	16	25	20
5.0	8	12	5	64	25	40
5.3	9.5	12	5	90.25	25	47.5
4.2	5.5	13	7	30.25	49	38.5
2.4	1	17	9	1	81	9
3.7	2	17	9	4	81	18
5.3	9.5	17	9	90.25	81	85.5
5.8	12	18	11.5	144	132.25	138
6.0	14	18	11.5	196	132.25	161
5.9	13	21	14.5	169	210.25	188.5
6.3	15	21	14.5	225	210.25	217.5
5.5	12	20	13	121	169	143
6.5	16	24	16	256	256	256
	$\sum u = 136$		$\sum v = 136$	$\sum u^2 = 1495$	$\sum v^2 = 1490.5$	$\sum uv = 1396.25$

$$SS_{uv} = \sum uv - \frac{\left(\sum u\right)\left(\sum v\right)}{n} = 1396.25 - \frac{136(136)}{16} = 240.25$$

$$SS_{uu} = \sum u^2 - \frac{\left(\sum u\right)^2}{n} = 1495 - \frac{136^2}{16} = 339$$

$$SS_{vv} = \sum v^2 - \frac{\left(\sum v\right)^2}{n} = 1490.5 - \frac{136^2}{16} = 334.5$$

a.   The ranks of the blood lactate levels are stored in the "u" column above.

b.   The ranks of the perceived recovery values are stored in the "v" column above.

c. $r_s = \dfrac{SS_{uv}}{\sqrt{SS_{uu}SS_{vv}}} = \dfrac{240.25}{\sqrt{339(334.5)}} = .713$

Since r_s is relatively close to 1, blood lactate level and perceived recovery are fairly strongly, positively, linearly rank related.

d.   Reject H_0 if $r_s < -r_{s,\alpha/2}$ or if $r_s > r_{s,\alpha/2}$ where $\alpha/2 = .05$ and $n = 16$;
Reject H_0 if $r_s < -.425$ or $r_s > .425$

e.   To determine if blood lactate level and perceived recovery are rank correlated, we test:

H_0: Blood lactate level and Perceived recovery are not rank correlated
H_a: Blood lactate level and Perceived recovery are rank correlated

The test statistic is $r_s = .713$.

Since the observed value of the test statistic falls in the rejection region, ($r_s = .713 > .425$), H_0 is rejected. There is sufficient evidence to indicate that blood lactate level and perceived recovery are rank correlated at $\alpha = .10$.

14.63   a.

Brand	Experts 1's Rank	Experts 2's Rank	d_i	d_i^2
A	6	5	1	1
B	5	6	−1	1
C	1	2	−1	1
D	3	1	2	4
E	2	4	−2	4
F	4	3	1	1
				$\sum d_i^2 = 12$

Notice that the observations are already in rank form.

$$r_s = 1 - \frac{6\sum d_i^2}{n(n^2 - 1)} = 1 - \frac{6(12)}{6(6^2 - 1)} = 1 - .343 = .657$$

b.   To determine if there is a positive correlation in the rankings of the two experts, we test:

$H_0: \rho_s = 0$
$H_a: \rho_s > 0$

The test statistic is $r_s = .657$

Since no α was given, we will use $\alpha = .05$. Reject H_0 if $r_s > r_{s,\alpha}$ where $\alpha = .05$ and $n = 6$:

Reject H_0 if $r_s > .829$ (from Table XIV, Appendix A)

Since the observed value of the test statistic does not fall in the rejection region ($r_s = .657 \not> .829$), H_0 is not rejected. There is insufficient evidence to indicate a positive correlation in the rankings of the two experts at $\alpha = .05$.

14.65 a.   Some preliminary calculations are:

Week	Number of Strikes	Rank u_i	Age of fish	Rank v_i	$d_i = u_i - v_i$	d_i^2
1	85	9	120	1	-8	64
2	63	8	136	2	-6	36
3	34	3	150	3	0	0
4	39	5	155	4	1	1
5	58	7	162	5	2	4
6	35	4	169	6	-2	4
7	57	6	178	7	-1	1
8	12	1	184	8	-7	49
9	15	2	190	9	-7	49

$$\sum d_i^2 = 208$$

$$r_s = 1 - \frac{6 \sum d_i^2}{n(n^2 - 1)} = 1 - \frac{6(208)}{9(9^2 - 1)} = 1 - 1.733 = -.733$$

b.   To determine whether the number of strikes and age are negatively correlated, we test:

H_0: $\rho_s = 0$
H_a: $\rho_s < 0$

The test statistic is $r_s = -.733$.

Reject H_0 if $r_s < -r_{s,\alpha}$ where $\alpha = .01$ and $n = 9$.

Reject H_0 if $r_s < -.783$ (From Table XIV, Appendix A)

Since the observed value of the test statistic does not fall in the rejection region ($r_s = -.733 \not< -.783$), H_0 is not rejected. There is insufficient evidence to indicate that the number of strikes and age are negatively correlated at $\alpha = .01$.

14.67   Some preliminary calculations are:

Patient	Recreation Time (Hours)	Rank (u_i)	Tranquilizers	Rank (v_i)	u_i^2	v_i^2	$u_i v_i$
1	16	6	3	4	36	16	24
2	22	7	1	2	49	8	14
3	10	3	4	5	9	25	15
4	8	2	9	7.5	4	56.25	15
5	14	5	5	6	25	36	30
6	34	9	2	3	81	9	27
7	26	8	0	1	64	1	8
8	13	4	10	9	16	81	36
9	5	1	9	7.5	1	56.25	7.5
		$\sum u_i = 45$		$\sum v_i = 45$	$\sum u_i^2 = 285$	$\sum v_i^2 = 284.5$	$\sum u_i v_i = 176.5$

$$SS_{uv} = \sum u_i v_i - \frac{\sum u_i \sum v_i}{n} = 176.5 - \frac{45(45)}{9} = -48.5$$

$$SS_{uu} = \sum u_i^2 - \frac{\left(\sum u_i\right)^2}{n} = 285 - \frac{45^2}{9} = 60$$

$$SS_{vv} = \sum v_i^2 - \frac{\left(\sum v_i\right)^2}{n} = 284.5 - \frac{45^2}{9} = 59.5$$

To determine if recreation is negatively correlated with the number of times a tranquilizer has to be given, we test:

H_0: $\rho_s = 0$
H_a: $\rho_s < 0$

The test statistic is $r_s = \dfrac{SS_{uv}}{\sqrt{SS_{uu}SS_{vv}}} = \dfrac{-48.5}{\sqrt{60(59.5)}} = -.812$

Reject H_0 if $r_s < -r_{s,\alpha}$ where $\alpha = .05$ and $n = 9$:

Reject H_0 if $r_s < -.600$ (from Table XIV, Appendix A).

Since the observed value of the test statistic falls in the rejection region ($r_s = -.812 < -.600$), reject H_0. There is sufficient evidence to indicate that recreation time is negatively correlated with the number of times a tranquilizer is given at $\alpha = .05$.

14.69   It is appropriate to use the t and F tests for comparing two or more population means when the populations sampled are normal and the population variances are equal.

14.71   a.   Some preliminary calculations are:

Pair	X	Rank u_i	Y	Rank v_i	u_i^2	v_i^2	$u_i v_i$
1	19	5	12	5	25	25	25
2	27	7	19	8	49	64	56
3	15	2	7	1	4	1	2
4	35	9	25	9	81	81	81
5	13	1	11	4	1	16	4
6	29	8	10	2.5	64	6.25	20
7	16	3.5	16	6	12.25	36	21
8	22	6	10	2.5	36	6.25	15
9	16	3.5	18	7	12.25	49	24.5
		$\sum u_i = 45$		$\sum v_i = 45$	$\sum u_i^2 = 284.5$	$\sum v_i^2 = 284.5$	$\sum u_i v_i = 248.5$

$$SS_{uv} = \sum u_i v_i - \frac{\sum u_i v_i}{n} = 248.5 - \frac{45(45)}{9} = 23.5$$

$$SS_{uu} = \sum u_i^2 - \frac{\left(\sum u_i\right)^2}{n} = 284.5 - \frac{45^2}{9} = 59.5$$

$$SS_{vv} = \sum v_i^2 - \frac{\left(\sum v_i\right)^2}{n} = 284.5 - \frac{45^2}{9} = 59.5$$

To determine if the Spearman rank correlation differs from 0, we test:

$H_0: \rho_s = 0$
$H_a: \rho_s \neq 0$

The test statistic is $r_s = \dfrac{SS_{uv}}{\sqrt{SS_{uu}SS_{vv}}} = \dfrac{23.5}{\sqrt{59.5(59.5)}} = .40$

Reject H_0 if $r_s < -r_{s,\alpha/2}$ or if $r_s > r_{s,\alpha/2}$ where $\alpha/2 = .025$ and $n = 9$:

Reject H_0 if $r_s < -.683$ or if $r_s > .683$ (from Table XIV, Appendix A)

Since the observed value of the test statistic does not fall in the rejection region ($r_s = .40$ $\not> .683$), H_0 is not rejected. There is insufficient evidence to indicate that Spearman's rank correlation between x and y is significantly different from 0 at $\alpha = .05$.

b. Use the Wilcoxon signed rank test. Some preliminary calculations are:

Pair	X	Y	Difference	Rank of Absolute Difference
1	19	12	7	3
2	27	19	8	4.5
3	15	7	8	4.5
4	35	25	10	6
5	13	11	2	1.5
6	29	10	19	8
7	16	16	0	(eliminated)
8	22	10	12	7
9	16	18	−2	1.5
				$T_- = 1.5$

To determine if the probability distribution of x is shifted to the right of that for y, we test:

H_0: The probability distributions are identical for the two variables
H_a: The probability distribution of x is shifted to the right of the probability distribution of y

The test statistic is $T = T_- = 1.5$

Reject H_0 if $T \leq T_0$ where T_0 is based on $\alpha = .05$ and $n = 8$ (one-tailed):

Reject H_0 if $T \leq 6$ (from Table XIII, Appendix A).

Since the observed value of the test statistic falls in the rejection region ($T = 1.5 \leq 6$), reject H_0 at $\alpha = .05$. There is sufficient evidence to conclude that the probability distribution of x is shifted to the right of that for y.

14.73 Some preliminary calculations are:

Block	1	Rank	2	Rank	3	Rank	4	Rank	5	Rank
1	75	4	65	1	74	3	80	5	69	2
2	77	3	69	1	78	4	80	5	72	2
3	70	4	63	1.5	69	3	75	5	63	1.5
4	80	3.5	69	1	80	3.5	86	5	77	2
		$R_1 = 14.5$		$R_2 = 4.5$		$R_3 = 13.5$		$R_4 = 20$		$R_5 = 7.5$

To determine whether at least two of the treatment probability distributions differ in location, use Friedman F_r test.

H_0: The five treatments have identical probability distributions
H_a: At least two of the populations have probability distributions that differ in location

The test statistic is $F_r = \dfrac{12}{bp(p+1)} \sum R_j^2 - 3b(p+1)$

$$= \frac{12}{4(5)(6)} [(14.5)^2 + (4.5)^2 + (13.5)^2 + (20)^2 + (7.5)^2] - 3(4)(6) = 14.9$$

The rejection region requires $\alpha = .05$ in the upper tail of the χ^2 distribution with df $= p - 1$ $= 5 - 1 = 4$. From Table VII, Appendix A, $= 9.48773$. The rejection region is $F_r > 9.48773$.

Since the observed value of the test statistic falls in the rejection region ($F_r = 14.9 > 9.48773$), H_0 is rejected. There is sufficient evidence to indicate that at least two of the treatment means differ in location at $\alpha = .05$.

14.75

Test A	Rank	Test B	Rank
90	12	66	4
71	6	78	8
83	11	50	1
82	10	68	5
75	7	80	9
91	13	60	2
65	3		$T_2 = 29$
	$T_1 = 62$		

H_0: The probability distributions of scores for tests A and B are identical
H_a: There is a shift in the locations of the probability distributions of scores for tests A and B

The test statistic is $T_2 = 29$.

The null hypothesis will be rejected if $T_2 \leq T_L$ or $T_2 \geq T_U$ where $\alpha = .05$ (two-tailed), $n_1 = 6$ and $n_2 = 7$. From Table XII, Appendix A, $T_L = 28$ and $T_U = 56$.

Reject H_0 if $T_2 \leq 28$ or $T_2 \geq 56$.

Since $T_2 = 29 \nleq 28$ and $T_2 = 29 \ngeq 56$, do not reject H_0. There is insufficient evidence to indicate a shift in location for tests A and B at $\alpha = .05$.

14.77 a. Some preliminary calculations are:

Door-to-Door	Rank	Telephone	Rank	Grocery Store Stand	Rank	Department Store Stand	Rank
47	14	63	18	113	22	25	5
93	21	19	2	50	15	36	10
58	16	29	7	68	19	21	3
37	11.5	24	4	37	11.5	27	6
62	17	33	9	39	13	18	1
				77	20	31	8
$R_A = 79.5$		$R_B = 40$		$R_C = 100.5$		$R_D = 33$	

To determine if the probability distributions of number of sales differ in location for at least 2 of the four techniques, we test:

H_0: The four population probability distributions are identical
H_a: At least two of the four probability distributions differ in location

The test statistic is $H = \dfrac{12}{n(n+1)} \sum \dfrac{R_j^2}{n_j} - 3(n+1)$

$= \dfrac{12}{22(23)} \left[\dfrac{79.5^2}{5} + \dfrac{40^2}{5} + \dfrac{100.5^2}{6} + \dfrac{33^2}{6} \right] - 3(23) = 81.7927 - 69 = 12.7927$

The rejection region requires $\alpha = .10$ in the upper tail of the χ^2 distribution with df $= p - 1 = 4 - 1 = 3$. From Table VII, Appendix A, $\chi^2_{.10} = 6.25139$. The rejection region is $H > 6.25139$.

Since the observed value of the test statistic falls in the rejection region ($H = 12.7927 > 6.25139$), H_0 is rejected. There is sufficient evidence to indicate the probability distributions of number of sales differ for at least two of the four techniques at $\alpha = .10$.

14.79 Some preliminary calculations are:

Child	Paranoia	Aggression	d_i	d_i^2
1	7	5	2	4
2	3	1	2	4
3	6	4	2	4
4	1	2	−1	1
5	2	8	−6	36
6	4	7	−3	9
7	10	9	1	1
8	8	3	5	25
9	5	6	−1	1
10	9	10	−1	1
				$\sum d_i^2 = 86$

To determine if there is a relationship between aggression and paranoia, we test:

H_0: $\rho_s = 0$
H_a: $\rho_s \neq 0$

The test statistic is $r_s = 1 - \dfrac{6\sum d_i^2}{n(n^2-1)} = 1 - \dfrac{6(86)}{10(10^2-1)} = 1 - .521 = .479$

Reject H_0 if $r_s < -r_{s,\alpha/2}$ or $r_s > r_{s,\alpha/2}$ where $\alpha/2 = .025$ and $n = 10$:

Reject H_0 if $r_s < -.648$ or $r_s > .648$ (from Table XIV, Appendix A).

Since the observed value of the test statistic does not fall in the rejection region, ($r_s = .478 \not>$ .648), do not reject H_0. There is insufficient evidence to indicate a relationship between aggression and paranoia in children as judged by this psychologist at $\alpha = .05$.

14.81 Some preliminary calculations are:

Type A	Rank	Type B	Rank
95	1	110	6
122	10	102	4
102	3	115	8
99	2	112	7
108	5	120	9
	$T_1 = 21$		$T_2 = 34$

To determine if print type A is easier to read, we test:

H_0: The two sampled populations have identical probability distributions
H_a: The probability distribution for print type A is shifted to the left of that for print type B

The test statistic is $T_1 = 21$.

The rejection region is $T_1 \leq 19$ from Table XII, Appendix A, with $n_1 = 5$ and $n_2 = 5$, and $\alpha = .05$.

Since the observed value of the test statistic does not fall in the rejection region ($T_1 = 21 \not\leq$ 19), H_0 is not rejected. There is insufficient evidence to indicate print type A is easier to read at $\alpha = .05$.

14.83 Since the data are already ranks, all that is needed are the rank sums:

$R_1 = 51$ $R_2 = 71$ $R_3 = 24$ $R_4 = 51$ $R_5 = 28$

To determine if there is a difference in the amount of prestige the public attaches to the five professions, we test:

H_0: The probability distributions of prestige rankings are identical for the five professions
H_a: At least two of the probability distributions differ in location

The test statistic is $F_r = \dfrac{12}{bp(p+1)} \sum R_j^2 - 3b(p+1)$

$$= \dfrac{12}{15(5)(5+1)} [51^2 + 71^2 + 24^2 + 51^2 + 28^2] - 3(15)(5+1) = 309.413 - 270 = 39.413$$

The rejection region requires $\alpha = .025$ in the upper tail of the χ^2 distribution with $df = p - 1 = 5 - 1 = 4$. From Table VII, Appendix A, $\chi^2_{.025} = 11.1433$. The rejection region is $F_r > 11.1433$.

Since the observed value of the test statistic falls in the rejection region ($F_r = 39.413 > 11.1433$), reject H_0. There is sufficient evidence to indicate a difference in the amount of prestige the public attaches to the five professions at $\alpha = .025$.

14.85   Some preliminary calculations are:

List 1	Rank	List 2	Rank	List 3	Rank
48	19	41	14	18	2
43	16	36	10	42	15
39	12	29	5	28	4
57	20	40	13	38	11
21	3	35	9	15	1
47	18	45	17	33	8
58	21	32	7	31	6
	$R_1 = 109$		$R_2 = 75$		$R_3 = 47$

To determine if there is a difference between at least two of the probability distributions of the numbers of word associates that subjects can name for the three lists, we test:

H_0:   The probability distributions of the numbers of word associates named are the same for the three lists

H_a: At least two of the three probability distributions differ in location

The test statistic is $H = \dfrac{12}{n(n+1)} \sum \dfrac{R_j^2}{n_j} - 3(n+1)$

$$= \dfrac{12}{21(21+1)} \left[\dfrac{(109)^2}{7} + \dfrac{(75)^2}{7} + \dfrac{(47)^2}{7} \right] - 3(21+1)$$

$$= 73.154 - 66 = 7.154$$

The rejection region requires $\alpha = .05$ in the upper tail of the χ^2 distribution with $df = p - 1 = 3 - 1 = 2$. From Table VII, Appendix A, $\chi^2_{.05} = 5.99147$. The rejection region is $H > 5.99147$.

Since the observed value of the test statistic falls in the rejection region ($H = 7.154 > 5.99147$), reject H_0. There is sufficient evidence to indicate a difference in location for at least two of the probability distributions of the numbers of word associates at $\alpha = .05$.

14.87   a.   Some preliminary calculations are:

Difference Highway 1 – Highway 2	Rank of Absolute Differences
−25	5
4	1
−23	4
−16	2.5
−16	2.5
	$T_+ = 1$

To determine if the heavily patrolled highway tends to have fewer speeders per 100 cars than the occasionally patrolled highway, we test:

H_0:   The two sampled populations have identical probability distributions
H_a: The probability distribution for highway 1 is shifted to the left of that for highway 2

The test statistic is $T_+ = 1$.

The rejection region is $T_+ \leq 1$ from Table XIII, Appendix A, with $n = 5$ and $\alpha = .05$.

Since the observed value of the test statistic falls in the rejection region ($T_+ = 1 \leq 1$), H_0 is rejected. There is sufficient evidence to indicate the probability distribution for highway 1 is shifted to the left of that for highway 2 at $\alpha = .05$.

   b.   Some preliminary calculations are:

Day	Difference Highway 1 – Highway 2
1	25
2	4
3	−23
4	−16
5	−16

$$\bar{x}_D = \frac{\sum x_D}{n} = \frac{-76}{5} = -15.2$$

$$s_D^2 = \frac{\sum x_D^2 - \frac{\left(\sum x_D\right)^2}{n}}{n-1} = \frac{1682 - \frac{(-76)^2}{5}}{5-1} = 131.7$$

$$s_D = \sqrt{131.7} = 11.4761$$

To determine if the mean number of speeders per 100 cars differ for the two highways, we test:

H_0: $\mu_1 = \mu_2$
H_a: $\mu_1 \neq \mu_2$

The test statistic is $t = \dfrac{\bar{x}_D - 0}{s_D / \sqrt{n}} = \dfrac{-15.2}{\dfrac{11.4761}{\sqrt{5}}} = -2.96$

The rejection region requires $\alpha/2 = .05/2 = .025$ in each tail of the t distribution with df $= n - 1 = 5 - 1 = 4$. From Table VI, Appendix A, $t_{.025} = 2.776$. The rejection region is $t > 2.776$ and $t < -2.776$.

Since the observed value of the test statistic falls in the rejection region ($t = -2.96$ < -2.776), H_0 is rejected. There is sufficient evidence to indicate the mean number of speeders per 100 cars differ for the two highways at $\alpha = .05$.

We must assume that the population of differences is normally distributed and that a random sample of differences was selected.

14.89   Some preliminary calculations are:

Hours	Rank	Fraction Defective	Rank	d_i	d_i^2
1	1	.02	1	0	0
2	2	.05	3	−1	1
3	3	.03	2	1	1
4	4	.08	5	−1	1
5	5	.06	4	1	1
6	6	.09	6	0	0
7	7	.11	8	−1	1
8	8	.10	7	1	1
					$\sum d_i^2 = 6$

To determine if the fraction defective increases as the day progresses, we test:

H_0: $\rho_s = 0$
H_a: $\rho_s > 0$

The test statistic is $r_s = 1 - \dfrac{6 \sum d_i^2}{n(n^2 - 1)} = 1 - \dfrac{6(6)}{8(8^2 - 1)} = 1 - .071 = .929$

Reject H_0 if $r_s > r_{s,\alpha}$ where $\alpha = .05$ and $n = 8$:

Reject H_0 if $r_s > .643$ (from Table XIV, Appendix A).

Since $r_s = .929 > .643$, reject H_0. There is sufficient evidence to indicate that the fraction defective increases as the day progresses at $\alpha = .05$.

14.91   a.   Some preliminary calculations are:

Victoria A	Rank	Texas	Rank	Russian	Rank
12	20	9	14.5	7	9
6	6.5	10	16.5	3	1
13	21	5	4.5	7	9
10	16.5	4	2.5	5	4.5
8	12	9	14.5	6	6.5
11	18.5	8	12	4	2.5
7	9	11	18.5	8	12
	$R_1 = 103.5$		$R_2 = 83.0$		$R_3 = 44.5$

To determine if the recovery times for one or more types of influenza tend to be longer than for other types, we test:

H_0: The probability distributions of the recovery times are the same for the three types of influenza

H_a: At least two of the probability distributions of the recovery times are different in location

The test statistic is $H = \dfrac{12}{n(n+1)} \sum \dfrac{R_j^2}{n_j} - 3(n+1)$

$$= \dfrac{12}{21(21+1)} \left\{ \dfrac{(103.5)^2}{7} + \dfrac{(83)^2}{7} + \dfrac{(44.5)^2}{7} \right\} - 3(21+1) = 6.66$$

The rejection region requires $\alpha = .05$ in the upper tail of the χ^2 distribution with df = $p - 1 = 3 - 1 = 2$. From Table VII, Appendix A, $\chi_{.05}^2 = 5.99147$. The rejection region is $H > 5.99147$.

Since the observed value of the test statistic falls in the rejection region (6.66 > 5.99147), reject H_0 at $\alpha = .05$. There is sufficient evidence to indicate a difference in the probability distributions of the recovery times for at least two of the influenzas at $\alpha = .05$.

b. Some preliminary calculations are:

Victoria A	Rank	Russian	Rank
12	13	7	7
6	4.5	3	1
13	14	7	7
10	11	5	3
8	9.5	6	4.5
11	12	4	2
7	7	8	9.5
	$T_1 = 71.0$		$T_2 = 34.0$

To determine if there is a difference in location of the distributions of recovery times for Victoria A and Russian types, we test:

H_0: The probability distributions of recovery times for Victoria A and Russian types are identical

H_a: The probability distributions of recovery time for Victoria A is shifted to the right or left of that for Russian type

Test statistic is $T = T_1 = 71.0$ (or $T_2 = 34.0$, since $n_1 = n_2 = 7$)

The rejection region is $T \leq T_L$ or $T \geq T_U$, where $T_L = 37$ is the lower value and $T_U = 68$ is the upper value given by Table XII in Appendix A for $\alpha = .05$ two-tailed, with $n_1 = 7$, $n_2 = 7$. The rejection region is $T \leq 37$ or $T \geq 68$.

Since the observed value of the test statistic falls in the rejection region ($T_1 = 71.0 > 68$), H_0 is rejected. There is sufficient evidence to conclude that there is a difference in the location of the distributions for the two influenzas at $\alpha = .05$.